CHICAGO PUBLIC LIBRARY
D0147616

POLLU

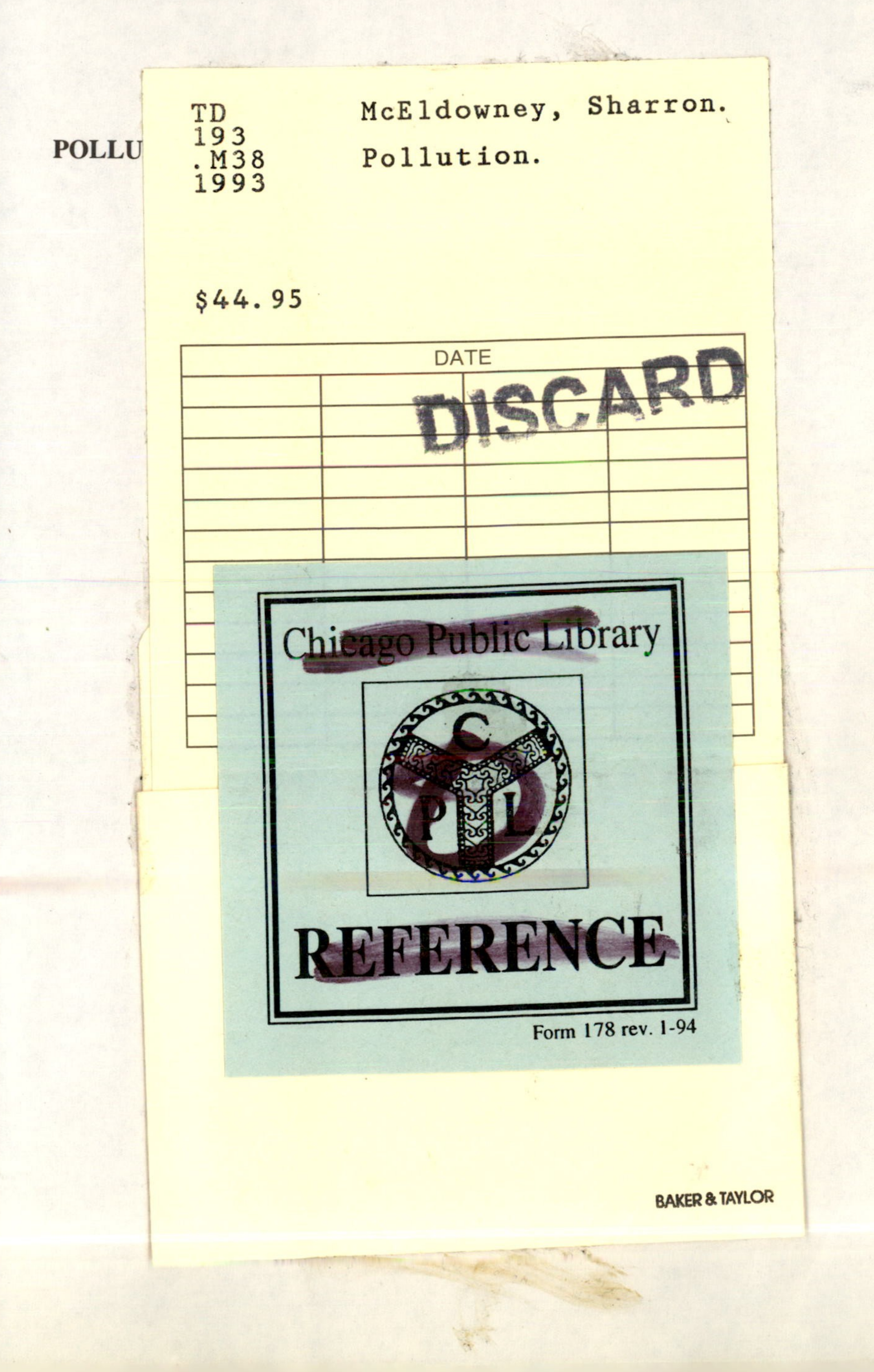
TD
193
.M38
1993
McEldowney, Sharron.
Pollution.
$44.95
DATE
DISCARD
Chicago Public Library
C
P L
REFERENCE
Form 178 rev. 1-94
BAKER & TAYLOR

POLLUTION: ECOLOGY AND BIOTREATMENT

Sharron McEldowney
Division of Biotechnology, University of Westminster

David J. Hardman
The International Institute of Biotechnology

Stephen Waite
Department of Pharmacy, University of Brighton

Longman Scientific & Technical
Longman Group UK Limited,
Longman House, Burnt Mill, Harlow,
Essex CM20 2JE, England
and Associated Companies throughout the world.

First published 1993

British Library Cataloguing in Publication Data

A catalogue record for this book is available from the British Library

ISBN 0–582–08655–8

Library of Congress Cataloging-in-Publication Data

McEldowney, Sharron.
Pollution : ecology and biotreatment / Sharron McEldowney, David J. Hardman, Stephen Waite.
p. cm.
"Copublished in the United States with John Wiley & Sons, Inc., New York."
Includes bibliographical references and index.
ISBN 0–582–08655–8
1. Environmental chemistry. 2. Bioremediation.
I. Hardman, David J. II. Waite, Stephen. III. Title.
TD193.M38 1993
628.5—dc20

Set by 5 in 10/12pt Times Roman

Produced by Longman Singapore Publishers (Pte) Ltd.
Printed in Singapore

To our families for their support, encouragement and above all understanding.

To our families for their support, encouragement and above all understanding

CONTENTS

PREFACE

The world's ever-growing population and its progressive adoption of an industrially-based lifestyle has inevitably led to an increased anthropogenic impact on the biosphere. The types and sources of pollutant are as diverse as their potential effects and fates within the environment. Opportunities exist for the release of potentially environmentally hazardous compounds at every stage of product manufacture, use and disposal. In addition to the manufacturing industries, pollutants may routinely enter the environment as a result of agricultural practices, food processing and transport. Some releases are intentional, such as the use of pesticides or discharges of waste effluent produced during manufacturing processes, others are unintentional, for example, releases resulting from accidents. In both cases the environmental consequences of the emissions are frequently not fully foreseen or completely understood.

Historically, pollution with the resultant human mortalities and disruption to the environment was largely confined to centres of resource extraction and product manufacture. Such centres of industry were also invariably centres of population which further added to the environmental impact. This is no longer the case. The potentially adverse effects of industrial development are no longer confined within national borders, the impacts of pollution are global. There are many reasons for this, but two are of paramount importance. Global weather systems and the biogeochemical cycling of elements mean that pollutants may be rapidly dispersed across national frontiers. Heavy metals and organochloride pesticide residues have been detected in the ice sheets of both polar caps, many thousands of miles from any significant centres of manufacture or use. Secondly, as newly industrializing nations attempt to emulate the lifestyle and economic wealth of the industrialized north, the number of manufacturing centres and waste disposal sites have increased and will continue to increase.

Many pollutants are totally artificial organic compounds, the products of an increasingly inventive chemical industry. The potential and actual impact of these xenobiotic compounds is extremely difficult to predict or assess. Relatively little is known about their environmental chemistry or of the ability of the biota to metabolize or degrade them. However, given the vast and largely untapped diversity of microorganisms and the versatility of their catabolic processes, it seems probable that many, if not all, organic pollutants could be amenable to biotreatment, by which they may be either degraded or transformed into compounds less toxic and environmentally damaging. Biotreatment of metals and other inorganic pollution is in some ways more limited. Metal elements cannot be degraded; once released into the environment they will persist indefinitely. The biotreatment of such pollutants is thus restricted to either 'locking them up' within the biota, thereby reducing their environmental mobility, and thus impact, or converting them into forms which no longer present a significant risk to the environment.

The scientific study of pollution and its impact on the environment, ecotoxicology, is in its infancy. In simple terms, the development of an applied scientific discipline or the study of a particular problem can often be viewed as involving three stages: (i) the identification, description and definition of the problem, (ii) analysis and identification of possible immediate practical solutions and (iii) the development of procedures to prevent the recurrence of the problem in the future. At present our knowledge is beginning to advance beyond the first stage of problem identification and description, towards the development and implementation of immediate remediable solutions. The ultimate objective of clean manufacturing technologies which minimize pollutant emissions has yet to be reached; we are still 'mopping up'!

With the realization of the extent of global pollution and the importance of chronic exposure to low levels of pollutants, national authorities have set legislative requirements for emissions and environmental quality. In future, multinational agreements and international protocols will become increasingly important and necessary. To be effective the consequences of this body of legislation must not only be politically and socially acceptable, but must be based on good environmental science and a realistic appreciation of the short- and long-term risks posed to the health of the planet and its occupants.

This book was written in the hope that it might stimulate the investigation, development and adoption of biotreatments during the initial mopping up stage, but also to consider applications in manufacturing processes as a means of reducing emissions. Given the scope of the material and topics relevant to this aim, this book is by necessity selective. We have chosen not to cover such major issues as greenhouse gases and global warming. The reasons for this are two-fold. Global climate change is an extremely complex issue, warranting more space than we could ever give it and, secondly, no practical biotreatment technologies capable of significantly reducing atmospheric levels of greenhouse gases currently exist. Those that have been suggested, e.g. increasing phytoplankton photosynthesis by the fertilization of seawater with iron or forest planting, are either of questionable value or at the very early stages of development. We have made a conscious effort not to exaggerate the possible applications and effectiveness of biotreatments. We are only too well aware that the public perception of science has suffered in the past from unrealistic and over-optimistic claims made of the ability of science and technology to solve environmental, social and technological problems. Thus, although all major groups of pollutants are covered within this book we have concentrated on those which are particularly amenable to biotreatment. For this reason consideration of the treatment of organic pollutants is largely focused on the amelioration of contaminated soils. All industrial countries are littered with a legacy of old industrial sites, waste dumps, etc., the soils of which are frequently contaminated with a cocktail of toxic carcinogenic compounds. These not only represent a health hazard to the local population but also act as diffuse sources of pollutants which may enter the atmosphere and leach into groundwaters. While the pollutants are effectively localized and concentrated within these sites, they are more likely to be amenable to biotreatment; once dispersed in the environment, treatment options become limited.

Equally damaging to science are exaggerated claims of impending global environmental disaster. In this book, we have attempted to provide as far as possible a balanced and objective account of the environmental issues and risks associated with particular groups of pollutants. The first section of the book provides a selective introduction to the fundamental aspects of the three topics in the title – pollution, ecology and biotreatment. The following sections deal with the major groups of pollutants. Each section contains chapters which discuss aspects of their environmental impact, the selection of possible biotreatment organisms and the development and application of biotreatment. We are convinced that the successful application of biotechnology to environmental problems requires an appreciation of both ecology and biotechnology. We hope this book will prove useful to ecologists and environmental managers and students with only little knowledge of biotechnology, and conversely to biotechnologists with little grounding in environmental science.

DAVID J. HARDMAN
SHARRON McELDOWNEY
STEPHEN WAITE
Spring 1993

ACKNOWLEDGEMENTS

We would like to record our appreciation for the continued help and support we have received from our friends and colleagues and especially to thank Jane Williams and Sarah Dines for their help in the preparation of this manuscript.

We are grateful to the following for permission to reproduce copyright material:

Academic Press Ltd. for Table 2.1 (Moriarty, 1990); Academic Press Inc. and the author Dr. W. Klöpffer for Fig. 2.6 (Klöpffer *et al.*, 1982); Academic Press and the author K. Yoshida for Fig. 2.5 (Yoshida *et al.*, 1983); American Chemical Society for Fig. 11.2 (Bennett & Hill, 1974) & Tables 1.1b (Keith & Telliard, 1979), 13.1 (Pearson, 1963) Copyright 1963, 1974 & 1979 American Chemical Society: American Society for Testing & Materials (ASTM) for Fig. 2.2 (Stern & Walker, 1978) Copyright ASTM; Botanical Society of America Inc. for Fig. 10.3 (Mendelssohn & Postek, 1982); Elsevier Applied Science Publishers Ltd. and the author, C.H. Walker for Fig. 2.7 (Walker, 1987); W.H. Freeman & Co. for Fig. 8.1b (Reklefs, 1990); the author, H.B.N. Hynes for Fig. 8.2 (Hynes, 1960); Institution of Water & Environmental Management for Fig. 10.2 (Bayes *et al.*, 1989); IOP Publishing Ltd. and the author, L.B. Wood for Fig. 8.3 (Wood, 1982); the author, Dr. D. Laxen for Fig. 13.4 (Laxen, 1983); Longman Group UK Ltd. for Fig. 11.4 & Table 1.1a (Mason, 1991); the author, Dr. A.M. Mannion for Fig. 11.1 (Mannion & Bowlby, 1992); Marcel Dekker, Inc. for Table 5.1 (Slater & Lovatt); National Center for Atmospheric Research/University Corporation for Atmospheric Research for Fig. 13.2; O.E.C.D. for Table 8.1 (Vollenweider & Kerekes, 1982); Pergamon Press Ltd. for Figs 2.4 (Neely, 1982), 10.1 (Reddy & De Busk, 1987) & Table 13.3 (Lantzy & Mackenzie, 1970) Copyright 1970, 1980 & 1987 Pergamon Press Ltd.; Plenum Publishing Corp. and the author, W.B. Neely for Fig. 2.3 (Neely & Blau, 1977); The Royal Society and the author Prof. A.T. Bull for Table 3.3 (Bull, 1992); the author, Prof. Dr. W. Stumm for Fig. 13.3 (Stumm & Bilinski, 1972); Water Research Council plc and the author,

J.F. Solbé for Fig. 2.1 (Solbé, 1988); John Wiley & Sons Ltd. for Fig. 8.1a (Etherington, 1975) Copyright © 1975 John Wiley & Sons Ltd.; John Wiley & Sons Inc. and the author Dr. M. Whitfield for Fig. 13.5 & Table 13.2 (Whitfield & Turner, 1987) Copyright © 1987 John Wiley & Sons Inc.

Whilst every effort has been made to trace the owners of copyright material, in a few cases this has proved impossible and we take this opportunity to offer our apologies to any copyright holders whose rights we may have unwittingly infringed.

PRINCIPLES AND PERSPECTIVES

Chapter 1

THE ENVIRONMENT AND POLLUTION

Introduction

Holdgate (1979) defined pollution as "the introduction by man into the environment of substances or energy liable to cause hazards to human health, harm to living resources and ecological damage, or interference with legitimate uses of the environment". Pollutants may be of numerous types, including certain metals, a diverse range of organic compounds and some gases. The importance attributed to a particular pollutant is normally linked to its perceived toxicity to humans. The European Economic Community (EEC) places the most dangerously toxic compounds on a 'Black List', the less dangerous pollutants occur on the 'Grey List'. Similarly the United States Environmental Protection Agency (EPA), lists 129 chemicals on its list of priority pollutants (Table 1.1). Chemicals placed on the European Economic Community 'Black List', and the EPA list of priority pollutants tend not only to be highly toxic but also persistent and capable of bioaccumulation (Ch. 2). The substance, or pollutant, may be a synthetic compound or a naturally occurring element or compound whose concentration is elevated by anthropogenic activity to levels which are either toxic or capable of disrupting the ecology of an area.

Depending on the activity responsible for the release, pollutants may be released from isolated point sources at high concentrations, e.g. metal smelters, or at low concentrations from many diffused sources, e.g. leachate from domestic disposal sites. In the case of point sources the impact of the pollutant may initially be restricted to the immediate locality of the source, the concentration of the pollutant decreasing rapidly away from the source as it is dissipated and diluted in the environment. However, the dilution and dissipation of a pollutant from a locality rarely, if ever, completely eliminates or protects the environment from possible adverse effects. Pollutants may accumulate within the biota and contaminate human food sources as well as affecting the fitness of an organism with a high body load and its predators.

Table 1.1 (a) 'Black List' and 'Grey List' compounds of the European Economic Community (Mason, 1991)

List no. 1 ('Black List')

1. Organohalogen compounds and substances which may form such compounds in the aquatic environment
2. Organophosphorus compounds
3. Organotin compounds
4. Substances, the carcinogenic activity of which is exhibited in or by the aquatic environment
(substances in List 2 which are carcinogenic are included here)
5. Mercury and its compounds
6. Cadmium and its compounds
7. Persistent mineral oils and hydrocarbons of petroleum
8. Persistent synthetic substances

List no. 2 ('Grey List')

1. The following metalloids/metals and their compounds

1. Zinc	2. Copper	3. Nickel	4. Chromium	5. Lead
6. Selenium	7. Arsenic	8. Antimony	9. Molybdenum	10. Titanium
11. Tin	12. Barium	13. Beryllium	14. Boron	15. Uranium
16. Vanadium	17. Cobalt	18. Thalium	19. Tellurium	20. Silver

2. Biocides and their derivatives not appearing in List 1
3. Substances which have a deleterious effect on the taste and/or smell of products for human consumption derived from the aquatic environment compounds liable to give rise to such substances in water
4. Toxic or persistent organic compounds of silicon and substances which give rise to such compounds in water, excluding those which are biologically harmless or are rapidly converted in water to harmless substances
5. Inorganic compounds of phosphorus and elemental phosphorus
6. Non-persistent mineral oils and hydrocarbons of petroleum origin
7. Cyanides, fluorides
8. Certain substances which may have an adverse effect on the oxygen balance, particularly ammonia and nitrites

Table 1.1 (b) EPA list of 129 Priority Pollutants (in Keith & Telliard, 1979)

31 are purgeable organics

Acrolein	1,2-Dichloropropane
Acrylonitrile	1,3-Dichloropropene
Benzene	Methylene chloride
Toluene	Methyl chloride
Ethylbenzene	Methyl bromide
Carbon tetrachloride	Bromoform
Chlorobenzene	Dichlorobromomethane
1,2-Dichloroethane	Trichlorofluoromethane
1,1,1-Trichloroethane	Dichlorodifluoromethane
1,1-Dichloroethane	Chlorodibromomethane
1,1-Dichloroethylene	Tetrachloroethylene
1,1,2-Trichloroethane	Trichloroethylene

Table 1.1 (b) Continued.

1,1,2,2-Tetrachloroethane	Vinyl chloride
Chloroethane	1,2-*trans*-Dichloroethylene
2-Chloroethyl vinyl ether	*bis*(chloromethyl) ether
Chloroform	
46 are base/neutral extractable organic compounds	
1,2-Dichlorobenzene	Fluorene
1,3-Dichlorobenzene	Fluoranthene
1,4-Dichlorobenzene	Chrysene
Hexachloroethane	Pyrene
Hexachlorbutadiene	Phenanthrene
Hexachlorobenzene	Anthracene
1,2,4-Trichlorobenzene	Benzo(a)anthracene
bis(2-Chloroethoxy)methane	Benzo(b)fluoranthene
Naphthalene	Benzo(k)fluoranthene
2-Chloronaphthalene	Benzo(a)pyrene
Isophorone	Indeno(1,2,3-c,d)pyrene
Nitrobenzene	Dibenzo(a,h)anthracene
2,4-Dinitrotoluene	Benzo(g,h,f)perylene
2,6-Dinitrotoluene	4-Chlorophenyl phenyl ether
4-Bromophenyl phenyl ether	3.3′-Dichlorobenzidine
bis(2-Ethylhexyl) phthalate	Benzidine
Di-*n*-octyl phthalate	*bis*(2-Chloroethyl)ether
Dimethyl phthalate	1,2-Diphenylhydrazine
Diethyl phthalate	Hexachlorocyclopentadiene
Di-*n*-butyl phthalate	N-Nitrosodiphenylamine
Acenaphthylene	N-Nitrosodimethylamine
Acenaphthene	N-Nitrosodi-*n*-propylamine
Butyl benzyl phthalate	*bis*(2-Chloroisopropyl) ether
11 are acid extractable organic compounds	
Phenol	*p*-Chloro-*m*-cresol
2-Nitrophenol	2-Chlorophenol
4-Nitrophenol	2,4-Dichlorophenol
2,4-Dinitrophenol	2,4,6-Trichlorophenol
4,6-Dinitro-*o*-cresol	2,4-Dimethylphenol
Pentachlorophenol	
26 are pesticides/PCBs	
α-Endosulfan	Heptachlor
β-Endosulfan	Heptachlor epoxide
Endosulfan sulfate	Chlordane
α-BHC	Toxaphene
β-BHC	Aroclor 1016
δ-BHC	Aroclor 1221
γ-BHC	Aroclor 1232
Aldrin	Aroclor 1242
Dieldrin	Aroclor 1248
4,4′-DDE	Aroclor 1254
4,4′-DDD	Aroclor 1260

Table 1.1 (b) Continued.

4,4′-DDT	2,3,7,8-Tetrachlorodibenzo-*p*-dioxin (TCDD)
Endrin	
Endrin aldehyde	
13 are metals	
Antimony	Mercury
Arsenic	Nickel
Beryllium	Selenium
Cadmium	Silver
Chromium	Thallium
Copper	Zinc
Lead	
Miscellaneous	
Total cyanides	Asbestos (fibrous)
	Total phenols

Metal smelting: an example of point source pollution

The metal concentrates produced by the milling of mined ores require further refinement before the metal product can be successfully extracted and processed. The first stage of this process occurs at a primary smelter, the waste products of which include molten material (slag which on cooling is disposed of on land), atmospheric emissions of metal-contaminated particulates and various gaseous pollutants, particularly SO_2. Slag dumps present similar problems to those associated with mine tailings, namely high metal concentrations, poor physical structure and extreme pH. The environmental impact of metal smelters and refinery works are normally substantial. Features typically associated with a smelter works are (i) localized contamination of surface soil and vegetation, (ii) an exponential decline in metal concentration with distance away from the point source, (iii) damaged ecosystem function and (iv) disruption of nutrient and carbon cycling.

Even modern smelters utilizing electrostatic precipitator and filter systems capable of removing more than 98% of metal-containing particulates, still emit substantial quantities of metal into the atmosphere. For example, a primary Zn, Pb and Cd smelter works at Avonmouth, south-west England, producing 100 000 tonnes of Zn, 40 000 tonnes of Pb and 300 tonnes of Cd a year releases some 6 kg of Zn, 4 kg of Pb, 0.4 kg of Cd and 0.1 kg of As per hour into the atmosphere as fine particles with a diameter of a few microns or less (Hopkin, 1989). The largest particles settle closest to the point of emission. Particles >10 μm in diameter are deposited within 1 km of the Avonmouth plant while smaller particles, ≤ 2.5 μm are transported considerable distances. For example, at the Avonmouth smelter there is evidence of metal contamination present in samples of soil, vegetation and invertebrates collected 25 km downwind of the works. Settlement rates are substantially increased by wet deposition occurring during rain events, when between 60 and 80% of airborne particulate material may be removed. Typically 50% of the material emitted by a smelter is deposited close to its source; as a result surface metal concentrations are normally extremely high within the vicinity of the works and decline rapidly with distance. For example, at a long established brassworks in Gusum, Sweden, surface soil concentrations of Zn (organic soil fraction) for samples collected within 0.3 km of the works ranged between 16 000 and 20 000 ppm. Samples collected 7–9 km away contained concentrations of around 200 ppm (Freedman, 1989).

The extent of ecological impact will reflect the pattern of deposition. Concentric zones of disruptions are frequently observable. Close to the source ecosystem function is severely disrupted. Species diversity and abundance is low. Vegetation is sparse, restricted to a few poorly growing, metal-tolerant species. The abundance and/or activity of other components of the ecosystem, e.g. invertebrate, microbial and fungal communities, will also be adversely affected.

Surface contamination of terrestrial ecosystems with heavy metals (Ch. 13) disrupts the cycling and decomposition of dead and senescent plant material (Ch. 5). In contaminated woodlands, populations of macroinvertebrates (e.g. earthworms, isopods and millipedes) which aid decomposition by shredding material into small fragments facilitating its subsequent decomposition by microorganisms, are severely depressed causing leaf litter to accumulate on the soil surface. Although total fungal and bacterial population sizes frequently appear relatively unaffected by metal contamination, their ability to decompose organic material is severely inhibited. Studies indicate that exposure to low levels of contamination can adversely affect microbial physiology, particularly respiration, nitrogen immobilization and rates of enzyme production. Significant reductions in the overall numbers of soil bacteria, fungi and actinomycetes compared with control sites are rarely found except at severely contaminated sites where Cu, Zn and Pd levels exceed 1% of the dry weight of the soil. In such situations undecomposed organic material may constitute 75% by dry weight of the soil. As levels of contamination increase, metal-tolerant ecotypes and species become dominant. Fungi, when compared to bacteria are generally more sensitive to metal pollutants. The density of fungal hyphae declines with increasing metal concentration. At moderate levels of contamination this decline is often associated with an increase in the abundance of bacteria which utilize nutrients released by dying hyphae and the resources no longer exploited by fungi.

Because of the abundance of cation exchange sites and ligands, deposited metals are largely retained in the humus and upper soil layers. Mobility differs between metals and is a complex function of many edaphic factors, including soil pH, cation exchange capacity and the microbial population of the soil–humus system (Ch. 13). In heavily contaminated sites, where microbial activity is severely inhibited, increased leaching of both essential plant nutrients and toxic metals occurs. Acid precipitation may also increase metal mobility (Ch. 11). Increased mobility of Cd, Zn and Pb within the soil profiles at Hallen Wood, which is subject to pollution from the Avonmouth works, has been attributed to increased emissions of SO_2 and a subsequent increase in the extent of acid precipitation at the site. Even in the absence of acidic precipitation, metal mobility at any given depth within the soil profile will tend to increase through time as organic material decomposes and becomes more acidic (largely due to the release and accumulation of organic acids). The decline in pH will increase the mobility of metals such as Cu and Pb, which in comparison with Cd and Zn, are initially relatively immobile (Hopkin, 1989; Ch. 13).

Vegetation may become contaminated by the direct uptake of metals from the soil and their subsequent translocation within the plant as soluble ionic and organic complexes, or by particulate contamination of leaf surfaces. Leaf surface contamination will be particularly marked among plant species with 'rough' or 'hairy' surfaces. It can account for up to 75% of the apparent metal content of plants close to sources of emission. Because of the essentially inert nature of this material, the bioavailability of metals associated with leaf surface particles will be limited in comparison to translocated metal within the plant. It is clear that the metal exposure of sap-feeding and leaf-cutting invertebrates feeding on the same plants will differ substantially and will not necessarily be correlated with total plant metal content.

The location and mobility of accumulated metals in tolerant and non-tolerant plant species differs. Tolerant species and ecotypes accumulate high concentrations of toxic metals, much of which is associated with the roots. Relatively little is translocated to the above-ground portions of the plant, much is adsorbed onto and within the cell walls of the root. In contrast, non-tolerant plants readily take up and translocate toxic metals to the shoots in which they accumulate. Thus, the low mobility of heavy metals within tolerant plants will tend to limit the transfer of metals to herbivores and favour the retention of metals within the soil–humus microcosm. Due to seasonal growth, variation in the concentrations of toxic metals in plant tissues occurs. In contaminated grassland dominated by tolerant ecotypes of the grass *Agrostis stolonifera* (creeping bent), marked winter peaks were observed in above-ground Cd and Cu tissue concentrations. This winter peak may be attributed to the mobilization and translocation of metals into older shoots and leaves prior to senescence and 'dilution' of the metals by new growth during the spring.

Invertebrates feeding on contaminated vegetation and leaf litter play a central role in the mobilization of metals from the soil/litter and plant components of terrestrial ecosystems. The movement of metals within a grassland system receiving significant amounts of Cu and Cd from a Merseyside smelter has been extensively studied (Hunter *et al.*, 1987a,b,c, 1989). Both Cd and Cu became concentrated during biotransfer from soil surface layers to plants, herbivores and their predators. At each stage of this transfer, the ratio of Cu : Cd decreased, the mobility and bioavailability of Cd greatly exceeding that of Cu. The ratio of total soil Cu to Cd in the heavily contaminated site was 716 : 1. Total soil concentrations are not a good indicator of the availability of toxic metals to plants; plant uptake and accumulation is more closely related to the concentration of metals present in the water-extractable fraction. Within this fraction the ratio of Cu : Cd was 186 : 1. In grasses growing in these soils and subject to leaf surface contamination the ratio was 37 : 1 in live material and 65 : 1 in senescent material. For plants not subject to surface deposition of metals, i.e. where metal accumulation occurred via root uptake alone, the Cu : Cd ratio was 19 : 1. Both detritivorous and herbivorous invertebrates accumulated significant quantities of Cu and Cd. Among detritivorous soil macrofauna typical dietary concentration factors of 3 for Cu and 15 for Cd were obtained. Among invertebrate herbivores concentration factors (Ch. 2) of 3 for Cu and 4 for Cd were typical. In both groups of organisms the ratio of Cu : Cd was substantially below that present in their dietary intake, reflecting the greater bioavailability of Cd over Cu. Marked differences in the ability of carnivorous invertebrates to accumulate and assimilate Cu and Cd were observed. Although both beetle and spider predators tended to concentrate the metals, Cu accumulation was most marked among beetle species in contrast to spiders which appear to preferentially accumulate Cd. Overall the bioavailability of Cd through the invertebrate food web exceeded that of Cu by between three and seven times. High tissue contents of Cu and Cd were found in small mammals feeding on contaminated invertebrates and plant material. However, with the exception of the insectivore *Sorex araneus* L. (common shrew), for which the Cu diet concentration factor equalled 1.75, small mammal concentration factors for Cu and Cd did not exceed 1.0. Thus, within this system the principal route of transfer responsible for the mobilization of toxic metals would appear to be from the soil to vegetation.

The behaviour of metals within terrestrial communities should be broadly similar, whether they are polluted by atmospheric deposition or have developed on abandoned mine wastes, where soil surface metal concentrations may have been reduced as a result of weathering and leeching of metals, or occur on uncontaminated soil used to cap toxic waste (i.e. reclamation). However, important differences do exist; where aerial contamination predominates the accumulation and transfer of metals will be largely governed by the extent and productivity of shallow-rooting herbaceous plants and the extent and characteristics of the soil/litter component. In contrast, where the source of contamination is confined to the subsoil, biotransfer will be dependent on the mobilization of metals from the subsoil by tolerant deep-rooted shrubs and trees able to take up and translocate metals.

Non-point source pollution

Unlike point source pollutants, non-point source pollutants typically occur in the environment at extremely low concentrations; their effects only become apparent after they become concentrated in the biota. The pollution of the Baltic Sea provides a good example of the effects of non-point sources. The Baltic Sea is a shallow, largely enclosed water body with a maximum depth of 18 m which, because of the restricted exchange of water through the Kattegat and Skergarrak to the North Sea, has a long water renewal time of between 20 and 50 years. Pollutants enter the Baltic from numerous minor sources. Many small rivers drain into it, and land run-off and discharges originate from many small industrial and population centres which occur along the coasts of Sweden and Finland. The restricted exchange of water, coupled with the short growth season, the low productivity and typically short food chains, mean that the Baltic is particularly sensitive to disturbance. The region has a history of pollution problems. In the 1960s organochloride pesticides (e.g. DDT) substantially reduced populations of fish and bird carnivores; more recently the eutrophication (Ch. 8) has become a problem. Currently concern centres on the polychlorinated biphenyls (PCBs). They were first synthesized around 1880 with large-scale commercial production beginning around 1929. They are very stable compounds and found widespread industrial use as solvents, coolants and sealants, particularly in the paint, print and electrical industries where they were extensively used in transformers. By the late 1960s their potential toxicity to humans was well documented and concern was mounting over the apparent relationship between the decline of Baltic seal populations and elevated levels of PCBs in fish and sea birds. At the beginning of the century Baltic seal populations, estimated at grey seals 100 000, ringed seals 400 000 and common seals 2000–3000, were sufficient to sustain their commercial exploitation for skin and oil products. From the early 1950s, populations began to decline and the incidence of malformations increased. Abnormalities recorded include occlusion and narrowing of the uterus causing sterility, deformation of the skull, malformation of the flippers, hardening of the arteries, kidney damage and brittle bones (Lothigius, 1991). Of the current estimated seal population of 1500 grey, 6000 ringed and 100–200 common seals, all female seals over 20 years have been found to be sterile; normal healthy seals would be expected to remain fertile up to an age of 40 years. The exact mechanism of toxicity is unknown. Laboratory studies show that PCBs can disrupt the reproductive physiology of rats and minks. However, the situation is complicated by the fact that not all PCBs are equally toxic and by the presence of other organic toxins, e.g. dioxins (dibenzofurans), which were present as contaminates of the original PCBs. Since the early 1970s the manufacture and use of PCBs in Sweden and most other countries has been prohibited. Following this PCB levels in Baltic fish and birds fell, but stabilized in 1984 and have since remained at an elevated level. One reason for this is the long-term persistence of PCBs, another is the continued environmental input of PCBs from small diffused sources, e.g. leachate from domestic waste tips containing discarded electrical equipment. Unlike pesticides, PCBs were never intended to be released directly into the environment, but given their widespread use and the way in which products containing them were used and disposed of, it was inevitable that they would, and will continue, to leak into the environment (Mason, 1990; Lothigius, 1991). Despite 20 years of investigation and concern over the potential environmental hazard posed by PCBs, there is little direct evidence linking them with the decline of seals or other populations. It is essentially a prudent and necessary interpretation of the correlative evidence which implicates them as environmentally hazardous compounds. This is not surprising when you consider the resources used, the numerous investigations and countless publications which were apparently required to establish an unambiguous link between cigarette smoking and lung cancer. The chain of causative interactions is considerably more complex and less easily defined when considering the potential impact of chemical pollutants on ecosystems.

Ecological considerations

The likely impact of a pollutant will depend to a large extent on its bioavailability, toxicity and concentration within the environment. The pollutant load that an organism acquires is a complex function of environmental concentrations, length of exposure and the form in which the pollutant occurs, i.e. whether associated with particulates, present as a dissolved species or as an inorganic or organic complex.

The sources, chemistry, environmental fate and impact of the major groups of pollutants are considered separately in this book. Although the general processes involved in determining the fate of a pollutant in the environment are similar (Ch. 2), each group and to a certain extent each pollutant within a group, behaves in a unique way. The situation may be further complicated by the emissions and discharges containing complex and sometimes totally unknown cocktails of toxic chemicals each with their own unique effects. In fact it is rare that an area is subject to the effects of a single contaminant. Similarly the form and extent of impact a pollutant may have on an ecosystem is best considered separately for each group of pollutants. Gross generalizations rarely prove to be either useful or reliable. However, it is appropriate to briefly comment on the nature of ecological communities and systems. Ecological systems are maintained by the flow and cycling of nutrients and energy (Fig. 1.1). Energy may be considered to flow through an ecosystem while nutrients tend to cycle within the system. Energy is trapped and fixed into organic compounds by autotrophic organisms (mainly plants). This material is then consumed by heterotrophic organisms (animals), saprophytes (e.g. fungi) and ultimately by the detritivores (a diverse group of organisms feeding on dead and decaying material including fungi and bacteria). It has been an established tradition to group these organisms into 'trophic' levels, each successive level feeding off the previous level. Since the efficiency of energy transfer between successive levels is rarely more than 10%, the biomass and energy associated with each trophic level decreases. This has several important ecological consequences. The maximum number of trophic levels or successive links in a food chain rarely exceeds five. The amount of biomass within each trophic level in which a pollutant may accumulate decreases from one trophic level to the next. The total amount of biomass which can be supported is determined by the number and activity of the primary producers. However, trophic structures are very artificial constructs. Since many organisms feed at more than one trophic level, a better picture of the pattern of energy and nutrient flow may be obtained from the consideration of food web diagrams which summarize the feeding behaviour of a community (Fig. 1.2). Although knowledge of a food web provides a more precise guide to the possible biological routes along which pollutants may pass and bioaccumulate, it is at best a very superficial and limited depiction of a community. The structure and composition of a community is only partially the result of the prey–predator and feeding interactions summarized in food web diagrams. The species composition and abundance are also strongly influenced by environmental conditions and a range of interspecies interactions including symbiosis and competition. The structure depicted in a food web will partly reflect the effect of species–species competitive interaction for resources other than food, e.g. nesting sites, refuge sites, etc. Most species, especially those within species-rich communities, may be considered to be specialists, having evolved often complex and unique characteristics to exploit a very restricted and well-defined

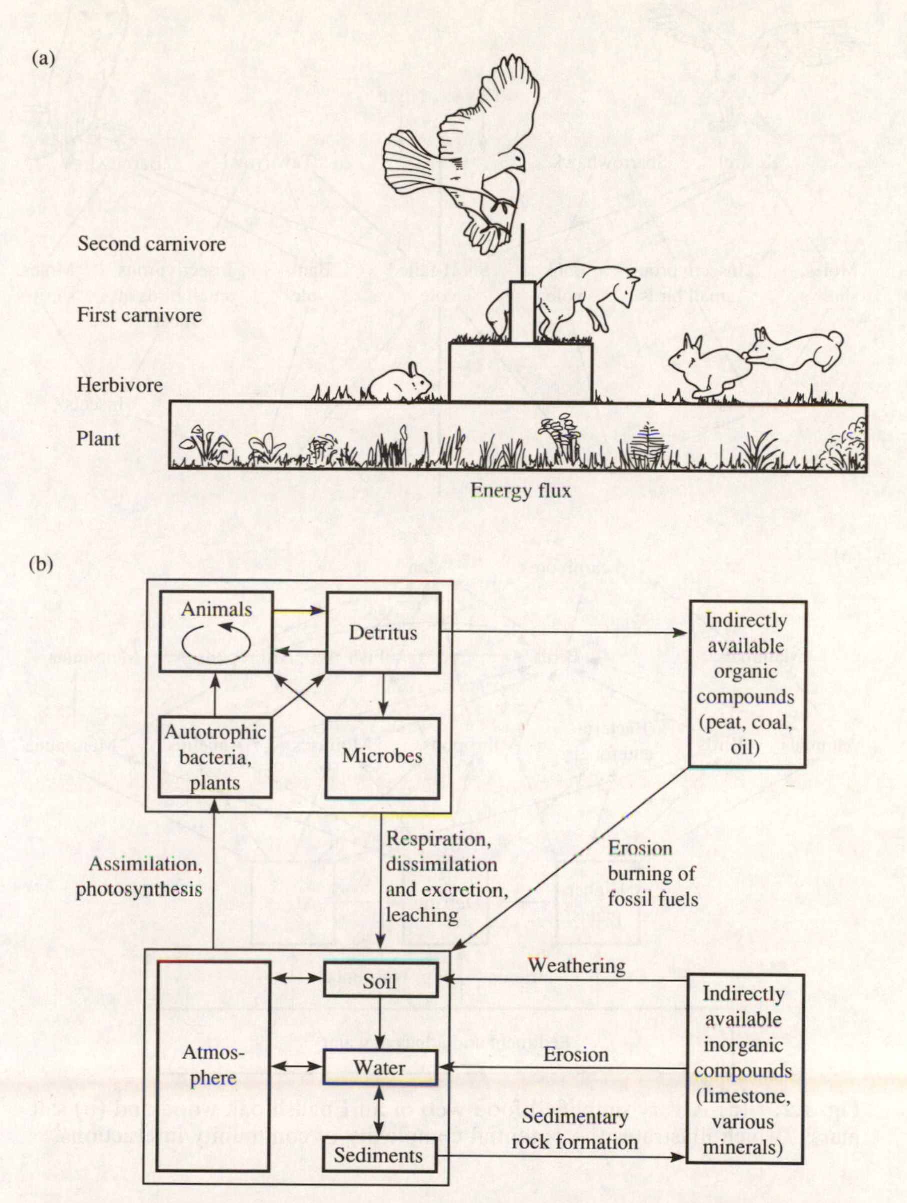

Fig. 1.1. (a) Trophic structure: an ecological pyramid in which the breadth of each bar represents the net productivity of each trophic level in the ecosystem. For this particular system ecological efficiencies are 20, 15 and 10% between trophic levels. (b) A generalized compartment model of the ecosystem, showing the relationship between energy and nutrient cycling.

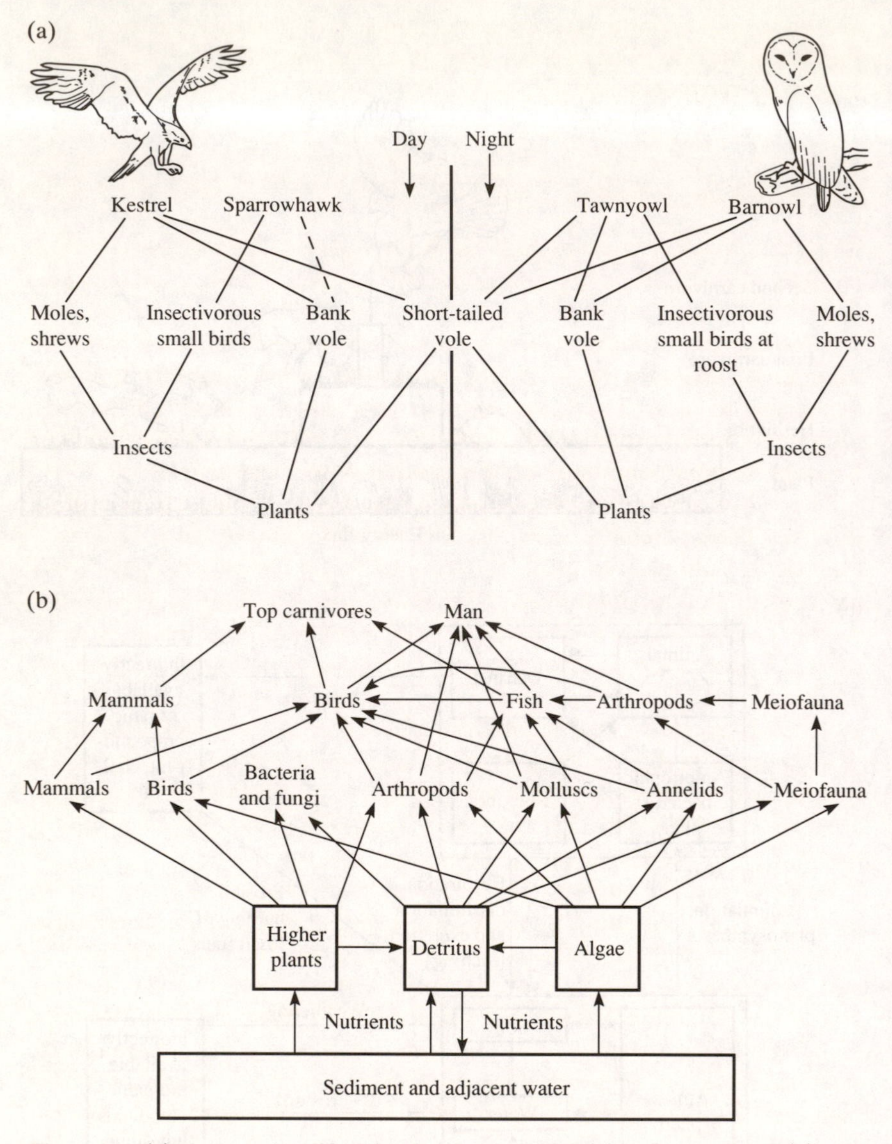

Fig. 1.2. (a) A very simplified food web of an English oak wood and (b) salt marsh, which illustrates the potential complexity of community interactions.

subset of the available resources. It cannot be assumed that the relative position, role and resources vacated by a species will necessarily become available to another species which appears capable of exploiting the same prey or resource item. Thus, since toxic pollutants will affect the fitness and competitive ability of species differentially, a knowledge of the pre-impact community ecology is only of limited use in predicting the likely impact of a pollutant and assessing the ability of a disrupted system to recover.

Environmental law

Many countries have responded to the environmental problems caused by industrial development and population growth by introducing significant changes in environmental legislation. National laws relating to environmental protection are normally dispersed between a large number of different Acts dealing with land, water and air. It is not necessary to describe these Acts in this book. Those particularly interested in environmental legislation are referred to the Further Reading section containing recommended books and articles.

There are a number of key issues and trends which are useful to discuss. These include consideration of national pollution agencies, integrated pollution control and the philosophy behind setting limits on environmental discharges. The discussion will largely be limited to the United States and to the United Kingdom as a member state of the European Community. The European Community influences environmental legislative initiatives of member states through the issue of European Community Directives based on the Community's 'Environmental Action Programmes'. The Single European Act, which amended the Treaty of Rome in 1987, and the Maastricht Summit in 1991 raise the prospect for increased and more demanding legislation originating in the European Community.

Environmental problems do not respect national boundaries, e.g. acid rain (Ch. 11), and may even have a global effect, e.g. greenhouse gases and chlorofluorocarbons (CFCs) emissions. International treaties and conventions, often developed under the auspices of the United Nations, e.g. United Nations Conference on Environment and Development (the Earth Summit) in Brazil, June 1992, may play an increasingly significant role in environmental protection. Some discussion of International Law relating to environmental issues will be given.

Integrated pollution control and national agencies

Integrated pollution control (IPC) recognizes that air, land and water are interrelated systems. Thus, controls on particular pollution sources must recognize the cross-media dimension. In reality this means that pollution control requires a balance to be struck between the three receiving media, e.g. lower water quality against cleaner air. A practical example of the conflict that may arise may be found in the use of gas scrubbing equipment to reduce atmospheric emissions from an industrial chimney which produces liquid waste capable of polluting water (see Ch. 9).

United States The United States was in the forefront of adopting laws tackling environmental problems in an interrelated way. In 1969 and 1970 the United States addressed fully the concept of integrated pollution control by enacting a National Environmental Policy (1969) followed by the Environmental Protection Agency (EPA) (1970). In addition to this federal framework for pollution control, individual states legislate for environmental standards and are required to implement federal environmental regulations.

The EPA is involved in both the development and implementation of regulations to control pollution. For example, in 1984 the EPA issued a policy statement indicating that permit levels on toxicity could be imposed together with the requirement for toxicity reduction evaluation (TRE) (see below) to be performed. States were encouraged to require industrial plants suspected of discharging highly toxic waste to undertake TREs without necessarily having a permit limit on toxicity. In 1990 the EPA proposed that these recommendations should become binding regulations.

The EPA also has statutory responsibilities within acts dealing with environmental protection. Three examples are set out below.

1 Under the Federal Water Pollution Act (1977, reauthorized 1987), more commonly known as the Clean Water Act (CWA), the EPA is required to produce and publish recommended water quality criteria.

2 Under 1984 amendments to the Resource Conservation and Recovery Act (RCRA) (1976) the EPA is required to evaluate and if necessary prohibit land disposal of certain hazardous wastes. The regulations (40 CFR 268) became effective in 1986, and the EPA has (or will) establish treatment standards for land disposal of RCRA hazardous wastes.

3 The Clean Air Act (CAA) (1982) and its amendments require the EPA to issue two forms of air quality standards for the criteria pollutants: primary standards aimed at protecting public health and secondary standards protecting the public welfare from any known or anticipated adverse effects on environmental resources, i.e. water, wildlife, vegetation and property. A clear example of IPC.

United Kingdom The United Kingdom is a relative newcomer to the concept of IPC. New powers have recently been established under the Environmental Protection Act (1990) to allow Her Majesty's Inspectorate of Pollution (HMIP) to implement a framework for the purposes of IPC. HMIP has become a statutory pollution regulation agency covering those industrial processes harmful to the environment. Industrial plants releasing pollutants to the environment must obtain authorization from HMIP. This requires the use of 'best available techniques not entailing excessive cost' (BATNEEC) (see below) to control discharges. Consideration must be given to emission to land, air and water when determining the impact of a process on the environment as a whole. The 1990 Act embodies the principle that the polluter must pay to meet the cost of determining IPC applications. The HMIP, as an agency implementing the new system of IPC, works alongside local authority environmental health departments. These control releases to the atmosphere from lesser polluting processes. The IPC relates to larger processes.

At present there is considerable discussion within the United Kingdom about the development of a national environmental agency consolidating HMIP and National Rivers Authority (NRA) functions. The NRA was established in 1989 and is responsible for policing discharges to water including sewage and granting abstraction licences. The waste regulation functions currently held by local authorities would also be incorporated within the remit of the new agency.

Undoubtedly the role of IPC in the control of discharges to the environment will continue to grow. The development of any new biological technology for the treatment of potentially polluting solid waste, liquid effluent or gas emission must be constrained by the limitations imposed by IPC. It is unacceptable if pollution reduced for one of the receiving media results in a substantial increase in releases to another medium.

Concepts underlying the establishment of discharge limits

United Kingdom Led by European Community environmental policy and Directives, the United Kingdom is adopting precautionary, uniform and technology-based standards for pollution control. This is well-illustrated in the precepts which underlie the operation of HMIP.

In the Environmental Protection Act (1990) the control of air pollution from potentially polluting processes is based on the requirement of BATNEEC. This is a European Community concept introduced in the Directive on the combatting of air pollution from industrial plants (Framework Directive) of 1984 (84/360/EEC-OJ L188 16.7.84). This Directive requires that authorization for emissions can only be given if "all appropriate preventative measures against air pollution have been taken, including the application of best available technology, provided the application of such measures does not entail excessive costs". BATNEEC may be expressed in terms of emission standards or of specified hardware. HMIP normally express BATNEEC in terms of a performance standard, i.e. a technique which produces release levels of *x* or better, where *x* values are obtained from identified BATNEEC.

In view of their role in IPC a new concept has been introduced by HMIP, that is, the best practicable environmental option (BPEO). BPEO requires consideration of the potential effects of given industrial processes as a whole on air, water and land (health, flora, fauna, buildings, etc.). The impact of possible accidents must be considered together with eventual plant decommissioning, plant disposal and land restoration. Not only must assessment of risks be defined by scientists but they should take account of the public's perception of risk. The option chosen to minimize pollution is not dictated, but is rather an examination of alternative ways of reducing all possible forms of pollution from a plant. BPEO is relatively open-ended. Changes in scientific knowledge and practical experience mean that BPEO must be kept under review and the effects of decisions must be monitored.

United States In the United States, the EPA defines federal air emission limitations based on the concept of best available control technology (BACT). BACT requirements are based on an evaluation of technology performance and have been set for a variety of pollutants. In effect then, BACT requirements set national technology-limited emission requirements. Similar criteria are used to set liquid effluent and other waste treatment standards for pollution reduction. The concept behind BACT is broadly similar to that of BATNEEC.

The United States also has similar types of requirements to BPEO. For example, under the CWA individual states are required to control the release of toxic (priority) pollutants. This may necessitate not only the reduction or even elimination of toxic materials, e.g. organics and heavy metals, but may require some industries to undertake a toxicity reduction evaluation (TRE) at particular sites. A TRE varies with state. In some it is not only considered to be the identification of potential pollution sources, but also an action-orientated programme providing and implementing solutions. In others TRE is regarded as a mechanism for documenting progress towards pollution reduction.

The above discussion clearly does not examine all the concepts behind environmental protection. It does illustrate some concepts in environmental law which biotreatment processes must accommodate and indicates the legal significance of improved scientific knowledge on the impact of environmental pollutants.

International law and environmental protection

The Stockholm Conference on the Human Environment (1972) saw the establishment of the United Nations Environment Programme

(UNEP). Since the foundation of UNEP the United Nations has taken an increasingly important role in the development of international initiatives and treaties covering environmental control. For example, United Nations initiatives in the area of CFC production and the protection of the ozone layer led to the Montreal Protocol in September 1988 and the agreement of more stringent measures at the London Conference on the Montreal Protocol in July 1989. The United Nations Conference on Environment and Development (Brazil, 1992) resulted in conventions on biodiversity and climate change.

The terms of these conventions and treaties, however, only form part of the law of individual countries when they are ratified or otherwise accepted into the law of that country or state. Until this has happened there are considerable difficulties in enforcement. This is particularly true in the absence of any international enforcement agency. Recourse to the courts is of limited value since the International Court of Justice is not really a practical option. Administrative mechanisms may offer the best route to achieve compliance by signatories. Some international treaties and conventions establish a permanent secretariat which undertakes a regular review of compliance. As the global issues of pollution control become more evident it is to be hoped that appropriate treaties will be developed and adhered to by signatory countries.

References

FREEDMAN, B. (1989) *Environmental Ecology. The Impacts of Pollution and Other Stresses on Ecosystem Structure and Function*. Academic Press, San Diego.

HOLDGATE, M.W. (1979) *A Perspective of Environmental Pollution*. Cambridge University Press, Cambridge.

HOPKIN, S.P. (1989) *Ecophysiology of Metals in Terrestrial Invertebrates*. Elsevier Applied Science, London, New York.

HUNTER, B.A., JOHNSON, M.S. and THOMPSON, D.J. (1987a) Ecotoxicology of copper and cadmium in a contaminated grassland ecosystem. I. Soil and vegetation contamination. *J. Appl. Ecol.*, **24(2)**, 573–586.

HUNTER, B.A., JOHNSON, M.S. and THOMPSON D.J. (1987b) Ecotoxicology of copper and cadium in a contaminated grassland ecosystem. II. Invertebrates. *J. Appl. Ecol.*, **24(2)**, 587–600.

HUNTER, B.A., JOHNSON, M.S. and THOMPSON D.J. (1987c) Ecotoxicology of copper and cadium in a contaminated grassland ecosystem. III. Small mammals. *J. Appl. Ecol.*, **24(2)**, 601–614.

HUNTER, B.A., JOHNSON, M.S. and THOMPSON D.J. (1989) Ecotoxicology of copper and cadium in a contaminated grassland ecosystem. IV. Tissue distribution and age accumulation in small mammals. *J. Appl. Ecol.*, **26(1)**, 89–100.

KEITH, L.H. and TELLIARD, W.A. (1979) ES & T special report: Priority pollutants I – a perspective view. *Environ. Sci. Techn.* **13**, 416–23.

LOTHIGIUS, J. (ed.) (1991) Toxic organic compounds in the Baltic. Special Issue. *Environment*, Vol. **10**. Publication of the Swedish Environmental Protection Agency.

MASON, C.F. (1991) *Biology of Freshwater Pollution*, 2nd edn. Longman Scientific and Technical, Harlow.

Further reading

BALL, S. and BELL, S. (1991) *Environmental Law*. Blackstone, London.

BARAM, M.S. and PARTAN, D. (1992) *Corporate Disclosure of Environmental Risks*. Butterworths, London.

BERGESEN, H.O, NORDERHAUG, M. and PARMANN G. (1992) *Green Globe Yearbook 1992*. Oxford University Press, Oxford.

CHURCHILL, R., WARREN, L. and GIBSON, J. (1991) *Law Policy and the Environment*. Blackwell, Oxford.

FREEDMAN, B. (1989) *Environmental Ecology. The Impacts of Pollution and Other Stresses on Ecosystem Structure and Function*. Academic Press, San Diego.

HM GOVERNMENT WHITE PAPER (1990) *This Common Inheritance. Britain's Environmental Strategy*. HMSO, London.

HOLDGATE, M.W. (1979) *A Perspective of Environmental Pollution*. Cambridge University Press, Cambridge.

HOPKIN, S.P. (1989) *Ecophysiology of Metals in Terrestrial Invertebrates*. Elsevier Applied Science, London, New York.

HUGHES, D. (1992) *Environmental Law*, 2nd edn. Butterworths, London.

LANKFORD, P.W. and ECKENFELDER JR, W.W. (eds) (1990) *Toxicity Reduction in Industrial Effluents*. Van Nostrand Reinhold, New York.

LOMAS, O. and McELDOWNEY, J. (eds) (1991) *Frontiers in Environmental Law*. Chancery, London.

LOTHIGIUS, J. (ed.) (1991) Toxic organic compounds in the Baltic. Special Issue. *Environment*, Vol. **10**. Publication of the Swedish Environmental Protection Agency.

MOGK, J. and LEPLEY JR., F.J. (1990) The evolving regulation of co-generation (CHP) in the United States. *Util. Law Rev.* **1**, 44–55.

SALTER, J. (1992) *Corporate Environmental Responsibility*. Butterworths, London.

VAUGHAN, D. (1992) *Environment and Planning Law in the EC*. Butterworths, London.

Chapter 2

ASSESSING THE ENVIRONMENTAL FATE AND POTENTIAL IMPACT OF POLLUTANTS

Introduction

Ecotoxicology is an interdisciplinary science concerned with investigating and quantifying the environmental fate and possible adverse effects of pollutants (Brouwer *et al.*, 1990). In many ways it may be viewed as a development or extension of traditional toxicology. Although environmental and ecosystem level responses and impacts are the primary focus of ecotoxicology, the assessment procedures commonly used to evaluate the potential environmental risks posed by a chemical are largely based on a consideration of its physico-chemical properties and toxicity in simple single-species toxicity tests (Moriarty, 1990). In part this reflects the nature of the subject and the demands placed on it. If assessments of the ecotoxicities of compounds are to be embodied effectively in regulatory legislation and used for the setting and monitoring of discharge consents, assessment procedures must be easily repeatable, reliable and allow unambiguous interpretation. However, such procedures, e.g. standard single-species toxicity tests, will not increase our understanding of the mechanisms and processes which determine the environmental fate and impact of a pollutant. The need for simple inexpensive assessment procedures can be readily understood when the scale of the problem is appreciated. It has been estimated that world-wide over 63 000 chemicals are in common use and that between 200 and 1000 new synthetic chemicals are marketed each year (Moriarty, 1990). For many of these substances little, if anything, is know of their possible ecological effects. Our experience of organochloride pesticides and more recently with polychlorinated biphenyls (PCBs) illustrates the need to assess chemicals more closely prior to their widespread use or release into the environment. (Ch. 1).

To be useful, prior market or pre-release screening must take account, not only of the toxicity of a compound to humans and other species, but also of the quantity likely to be manufactured, the proportion of this which may leak or be directly released into the environment and the probable environmental fate of a compound once in the environment. The extent of releases and their possible ecological effects must be considered throughout the foreseeable life cycle of the product. This should include an assessment of the threats posed by degradation products and the ultimate disposal of the original product.

The underlying rationale of current ecotoxicity assessments is deceptively simple (Fig. 2.1, outline of product cycle in Solbé, 1988). Estimates are made of the quantities of a compound likely to enter the environment. Then based on a knowledge of the chemical properties of the compound, which in some cases can be supplemented by knowledge of the environmental behaviour of chemically-similar compounds and

Table 2.1 Effect of amount of chemical within the European market on the initial amount of information needed for hazard assessment (Moriarty, 1990)

Quantity marketed (tonnes) *Per annum*	 *Total*		*Amount of* *information needed*
<1	–		Limited announcement
1 or more	<50		Base set
10 or more	50 or more	Full notification	Base set *or* Level I
100 or more	500 or more		Level I
1000 or more	5000 or more		Level II

Base set:

(1) The identity of the chemical, e.g. structural formula, spectral data. Impurities are usually ignored, both at this stage and in any subsequent assessments.

(2) Information about the chemical, i.e. uses, amounts and how to handle.

(3) Physico-chemical properties, including the partition coefficient between *n*-octanol and water.

(4) Toxicology.

(5) Ecotoxicology.

(a) The acute toxicity (LC_{50}) for one species of fish.

(b) The acute toxicity (LC_{50}) for one species of *Daphnia*.

(c) The rate of biotic and abiotic degradation processes, measured in standardized aerobic conditions. Anaerobic tests may be more appropriate for some chemicals, and specified by some national authorities.

Level I tests

(a) A longer toxicity study, of at least 14 days, for one species of fish.

(b) A longer toxicity test, of at least 21 days, for one species of *Daphnia*.

(c) A prolonged biodegration study.

(d) A test of 'growth inhibition', i.e. of effect on the rate of cellular division, in one algal species.

(e) A test of bioaccumulation, usually with a species of fish.

(f) A toxicity test with one species of earthworm.

(g) A toxicity test with a 'higher' plant. *Lemna*, a small free-floating aquatic species, is commonly used. This meets the requirement, but cannot be regarded as a very representative genus.

the results of experimental studies (e.g. degradation studies), predictions are made on the likely environmental fate of the compound, i.e. its distribution between, and concentration within, the major environmental compartments or phases – water, atmosphere and soil. The possible hazards posed by a given concentration of the compound in a particular compartment are assessed in the light of results of single- or multispecies toxicity tests.

An example of this approach is the system of hazard assessment ordained by current European Community legislation (Council Directive, 79/831/EEC). This Directive specifies the amount of information required from a manufacturer during product notification prior to marketing (Table 2.1). As the quantity of the compound to be marketed increases, so the scope of information required increases. The use of aquatic species and systems is rationalized on the basis that sewage effluent and direct industrial discharges into aquatic systems frequently represent major routes of pollutant entry into the environment (Fig. 2.1). For aquatic systems predicted environmental concentrations are compared with the lowest LC_{50} value, (i.e. the lowest concentration of the compound to cause the death of 50% of the test population). If there is found to be an adequate safety margin between this and the predicted

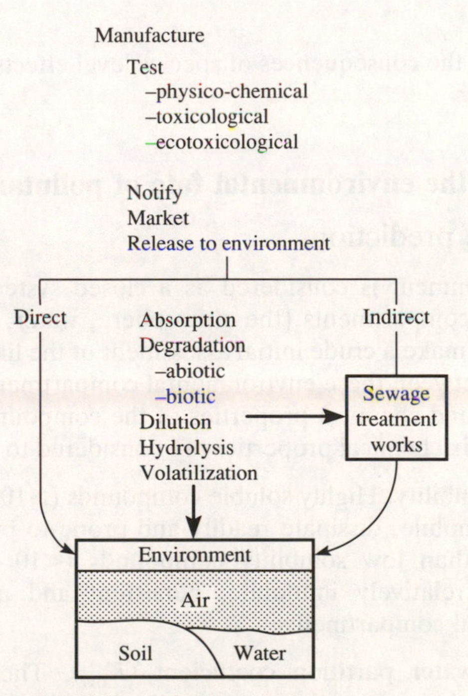

Fig. 2.1. Fate of a chemical (from Solbé, 1988).

concentration (frequently taken to be a 100- or 1000-fold difference), the proposed uses and quantities of the compound are deemed not to present a significant environmental hazard. If the chemical is considered to present a major risk to terrestrial systems, either because of its mode of use or because of its physico-chemical properties, supplementary test species may be specified. For example, the United Kingdom Control of Pesticides Regulations (1986), require that for pesticides used outdoors when crops, fruit trees and weeds are in flower data on the acute oral toxicity of the compound to a species of bird (not domestic hens) and honey bees must be obtained. Although Level I considerably extends the base line of ecotoxicological tests (Table 2.1), the quantity and quality of information obtained is still insufficient to make reliable or accurate predictions of the possible ecological effects of a compound.

In order to predict the ecological consequences of a particular pollutant, four fundamental questions need to be answered (Holdgate, 1979; Ramade, 1987; Moriarty, 1990).

1 Where and at what concentration in the environment is the pollutant likely to occur?

2 What is the relationship between these concentrations and the amounts taken up and accumulated by organisms?

3 What are the effects of accumulated pollutant on individual organisms?

4 What are the consequences of species level effects on the ecosystem as a whole?

Predicting the environmental fate of pollutants

Qualitative predictions

If the environment is considered as a closed system containing four basic linked compartments (the atmosphere, water, soil and biota), it is possible to make a crude initial assessment of the likely distribution of a chemical between these environmental compartments based solely on the physical and chemical properties of the compound (Ney, 1990). In this context six chemical properties are considered to be of importance.

1 Water solubility. Highly soluble compounds (>1000 ppm) are likely to be more mobile, dissipate readily and prone to biodegradation and metabolism than low solubility compounds (<10 ppm), which will tend to be relatively immobile, persistent and accumulate within environmental compartments.

2 Octanol–water partition coefficient (K_{ow}). The K_{ow} value of a compound is the ratio of the concentration of the chemical in *n*-octanol to that in water at equilibrium for a system containing octanol and an aqueous solution of the pollutant. It is effectively a measure of the 'lipophilicity' of the chemical, i.e. the lipid solubility of the compound relative to its solubility in water. The rate of passive transport across cell membranes for organic chemicals is positively correlated with their lipid solubilities. In addition, organisms frequently contain large amounts of relatively metabolically inert lipid, within which lipid soluble compounds may accumulate (Moriarty, 1990). Thus, octanol : water partition coefficients are potentially valuable for assessing the likelihood that a compound will become concentrated within the biota. Numerous studies have demonstrated positive relationships between the K_{ow} of a compound and its toxicity (Samiullah, 1990). Ney (1990) suggests that chemicals with $K_{ow} < 500$ are unlikely to bioaccumulate, while those with values ≥ 1000 are likely to be bioaccumulative, be sorbed in soils and be environmentally persistent.

3 Rates of hydrolysis. Hydrolysis is frequently the major mechanism involved in the breakdown and degradation of chemical pollutants. It may be predicted on the basis of chemical structure, in practice, because it is affected by many factors including temperature, pH, solubility and volatility. It is best determined experimentally. The rate of hydrolisation is normally defined as the length of time required for the hydrolysis of half of the original amount (t_{50}) to occur. It is thus effectively a measure of the potential persistence of a pollutant. Compounds with calculated or empirically determined hydrolytic t_{50}s of less than 30 days are unlikely to pose a significant environmental hazard due to their relatively short environmental lifespan unless they occur at sufficiently high concentrations to be acutely toxic or the products of hydrolysis are hazardous.

4 Photolysis. A chemical capable of sorbing light energy may be subject to phototransformation. Phototransformation, e.g. photolysis, particularly in aquatic systems, can act as an important sink for the pollutant (Samiullah, 1990). The rate of photolysis, i.e. the time required for half of the original compound to be transformed, as with the rate of hydrolysis, may be used as a measure of the likely persistence of the compound.

5 Volatilization. The volatility of a chemical may be crudely assessed from its partial pressure under standard conditions. Compounds with high vapour pressures (>0.01 mmHg), i.e. volatile compounds, are likely to be rapidly dispersed by and accumulate within the atmospheric component of the environment.

6 Soil sorption and leaching. The tendency of a chemical to become adsorbed within soils or sediments can be assessed experimentally by determining the equilibrium concentration of the chemical in the solid and liquid phases. The soil sorption or partition coefficient (K_a) is equal to the ratio of the concentration in the soil (sediment) to that in water. The leaching potential of a chemical once associated/adsorbed within a soil or sediment may be assessed from simple extraction experiments or thin layer soil chromatography (Helling and Turner, 1968).

The potential use of this type of information is illustrated by the scheme shown in Fig. 2.2, which allows a rapid qualitative assessment of the likely fate of a chemical to be made.

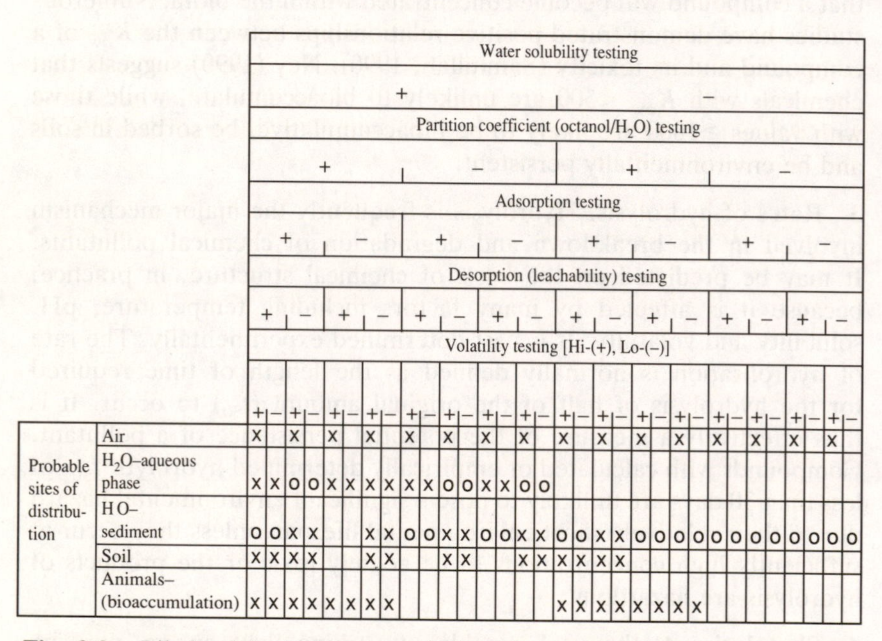

Fig. 2.2. Likely sinks for compounds based on their chemical affinities for air, water organic solvents and sediment (from Stern and Walker, 1978).

Quantitative predictions

Quantitative predictions depend on the development and use of models. Because of the scarcity of information these are by necessity extremely simple representations of the environment. The modelling process involves the construction of a simplified representation of the real world. In any modelling process there is a potential conflict between the need to produce a realistic (mechanistic) representation of the system being modelled and the need to minimize complexity so that the model is amenable to analysis and the results capable of unambiguous interpretation. A balance must be struck between complexity and realism. The level of complexity or realism incorporated into a model should be governed by the quality of data available, the use to which the model is to be put, e.g. how detailed and what type of predictions are required, and how the model is to be validated (Jeffers, 1982; Keen and Spain, 1992). The complexity of any model, be it a physical or mathematical representation, is largely governed by the number of processes and compartments included within the model boundary. As the number of compartments increases, the number of possible interactions, feedback and forward loops increases dramatically, as does the amount of information required to quantify the various fluxes and compartment sizes (Samiullah, 1990; Farmer and Rycroft, 1991). A realistic, but complicated multicompartmental model which is inadequately parameterized, i.e. if transfer rates, fluxes and compartment capacities are poorly defined, is of less value than a simple well-defined model, the behaviour of which can be adequately described and understood.

In the present context where the aim is to predict the environmental fate of chemicals prior to the occurrence of significant emissions, initial models will be by necessity extremely simple. The modeller will have access to a very limited set of data, often containing little reliable information on the environmental behaviour of the compound. The primary aim of the modelling strategies discussed in this chapter is to predict the potential environmental distribution (PED) and the potential environmental concentration (PEC) from data available prior to the release and marketing of a compound. Once this is achieved it then becomes feasible to estimate the exposure of a target organism.

When developing compartmental models the material balance of the system should fulfil the conditions that:

Total amount of material in compartments	=	Material added/ entering compartments (sources)	=	Material dissipated/ leaving compartments (sinks)

Thus, at equilibrium or steady state the concentration of the material (pollutant) in each compartment and the total amount of material (including derivatives if degradation is incorporated into the model), will, in a closed system, be constant. In an open system the total amount need not be constant, although at steady state the concentration within individual compartments should again be constant. Three distinct but related approaches are commonly used during the mathematical formulation of compartmental models.

1 Equilibrium or partition models. In this approach the distribution of the pollutant between the various compartments at equilibrium is quantified in terms of partition coefficients (e.g. K_a). If A and B represent two compartments, e.g. sediment and water, f_1 and f_2 represent the flow of material between the two, e.g. if A represents sediment, then f_1 and f_2, respectively, represent the rates of desorption and sorption of the compound.

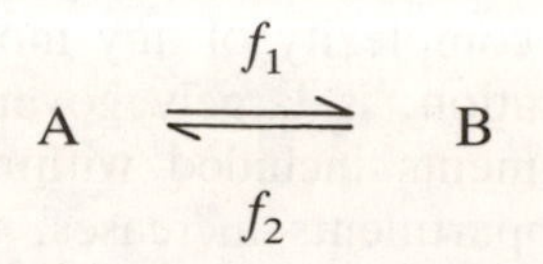

At equilibrium the concentration of pollutant in A and B will be constant, implying that $f_1=f_2$. If these fluxes are considered as being a direct function of the pollutant concentrations, i.e. show first-order kinetics, then:

$f_1 = k_1[A]$ and $f_2 = k_2[B]$

since $f_1 = f_2$ and $k_1[A] = k_2[B]$

$$\frac{k_2}{k_1} = \frac{[A]}{[B]} = K$$

where k_1 and k_2 are rate constants and K is the partition coefficient, which in this case, as we have seen before, is equal to the equilibrium concentration of the compound in the sediments divided by the concentration in the water.

2 Kinetic models may be applied to both open and closed systems where the total amount of the compound present need not be constant. Material may be transported out of the system or lost as a result of degradation (Walker, 1987). In kinetic models, flows are formulated in terms of rate equations. In most circumstances first-order kinetics have proved adequate (Baughman and Lassiter, 1978; Samiullah, 1990). Using this approach Neely and Blau (1977) successfully modelled the distribution of chlorpyrifos, an organophosphorus insecticide, between three environmental compartments – fish, water and soil plus plants – following a single application of the insecticide to a simple pond system (Fig. 2.3). In models of this type, concentration (C) of the pollutant in any compartment (x) is obtained from equations with the general form:

$$\frac{dc_x}{dt} = I_x + \sum_{i}^{n} k_i C_i - C_x \sum_{j}^{n} k_j$$

where I_x is the direct input of compound into compartment x, n the number of compartments, k_i the rate constant for flux/process i, $\sum_{i}^{n} k_i C_i$

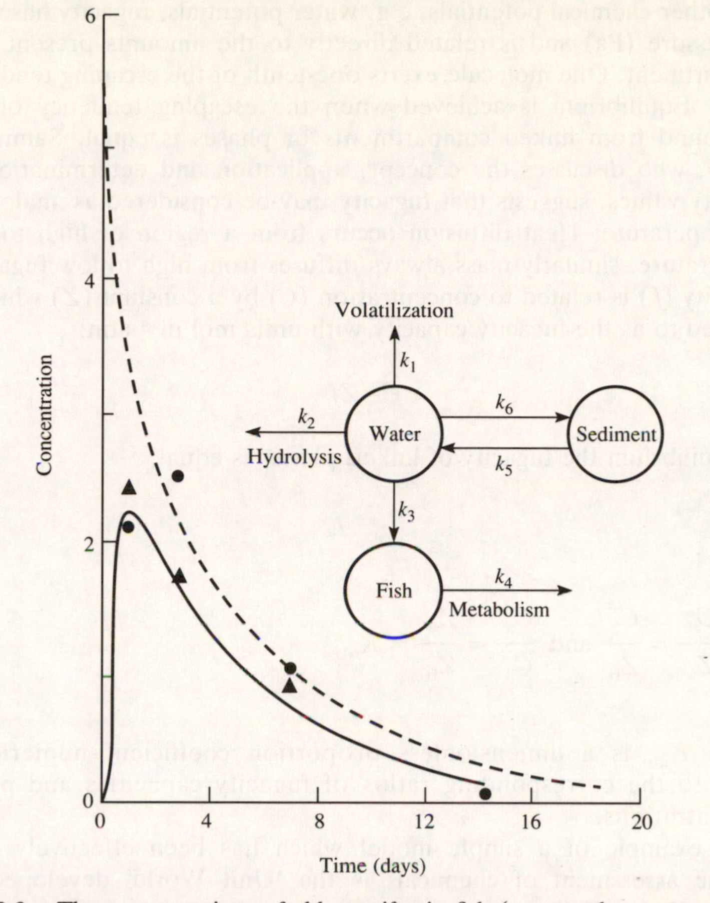

Fig. 2.3. The concentrations of chlorpyrifos in fish (expressed as parts × 10^{-6} and water (expressed as parts × 10^{-9} during the first 3 weeks after application. (●) Observed concentrations in fish; (▲) observed concentrations in water; (—) estimated concentrations in fish; (– –) estimated concentrations in water. These estimates are derived from the model shown (from Neely and Blau, 1977).

the sum of inputs into compartment x and $C_x \sum_{j}^{n} k_j$ the sum of outputs from compartment x.

Steady state conditions, where compartment concentrations are constant, i.e. $dC_x/dt = 0$, only occur when the flows into and out of a compartment are equal.

3 Fugacity models are based on consideration of the chemical potential of the substance present in each phase or compartment. In simplistic terms fugacity may be viewed as a measure of the 'escaping' tendency of the compound from one particular phase. As with other chemical potentials, e.g. water potentials, fugacity has units of pressure (Pa) and is related directly to the amounts present in a compartment. One molecule exerts one-tenth of the escaping tendency of 10. Equilibrium is achieved when the escaping tendency of the compound from linked compartments or phases is equal. Samiullah (1990), who discusses the concept, application and determination of fugacity values, suggests that fugacity may be considered as analogous to temperature. Heat diffusion occurs from a region of high to low temperature; similarly mass always diffuses from high to low fugacity. Fugacity (f) is related to concentration (C) by a constant (Z) which is referred to as the fugacity capacity with units mol m^{-3} atm:

$$C = Zf$$

At equilibrium the fugacity of linked phases is equal,

$$f_a = f_a$$

$$\text{thus } \frac{C_a}{Z_a} = \frac{C_b}{Z_b} \text{ and } \frac{C_a}{C_b} = \frac{Z_a}{Z_b} = K_{ab}$$

where K_{ab} is a dimensionless proportion coefficient numerically equal to the corresponding ratios of fugacity capacities and phase concentrations.

An example of a simple model which has been effectively used for the assessment of chemicals is the 'Unit World' developed by Neely (1982). The model contains six compartments, the volumes of which were selected to mimic the real environment as far as possible (Fig. 2.4). The mathematical formulation is based on estimated partition coefficients:

$$\frac{C_a}{C_w} = H \text{ (Henry's constant)}$$

$$\frac{C_s}{C_w} = K_a \text{ (soil : water sorption partition coefficent)}$$

$$\frac{C_f}{C_w} = BCF \text{ (bioconcentration factor, i.e. partition coefficient between biota and water)}$$

where C is equal to the concentrations in water (w), soil (s), fish (f) and air (a). The partition coefficients may be empirically determined or

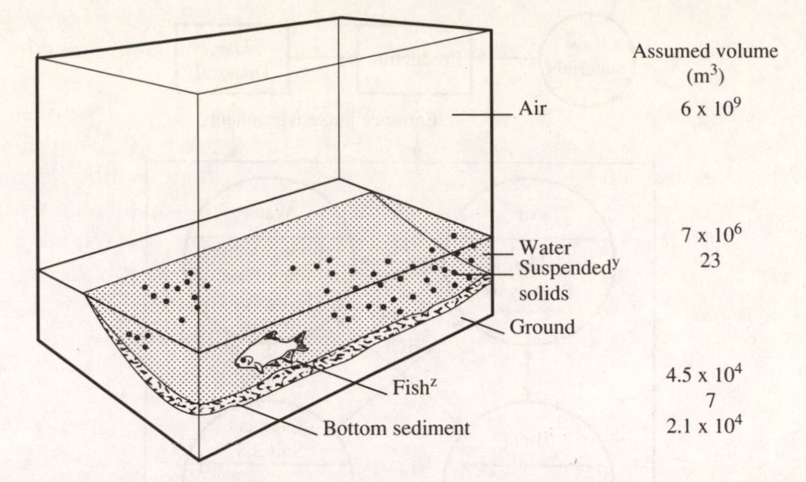

Fig. 2.4. A schematic representation of the 'Unit World' used to model the major environmental compartments (from Neely, 1982).

calculated directly from molecular weights, vapour pressures and water solubilities using the equations set out below.

$$H = PM\ 16.04/TS$$

$$\log BCF = 0.85 \log K_{ow} - 0.70$$

$$K_a = \%\text{organic}(0.6 K_{ow})$$

$$\log K_{ow} = 6.5 - 0.89(\log S/M) - 0.015(mp)$$

where P is the vapour pressure (mmHg), M the molecular weight, T the absolute temperature, S the water solubility in g m^{-3}, K_{ow} the octanol : water partition coefficient, mp the melting point in °C, K_a the soil : water partition coefficient and %organic is the percentage of organic carbon in soils, assumed to be 2% in soils, and 40% for bottom sediments.

Note that both the bioconcentration factors and the soil : water partition coefficient are considered to be functions of the octanol : water partition coefficient, which in turn may be estimated from the water solubility of the compound, its molecular weight and melting point. The calculation of partition coefficients from chemical data alone is discussed by Lyman *et al*. (1982). Although the 'Unit World' model has been successfully used to assess the water contamination potential of 65 classes and 129 specific chemicals, the model is only applicable to organic compounds with certain well-defined chemical structures. It is not appropriate for inorganics, polymers or formulations of chemicals

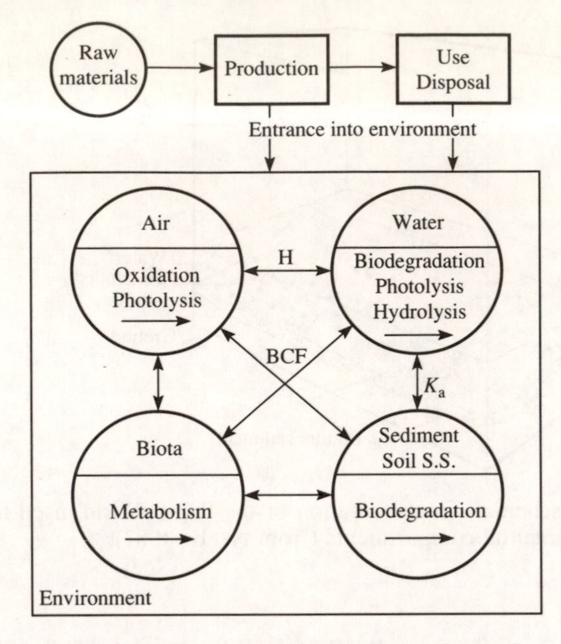

Fig. 2.5. Schematic diagram of transfer and transformation processes for chemicals in the environment. *H*, Henry's constant; *BCF*, bioconcentration factor in fish; K_a, soil sorption coefficient (Yoshida *et al.*, 1983).

(Neely, 1982). The 'Unit World' model is only one of several, with a similar basic structure outlined in Fig. 2.5. While such models have proved useful during the initial stages of hazard assessment, their resolution and precision is clearly limited. Precision can only be achieved by the development of more specific, complicated and non-generic models. For example, in the type of generic environmental models described above, six major compartments and ten major sink processes are sufficient to describe the general environmental fate of a compound (Fig. 2.5). A more detailed but still restricted model, dealing only with the fate of a pollutant in the aquatic system, requires many more compartments (Fig. 2.6).

Biological models

In addition to mathematical models, a diverse range of physical models has been used to investigate the potential environmental fates and impacts of pollutants. Pritchard (1982) distinguishes two broad classes of model ecosystem studies: microcosm studies in which

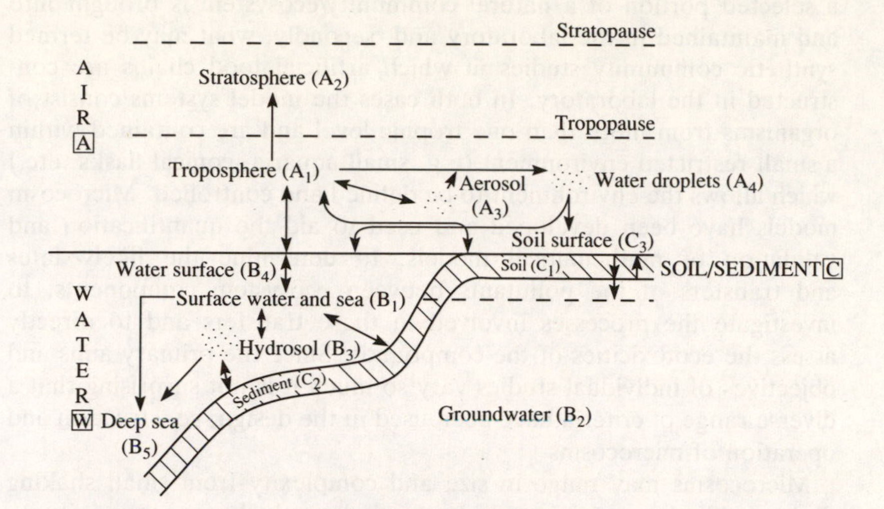

Major pathways of chemicals

	Pathways between sectors[a]	Mechanism	*pc or env.*[b]
	$A_1 \rightarrow A_2$	Turbulent diffusion	env.
	$A_1 \rightarrow A_4$	Dissolution gas → water	pc
	$A_4 \rightarrow A_1$	Volatility from aqueous solution	pc
A	$A_1 \rightarrow A_3$	Adsorption gas → aerosol	pc
	$A_3 \rightarrow A_1$	Volatility from adsorbed state	pc
	$A_3 \rightarrow A_4$	Condensation of water on aerosol	env.
A	$A_1 \rightarrow B_4$	'Dry deposition' (dissolution gas → water)	pc
	$A_1 \rightarrow C_3$	'Dry deposition' (adsorption)	pc
↓	$A_3 \rightarrow C_3 + B_4$	'Dry deposition' (of particles)	env.
B	$A_4 \rightarrow C_3 + B_4$	Rain	env.
+	and		
C	$A_3 \rightarrow A_4 \rightarrow C_3 + B_4$ (direct)	'Raining out'	pc
	$B_4 \rightleftharpoons B_1$	Turbulent diffusion and mixing	env.
	$B_1 \rightarrow B_5$	Turbulent diffusion	env.
B	$B_1 \rightarrow B_3$	Adsorption	pc
	$B_3 \rightarrow B_1$	Desorption	pc
	$B_3 \rightarrow B_5$	Sedimentation	env.
	$B_1 \rightarrow B_2$	Adsorption	pc
	$B_2 \rightarrow B_1$		
B	$B_4 \rightarrow A_1$	Volatility from aqueous solution	pc
↓	$B_4 \rightarrow A_2$	'White capping'	env.
A	$B_1 \rightarrow C_2$	Adsorption	pc
+			
C	$B_3 \rightarrow C_2$	Sedimentation	env.
	$C_3 \rightarrow C_1$	Adsorption/desorption ('leaching')	pc
C			
	$C_1 \rightarrow C_3$	Adsorption/desorption	pc
C	$C_1 \rightarrow B_2$	'Leaching'	pc
↓	$C_3 \rightarrow A_1$	Volatility from adsorbed state, aqueous solution (moist soil) and pure substance	pc
A			
+			
B			

[a]For notation of environmental media and sectors, reentry from stratosphere, deep sea and some other quantitatively less important pathways have been neglected.
[b]pc: Determined by physico-chemical properties of the substance and environmental limiting factors.
env.: Determined by environmental (e.g. meteorological) factors only: in some cases the assignment is still ambiguous.

Fig. 2.6. Important pathways for organic chemicals in the environment (from Klöpffer *et al.*, 1982).

a selected portion of a natural community/ecosystem is brought into and maintained in the laboratory and, secondly, what may be termed synthetic community studies in which artificial food chains are constructed in the laboratory. In both cases the model systems consist of organisms from more than one trophic level and are contained within a small restricted environment (e.g. small aquaria, conical flasks, etc.) which allows the environment to be defined and controlled. Microcosm models have been developed and used to aid the quantification and validation of mathematical models, to determine the likely fates and transfers of the pollutants between ecosystem components, to investigate the processes involved in these transfers and to directly assess the ecotoxicities of the compounds. Since the primary aims and objectives of individual studies vary so much it is not surprising that a diverse range of criteria have been used in the design, construction and operation of microcosms.

Microcosms may range in size and complexity from small shaking flasks (500 ml) containing seawater and a plankton community to systems containing 2 m high tree saplings, associated forest fauna and flora growing on intact natural blocks of forest floor and soil contained in large fibreglass boxes, and maintained in a greenhouse. In some cases, microcosm systems may be linked, for example, the leachate from a soil–plant system may be drained directly into an aquarium-based aquatic microcosm allowing the fate of a pollutant to be followed through both a terrestrial and aquatic system. The division between systems which are described as microcosm and those described as mesocosm or macrocosm studies is blurred, but, in general, microcosms are sufficiently small to be easily replicated, while mesocosm and macrocosm systems, because of their size, allow only limited replication and are often housed and constructed outside of the laboratory.

The use of microcosms enables, in theory, some of the structural complexity of real systems to be incorporated into the routine assessment of chemicals (Pritchard, 1982), partially bridging the gap between single-species experiments and field studies (Costello and Thrush, 1991). However, since they are, and can only be, simplified model representations of an ecosystem (or part of an ecosystem) it is important to appreciate their limitations. Pritchard (1982) argues forcefully for the need to calibrate laboratory microcosms against environmental conditions. It cannot be assumed *a priori* that a natural component removed from its original setting in the field and brought into the laboratory, will retain either its original structure or function. Ecosystems are normally considered to be essentially self-contained functional units capable of sustaining both their function (e.g. nutrient cycling and energy flow) and structure (e.g. species composition), within which a particular portion of the ecosystem is maintained by the processes of population growth, mortality, immigration and emigration. Few microcosms are capable of maintaining their structure or function in this way. Most require continued inoculation with organisms (i.e. artificial recruitment) or are maintained for only short periods of time. This is not surprising; contrary to popular belief, species diversity alone does not increase the inherent stability of a system. The results of a substantial body of both experimental and mathematical studies indicate that the conditions necessary to permit the stable co-existence of species become more exacting as the number of interacting species increases (Begon *et al.*, 1990). Clearly the ability of a microcosm to maintain itself must be considered when undertaking and interpreting the results of long-term microcosm-based toxicity studies. The results of such studies may well be affected by changes in the structure and functioning of the experimental system during the course of the study.

Because of the restriction on their size, microcosms cannot duplicate the complexity of real ecosystems or communities; some component species will always be excluded. In addition, due to their confinement, they will be subject to pronounced boundary effects. Similarly the ratios of the total amounts, volumes and contact surfaces between the various phases included in the model will be restricted by the size of the system to a limited and often unnatural range of values.

However, against these problems, which are largely associated with scaling, microcosms do allow the experimenter to work with multi-species systems which can be replicated. They enable processes and effects to be studied which cannot be observed in simpler single-species experiments or under field conditions, where because of the complexity and lack of control available information relevant to specific processes is difficult to isolate and extract. However, microcosms are best viewed not as functional models of natural systems but as distinct artificial systems which have characteristics analogous to and in common with undisturbed natural systems. With this in mind Pritchard (1982) convincingly argues that the primary use of microcosm studies should be as a verification tool for the validation of models used to predict exposure levels, toxicity testing protocols, for example, biodegradability tests and single-species bioassays, and to assess the potential pollutant assimilatory capacity of environmental components.

In contrast to microcosms, synthetic communities, such as that developed by Metcalf *et al.* (1971), have been specifically constructed and extensively used to assess the potential toxicities of a substantial number of organic pesticides (Metcalf, 1977; Pritchard, 1982; Moriarty, 1990). The basic aquatic/terrestrial Metcalf system consists of a sloping shelf of washed sand, 7 l of standard pond water containing the alga *Oedogonium cardiacum*, several planktonic species including *Daphnia magna*, *Culex pipiens* (mosquito larva) and *Physa* sp. (snail species) housed in a 10×10×10 inch glass aquarium. The upper exposed surface of the sand, representing the terrestrial component of the 'model ecosystem' is planted with *Sorghum vulgare* seeds. After 20 days, by which point the *Sorghum* leaves are approximately 10 cm high, a radiolabelled pesticide is added along with caterpillars (*Estigmene acrea*). The caterpillars serve two functions: they act as herbivores and their defecation products provide an additional route for the movement of the test compound into the soil and the aquatic portion of the model. After 30 days, three *Gambusis affinus* (fish), which feed on the *Culex*, are added and allowed to feed for a further three days, at which time the system is dismantled and the quantities of pesticides and derivatives are determined in each component of the model. Although this simple system allows at least eight distinct food chain links (Metcalf *et al.*, 1971), it is clearly an extremely artificial and ephemeral system, incapable of self-maintenance and has little in common with natural systems. It is best to consider synthetic communities of this type as multispecies toxicity test systems rather than as model ecosystems. Despite this, useful information on the degradation of pesticides and their tendency to become concentrated within the biota can be obtained. However, in most instances this information could have been readily obtained from simpler single-species studies. Moriarty (1990) concludes that synthetic community studies "appear to have the worst of both worlds: they are too complicated to give results that are easily interpreted, but too artificial to be of immediate relevance to field situations".

In mesocosm and macrocosm studies the larger volumes of natural ecosystems which may be enclosed, e.g. 1700 m^3 of seawater enclosed in plastic bags suspended from the water surface (Lundgren, 1985) or in outdoor pond systems (Crossland, 1988), ameliorate many of the problems associated with the restricted ability of microcosms to sustain and model natural ecosystem behaviour. However, with increased size and complexity, the ability of the experimenter to define, control and monitor the environment within the system is reduced. While their physical size and the costs of construction limit the degree of experimental replication, as a result their use is largely limited to the long-term investigation of ecological processes (Moriarty, 1990).

The relationship between exposure and the amount of pollutant accumulated by an organism

Three parameters are commonly used to describe the relationship between abiotic environmental concentrations and the amounts within organisms. Bioconcentration and bioaccumulation refer to the increase in pollutant concentration from water into aquatic organisms. The extent of bioconcentration or bioaccumulation is normally quantified as the concentration of the pollutant present in the organism on a wet weight basis, divided by the concentration in the surrounding water. This approach is essentially analogous to the concept of a physical partition coefficient. Although the terms bioconcentration and bioaccumulation are often used synonymously, they are technically distinct. Bioconcentration refers to the accumulation of the pollutant by direct uptake from water into the organism (invertebrate, fish, alga species, etc.), while bioaccumulation refers to the combined uptake of pollutants from food as well as water. Biomagnification refers to increases in pollutant concentration in the tissues of successive organisms along a food chain. Biomagnification is thus the only term applicable to both aquatic and terrestrial systems.

Compounds prone to bioaccumulation are frequently either lipophilic or analogues of essential nutrients. The tendency of many organic pollutants, e.g. organochloride pesticides and PCBs, to bioconcentrate is related to the rate at which the compound diffuses across the cell membrane. Diffusion across the outer membrane is partly a function of molecular size and the solubility of the compound in lipid, which may be assessed from the octanol : water partition coefficient (cf. p. 25). Chemical bioconcentration factors (*BCF*s) of lipophilic-persistent pollutants frequently correlate with their octanol : water partition constants. Similarly the body load of these compounds often correlates with the lipid content of the organism (Ramade, 1987; Samiullah, 1990; Walker, 1990). However the accumulation of extremely hydrophobic, i.e. strongly lipophilic compounds, may be limited by their tendency to become 'trapped' in external lipids (Moriarty, 1990). In addition to estimating *BCF*s from K_{ow} coefficients, they may be determined directly from field samples or more accurately during the course of acute toxicity tests. The latter procedure allows critical *BCF*s and tissue concentrations associated with mortality or other significant adverse affects (e.g. reduced reproductive output) to be determined.

In organisms where the uptake of the pollutant occurs via the ingestion of food and its subsequent transport across the gut wall (e.g. terrestrial animals), it is not appropriate to estimate chemical accumulation from K_{ow} values alone. Accumulation of the pollutant will be a function of the concentration of the pollutant in the food source, the rate and pattern of pollutant redistribution between body components and organs and the rate at which the pollutant is metabolized and excreted. Walker (1990) outlines two simple procedures which allow steady state *BCF*s of persistent pollutants in higher terrestrial animals (e.g. birds, mammals) to be readily estimated from either a knowledge of the biological half-life of the compound or from *in vitro* studies of the metabolism of liver preparations. Where the biological half-life (t_{50}) is known or can be experimentally determined,

*BCF*s may be estimated from the following proposed relationship (Walker, 1987):

$$BCF = \frac{2A_f(t_{50})}{N}$$

where A_f is the fraction of ingested pollutant which is absorbed, N the number of days required for an animal to consume a weight of food equal to its own body weight and t_{50} the initial biological half-life of a compound in an animal.

Many organic pollutants, particularly organochlorine compounds, bioaccumulate in the livers of predators (e.g. owls) and scavengers (e.g. corvids). The liver is frequently the primary site of pollutant detoxification and metabolism (Walker, 1990). Walker (1987) has shown that the ability of *in vitro* liver microsomal preparations to metabolize a dieldrin analogue (HCE) could be used to predict rates of *in vivo* HCE metabolism and excretion. If it is assumed that (i) the rate of excretion of unchanged pollutant is negligible and that the rate of loss approximates to the rate of primary metabolism and (ii) that the metabolism of the pollutant occurs primarily in the liver, then under these conditions the rate of adsorption will be equal to the rate of metabolism (Fig. 2.7). In this simple model, the amount

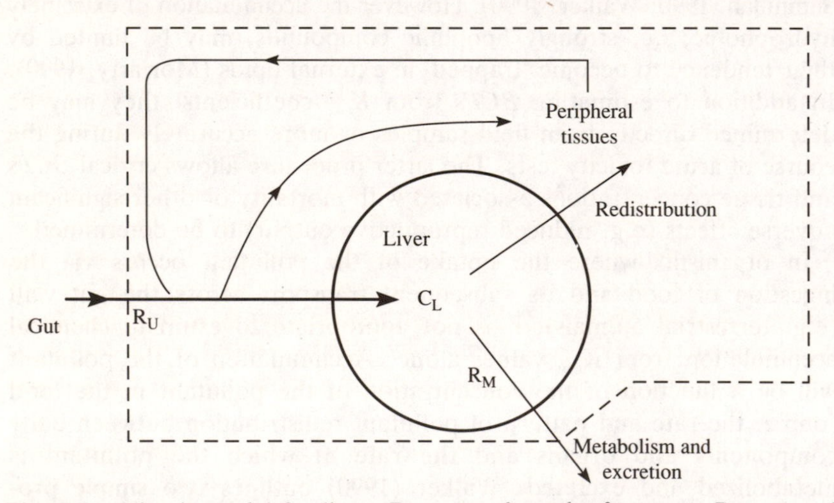

Fig. 2.7. Kinetic model for liver. R_U, rate of uptake from gut; R_M, rate of metabolism in liver; C_L, concentration of pollutant in liver. The arrows in the diagram indicate routes of transfer of pollutant within the animal. Rates of uptake and metabolism are expressed in terms of kg body weight. The final elimination of water soluble products (metabolites and conjugates) is in the bile and urine (after Walker, 1987).

of pollutant present in the body tissues depends on the extent of pollutant redistribution to tissues peripheral to the liver. The extent of redistribution is a function of the ability of the enzymes present in the liver to metabolize the pollutant, which can be assessed from *in vitro* studies. Pollutants which are readily metabolized will be excreted rather than accumulate in peripheral body tissues. A prediction of *BCF* for the liver can thus be made from a Lineweaver–Burke plot of *in vitro* liver preparation metabolism by determining the concentration of pollutant at which the rate of metabolism is equal to the rate of absorption, thus providing an estimate of the steady state concentration (C_i) in the intact liver. The *BCF* is then calculated as C_i/C_0, where C_0 is equal to the concentration of the pollutant in the food which would yield a relevant rate of absorption. If data on the distribution of the pollutant between body organs is available, *BCF*s may be subsequently estimated for other organs and the whole organism (Walker, 1987, 1990). Currently this approach has only been applied to a limited number of animals and compounds, but it has considerable potential, both as an experimental system to investigate the effects of pollutants on enzyme systems and as a means of estimating *BCF*s where the number of animals available precludes more traditional *in vivo* studies of bioaccumulation (Walker, 1987, 1990).

Application of bioconcentration and bioaccumulation factors

Once reliable or representative *BCF* values have been obtained they may be used to derive water quality criteria. Van der Kooij *et al.*(1991) describe a scheme by which the equilibrium partition approach may be used to derive a coherent set of water quality criteria for aquatic systems which set maximum standards for pollutant concentrations in water, suspended solids and sediments. Historically the role and importance of surface effects and sediment-bound chemical on the biological availability, mobility and toxicity of chemicals has been underestimated (Dickson *et al.*, 1987; Simkiss, 1990). Many species obtain a significant proportion of their pollution loads via surface adsorption and uptake of contaminated particulates (McEldowney and Waite, 1993). Dissolved water concentrations are frequently maintained in equilibrium with a sometimes substantial reservoir of pollutant bound or associated with particle surfaces (Moriarty, 1990; McEldowney and Waite, 1993). The toxicity of sediments to sediment-dwelling organisms is related to pore-water concentrations which are a function of the equilibrium partitioning of the pollutant between the solid and liquid phases.

If critical toxic tissue concentrations are known from toxicity testing or from product standard information, then using partition coefficients these levels may be translated into sediment and water quality standards. The critical toxic water and tissue concentrations are related by the *BCF* ($BCF = C_{organ}/C_w$; C_{organ} is toxic tissue concentration, C_w is concentration in water). We have seen that for organic pollutants, K_a may be deduced from the physical characteristics of the solid phase and the chemical properties of the pollutant. Similar general relationships are not available for inorganic pollutants; however, K_a values may be readily derived from field measurements of pollutant concentrations in filtered and unfiltered water:

$$K_a = \frac{C_{tot} - C_w}{SM \times C_w}$$

where C_{tot} is the total concentration in water before filtration (µg l^{-1}), C_w the concentration in water phase after filtration (µg l^{-1}) and *SM* the concentration of suspended matter in water (g l^{-1}).

Field studies indicate that the concentrations of pollutants in sediments differ systematically from the concentrations which occur in the suspended solids present in the water column above the bottom sediments. Empirical ratios of 1.5 for inorganic pollutants and 2.0 for organic chemicals have been obtained. Using these ratios the concentrations in the sediment and suspended solids may be obtained from the relationship below:

$$\frac{C_{sed}}{C_{sus}} = r$$

$$C_{sed} = \frac{C_{sus}}{r} = \frac{C_s}{r} = \frac{K_a}{r} \times C_w$$

where C_{sed} is the concentration in sediments (mg kg^{-1}), C_{sus} the concentration in suspended matter (mg kg^{-1}), C_s the concentration in the solid phase (mg kg^{-1}) and r is the empirically determined ratio of suspended matter : sediment concentrations. For metals r=1.5 and for organics r=2.

The values of r will vary between systems, depending on the chemical characteristics of the water, suspended solids and bottom sediments, and therefore if used, should be determined for the relevant system and conditions. The total waterborne concentration may subsequently be obtained from:

$$C_{tot} = C_w \times (1+K_a \times SM)$$

If BCF and K_a values are known along with the critical organism concentrations (often available from product standard and notification data) an alternative but related set of equations may be used:

$$C_w = \frac{C_{organ}}{BCF}$$

$$C_{sed} = \frac{C_{organ} \times K_a}{r \times BCF}$$

$$C_{tot} = \frac{C_{organ}\ (1+K_a \times SM)}{BCF}$$

Van der Kooij *et al.* (1991) and the references therein, discuss the basis and application of this scheme which has been proposed for the setting of national water quality criteria in The Netherlands. They also provide worked examples and further information on the use of product standard information.

Biomagnification

Biomagnification has been clearly shown to occur for many persistent organic pollutants, particularly those that are lipid soluble and for certain organometal compounds, e.g. methyl mercury, although in general inorganic metals do not biomagnify (Mason, 1991). Moriarty (1990) has questioned the general acceptance of the concept, pointing out the uncertainty and errors involved when comparing tissue bioaccumulation or bioconcentration factors obtained for taxonomically unrelated groups of organisms (e.g. invertebrate arthropods and vertebrate fish). He argues that since it is difficult, if not impossible, to identify analogous or equivalent body tissues in taxonomically disparate groups, bioconcentration and accumulation factors based on whole body burdens should be used for comparative purposes and not, as is often the case, whole body values for the smaller organisms and tissue based values for larger animals. In addition, although ecological communities may be described in term of trophic levels, pollutant transfer does not occur from trophic level to trophic level. Since many species feed on organisms from more than one trophic level and since trophic level species may differ considerably in their feeding behaviour and ecology, the exposure and body burdens of species from the same trophic level will vary greatly. Biomagnification, particularly in terrestrial systems, is most likely to occur along specific and restricted series of food chain links (Hopkin, 1989).

The uncritical use of *BCF*s should be guarded against. Bioaccumulation and *BCF*s are often perceived as taxon- (e.g. algae) or species-specific constants which they are not. Even within single-species populations individuals are not necessarily subject to the same level of exposure and the tendency of individuals to accumulate the pollutant will vary with age, size, physiological status and past history of exposure. For example, among marine macroalgae (seaweeds), the accumulation of heavy metals and radionuclides such as technetium varies between the major taxonomic groups (e.g. red, brown and green seaweeds), between the species of these taxons and between individuals within a population depending on their age, metabolic activity and position of the individual on the shore. It is also affected by a range of environmental factors including the nature of the sediments, the particulate load of the water body, water velocity, depth, pH, temperature and the presence and concentration of organic chelating ligands and other metal ions. Even within individual seaweed plants, which are structurally very simple organisms, tissue metal concentrations can differ considerably between different portions of the plant (McEldowney and Waite, 1993). Given the current state of knowledge it is difficult to justify the use of single *BCF*s derived from one species to characterize the bioaccumulative abilities of large groups of species which also happen to be fish, algae, etc., but have little else in common.

There is a vast literature of published *BCF* values, many derived from field studies; unfortunately the value of this information is often limited by the lack of information about the form in which the pollutant is present. Dissolved, particulate, organic and inorganic complexes of a pollutant differ in their bioavailability. Bioconcentration is not a direct function of total concentration. In laboratory studies, values are frequently obtained when the pollutant is present at unrealistically high concentrations. It should not always be assumed that body concentrations and water concentrations are linearly related over the range of concentrations actually occurring in the field; *BCF*s are not necessarily independent of concentration (Coughtrey and Thorne, 1983; Moriarty, 1990; McEldowney and Waite, 1993). An additional problem is that *BCF*s describe, and are only really relevant to, systems at equilibrium or steady state. Such a state may not be reached in the field where pollutant concentrations and environmental conditions may be continually changing.

Bioconcentration and bioaccumulation factors provide a clear and convenient way of summarizing the ability of the biota to accumulate pollutants, allowing environmental quality criteria to be derived, an approach given a veneer of theoretical credibility by the apparent similarity of *BCF*s to chemical partition constants. These predictions should only be viewed as crude but potentially useful approximations. The cynical, however, may wonder if the widespread use and calculation of *BCF*s simply reflects the tendency of biologists when given two numbers to divide one into the other.

The effects of accumulated pollutant on individual organisms

The effects of pollutants on individual organisms are routinely assessed using laboratory based single-species acute toxicity tests, detailed description and discussion of which may be found in numerous texts (Brown, 1976; Ramade, 1987; Moriarty 1990). Most procedures allow the concentration of pollutant which either reduces the growth or kills 50% of the test organisms, to be determined. The relevance of these results to the field situation is difficult to assess. To be effective the test requires the use of particularly sensitive species. Thus, the use of standard toxicity test results could lead to the setting of unnecessarily high environmental standards. In the test situation uniformly aged, sized and genetically similar individuals are exposed to a single pollutant at a constant concentration and under uniform and well-regulated conditions. This differs markedly from the situation in the field. Because of this there has been considerable interest and activity in the development of new tests and the use of multispecies toxicity tests, bioassays and bioindicators for the testing and assessment of water quality (Cairns and Pratt, 1987; McCahon and Pascoe, 1990; Spellerberg, 1991; Jeffry and Madden, 1991).

The extent and diversity of approaches adopted precludes detailed discussion here; however, it is worth noting that new tests and monitoring procedures need to be calibrated as far as possible with the existing established tests and conditions in the field. The use of caged test organisms, e.g. caged fish and fish eggs, placed in the field represents one potentially useful approach to calibration. The main advantage claimed of biological tests and bioassays over chemical analysis is the ability of organisms to integrate the effects of prolonged exposure to multiple pollutants and reflect the actual bioavailability of the pollutant. In addition, the collection and analysis of biological material can provide valuable information on the effects of unexpected releases. For example, Druehl *et al.* (1988) were able to establish the global extent and approximate time course of the radionuclide fallout resulting from the Chernobyl incident in 1986, by examining the concentration of radionuclides in samples of the brown algae *Fucus*, which acted as an 'in-place' indicator species. However, given the number of factors which can affect the accumulation of pollutants by organisms, great care is needed in the way the results of bioassays of any type are interpreted.

Impacts at the ecosystem level

It is dangerous to make general predictions of the likely effects of pollutants on ecosystems. The severity of effects will depend on the nature and concentration of pollutants present, the ecosystem and mode of release, for example, whether release is from a point source or a non-point source. Point source releases normally have a more immediate and obvious effect. Sensitive species are rapidly lost close to the source and distinctive zones of decreasing species diversity and increased ecosystem dysfunction may be recognized as the source is approached from regions not subject to significant contamination (Freedman, 1989; Hopkin, 1989). The effects of a non-point source are initially much less evident, and may become apparent only slowly as the population density of species which are particularly vulnerable due either to their sensitivity or their ecology, begin to decline, at which point, it is often not immediately clear what is the cause of the decline. Certain ecosystems are likely to be more vulnerable than others. The vulnerability of an ecosystem to disruption, i.e. its ability to re-establish and retain its structure and function during and following disturbance, is not directly related to complexity or productivity. The low diversity shallow sea communities of the Baltic, which have low productivities are particularly sensitive to disruption. Against this, the tropical rainforests which are both diverse and productive, are also extremely sensitive to disturbance. In each case the unique biological and geographical characteristics of the systems determine their sensitivity. The one generalization which probably can be made is that the potential impact of a pollutant will be greater if it significantly affects the productivity and growth of organisms at the base of food chains.

References

BAUGHMAN, G.L. and LASSITER, R.R. (1978) Prediction of environmental pollutant concentration. In Cairns, J., Dickson, K.L., Maki, A.V. (eds) *Estimating the Hazard of Chemical Substances to Aquatic Life*, pp. 35–54. Special Technical Publications, No. 657, American Society for Testing and Materials, Philadelphia.

BEGON, M., HANPEN, J.L. and TOWNSEND, C.R. (1990) *Ecology, Individuals, Populations and Communities*. Blackwell Scientific Publications, Oxford.

BROUWER, A., MURK, A.J. and KOEMAN, J.H. (1990) Biochemical and physiological approaches in ecotoxicology. *Function. Ecol.*, **4**, 275–281.

BROWN, V.M. (1976) Advances in testing the toxicity of substances to fish. *Chem. Indust.*, **21**, 143–149.

CAIRNS, J. JR. and PRATT, J.R. (1987) Ecotoxicological effect indices: A rapidly evolving system. *Water Sci. Technol.*, **19(11)**, 1–12.

COSTELLO, M.J. and THRUSH, S.F. (1991) Colonization of artificial substrate as a multi-species bioassay of marine environmental quality. In Jeffry, D.W., Madden, B. (eds) *Bioindicators and Environmental Management*. Academic Press, London.

COUGHTREY, P.J. and THORNE, M.C. (1983) *Radionuclide Distribution and Transport in Terrestrial and Aquatic Ecosytems. A Critical Review of Data*, Vol. 1. A.A. Balkema, Rotterdam.

CROSSLAND, N.O. (1988) Experimental design of pond studies. In Greaves, M.P., Greig-Smith, P.W., Smith, B.D. (eds) *Field Methods for the Study of Environmental Effects of Pesticides*. BCPC Monograph No. 40, BCPC Publications, Thornton Heath.

DICKSON, K.L., MAKI, A.W. and BRUNGS, W.A (eds) (1987) *Effects of Sediment-bound Chemicals in Aquatic Sediments*. SETAC Special Publication Series, Pergamon Press, New York.

DRUEHL, L.D., CACKETTE, M. and D'AURIA, J.M. (1988) Geographical and temporal distribution of codine-131 in the brown seaweed *Fucus* subsequent to the Chernobyl incident. *Marine Biol.*, **98,** 125–129.

FARMER, D.G. and RYCROFT, M.J. (1991) *Computer Modelling in Environmental Sciences*. Clarendon Press, Oxford.

FREEDMAN, B. (1989) *Environmental Ecology. The Impact of Pollution and Other Stresses on Ecosystem Structure and Function*. Academic Press, San Diego.

HELLING, C.S. and TURNER, B.C. (1968) Pesticide mobility: determination by soil thin-layer chromatography. *Science*, **162**, 562–563.

HOLDGATE, M.W. (1979) *A Perspective of Environmental Pollution*. Cambridge University Press, Cambridge.

HOPKIN, S.P. (1989) *Ecophysiology of Metals in Terrestrial Invertebrates*. Elsevier Applied Science, London.

JEFFERS, J.N.R. (1982) *Modelling*. Chapman and Hall, London.

JEFFRY, D.W. and MADDEN, B. (eds) (1991) *Bioindicators and Environmental Management*. Academic Press, London.

KEEN, R.E. and SPAIN, J.D. (1992) *Computer Simulation in Biology. A Basic Introduction*. Wiley-Liss, New York.

KLÖPFFER, W., RIPPEN, G. and FRISCHE, R. (1982) Physicochemical properties as useful tools for predicting the environmental fate of organic chemicals. *Ecotoxicol. Environ. Safety*, **6**, 294–301.

LUNDGREN, A. (1985) Model ecosystems as a tool in freshwater and marine research. *Arch. Hydrobiol.* 1 suppl., **7,2**, 157–196.

LYMAN, W.J., REEHL, W.F. and ROSENBLATT. (1982) *Handbook of Chemical Property Estimation Methods*. McGraw-Hill Book Company, London.

MASON, C.F. (1991) *Biology of Freshwater Pollution*, 2nd edn. Longman, Harlow.

McCAHON, C.P. and PASCOE, D. (1990) Episodic pollution: causes, toxicological effects and ecological significance. *Function. Ecol.*, **4(3)**, 375–384.

McELDOWNEY, S. and WAITE, S. (1993) *The Role of Microorganisms, Phytoplankton and Macrophyte Algae in Determining the Fate of Selected Radionuclides in the Marine Coastal Environment*. NIREX, Safety Series, Harwell (in press).

METCALF, R. L. (1977) Model ecosystem studies of bioconcentrations and biodegredation of pesticides. In Khan, M.A. (ed.) *Pesticides in Aquatic Environments*, pp. 127–144. Plenum Press, New York.

METCALF, R.L., SANGHA, G.K. and KAPOON, I.P. (1971) Model ecosystems for the evaluation of pesticide biodegradability and ecological magnification. *Environ. Sci. Technol.*, **5(8)**, 709–713.

MORIARTY. F. (1990) *Ecotoxicology, The Study of Pollutants in Ecosystems*, 2nd edn. Academic Press, London.

NEELY, W.B. (1982) Review: Organising data for environmental studies. *Environ. Toxicol. Chem.* **1**, 259–266.

NEELY, W.B. and BLAU, G.E. (1977) The use of laboratory data to predict the distribution of chlorpyrifos in a fish pond. In Khan, M.A. (ed.) *Pesticides in Aquatic Environments*, pp. 145–163. Plenum Press, New York.

NEY, R.E. (1990) *Where did that Chemical go? A Practical Guide to Chemical Fate and Transport in the Environment.* Van Nostrand Reinhold, New York.

PRITCHARD, P.H. (1982) Model ecosystems. In Conway, R.A. (ed.) *Environmental Risk Analysis*, pp. 257–473, Van Nostrand Reinhold, New York.

RAMADE, F. (1987) *Ecotoxicology*. John Wiley, Chichester.

SAMIULLAH, Y. (1990) *Prediction of the Environmental Fate of Chemicals*. Elsevier Applied Science, London.

SIMKISS, K. (1990) Surface effects in ecotoxicology. *Function. Ecol.*, **4**, 303–308.

SOLBE, J. (1988) Ecotoxicological testing and observations on pollutants in natural, semi-natural and artificial systems. *Chem. Ind.*, **1**, 16–22.

SPELLERBERG, I.F. (1991) *Monitoring Ecological Change*. Cambridge University Press, Cambridge.

STERN, A.M. and WALKER, C.R. (1978) Hazards assessment of toxic substances: environmental fate testing of organic chemicals and ecological effects testing. In Cairns, J.Jr., Dickson, K.L. and Maki, A.W. (eds) *Estimating the Hazards of Chemical Substances to*

Aquatic Life, pp. 81–131. STP 657. Amer. Soc. Testing Materials, Philadelphia, PA USA.

VAN DER KOOIJ, L.A., VANDE MEENT, D., VAN LEEUWEN, C.J. and BRUGGEMAN, A. (1991) Deriving quality criteria for water and sediments from the results of aquatic toxicity test and product standards: application of the equilibrium partition method. *Water. Res.*, **25(6)**, 697–705.

YOSHIDA, K., SHIGEOKA, T. and YAMAUCHI, F. (1983) Non-steady-state equilibrium model for the preliminary prediction of the fate of chemicals in the environment. *Ecotox. Environ. Safety*, **7**, 179–190.

WALKER, C.H. (1987) Kinetic models for predicting bioaccumulation of pollutants in ecosystems. *Environ. Pollu.*, **44**, 227–240.

WALKER, C.H. (1990) Kinetic models for predict bioaccumulation of pollutants. *Function Ecol*, **4**, 295–301.

Further reading

COUGHTREY, P.J. and THORNE, M.C. (1983) *Radionuclide Distribution and Transport in Terrestrial and Aquatic Ecosytems. A Critical Review of Data*. Vol. 1. A.A. Balkema, Rotterdam.

JEFFERS, J.N.R. (1982) *Modelling*. Chapman and Hall, London.

MORIARTY. F. (1990) *Ecotoxicology, The Study of Pollutants in Ecosystems*, 2nd edn. Academic Press, London.

SAMIULLAH, Y. (1990) *Prediction of the Environmental Fate of Chemicals*. Elsevier Applied Science, London.

Chapter 3

TREATMENT TECHNOLOGIES

Traditional approaches to pollution control

With the realization of the need for control of environmental pollution, both in terms of effluent treatment and remediation of polluted environments, has come a whole new industry, one dedicated to environmental clean-up.

The strategies employed in environmental pollution control have entered a period of change and rapid expansion. Many of the new technologies have been developed in the United States as a consequence of the Comprehensive Environmental Response, Compensation and Liability Act of 1980 (CERCLA) which resulted in the establishment of the Hazardous Waste Trust Fund, more commonly known as the 'Superfund'. The fund was established in order to clean up 1 500 sites designated on a National Priorities List (Ch. 1) from an estimated 20 000 hazardous waste sites in the United States.

The Superfund provided the impetus for the development and application of a wide variety of remediation strategies which represented technological developments for curing pollution. As concern for the environment has mounted, so interest has turned towards prevention rather than cure of environmental pollution. In parallel with the 'cure' strategies came a more general increase in emphasis on waste minimalization, with associated initiatives in recovery and recycling. The implementation of these strategies is directed towards reducing the environmental impact of human activity. The present technologies have been developed and employed in remediation, or as bolt-on, end-of-pipe processes to alleviate the environmental impact of intrinsically polluting manufacturing processes. More recently attention has turned to the so-called 'clean technologies', new production processes yielding a new generation of products that are jointly considered as environmentally sound.

There is a vast array of technologies which could be employed by the pollution control industry, but to date, many are unproven and those routinely employed are ones predominantly based on physico-chemical processes, in particular physical containment – *bury it* – or incineration – *burn it*.

Landfill

In the United Kingdom, much reliance has been placed on landfill as a means of disposal of hazardous wastes. However, in order for this approach to be operated to high environmental standards, strict controls must be maintained. In other European Community countries there are severe restrictions, even complete bans on landfill, because of the dangers of leachate contamination of soils and groundwaters. Landfilling is the controlled deposit of waste to land in such a way that no detrimental effects result. The controls regulating the use of landfills have been developed as a consequence of a substantial body of research, although over the years mistakes have been made and the controls flouted, resulting in major localized environmental damage. The design and construction of landfills requires consideration of means of deposition and containment of the wastes and also the short- and long-term control of the products of waste decomposition – landfill gases and leachates. In the United Kingdom landfill sites are considered as one of three generic types, depending upon the range of wastes accepted by the site. These are classified as *mono disposal*, where only one homogenous waste is deposited, *multi disposal*, where a range of wastes are accepted and *co-disposal* sites which accept general and controlled wastes. New European Community legislation is currently being considered which will require such sites to be classed as *inert waste sites* licensed for non-special wastes which are non-biodegradable and which have no potential to harm the environment, *household waste sites* licensed to take mainly household, commercial and biodegradable industrial wastes, and *hazardous waste sites* licensed to receive 'difficult' and 'special' industrial wastes.

When properly conducted, landfill offers an acceptable and economic way for the disposal of controlled wastes. Despite the application of new treatment technologies there will always be an irreducible amount of waste that requires disposal to landfill, be it the ash from incinerators or the residue from biological treatment works. Having said this, the availability of sites suitable for landfill operations is limited, hence there is an ever increasing need for technologies which can reduce the volume of waste entering such sites. This requires technologies which will, by their very nature, reduce the overall amount of solid waste. In terms of toxic wastes, processes involving destruction of the wastes rather than concentration through phase shifts are of particular interest.

Incineration

Incineration provides a destructive means of waste disposal and is used for municipal, clinical and toxic wastes. Effective incineration requires

(i) sufficient contact time between the waste and heat so as to ensure efficient combustion, a time usually measured in seconds, (ii) a sufficiently high temperature, which is dependent upon the type of waste – 850°C for municipal waste and 1200°C for chemical waste and (iii) turbulence, so as to ensure complete mixing in the combustion zone. Incineration has been considered as an effective treatment strategy which offers the 'nine-nines' efficiency required by the waste treatment industry, that is, it achieves a 99.9999999% destruction of the waste material. This has led to incineration being one of the favoured waste treatment technologies over the last decade. Incineration results in a reduction in the volume of waste requiring deposition in landfill sites, with only the non-combustible materials remaining after treatment. However, over the last few years increasing concerns have been expressed over the destructive efficiency of incinerators. Whilst it is recognized that this disposal technology is intrinsically safe, operational inadequacies have led to incomplete combustion of the waste materials. Emissions from smoke-stacks of compounds like dioxins, dibenzofuranes and polyaromatic hydrocarbons which result from incomplete combustion, and heavy metal pollution of the surrounding environment, have both been recorded. This has led to public concern over the siting of incinerators with the not-in-my-back-yard (NIMBY) syndrome making incineration a less attractive technology for waste treatment. Metals are also released from organometallic compounds during incineration and this can lead to greater problems of leaching from the ash and slag when it is sent to landfill.

Other technologies

The application of new technologies is seen as the way to solve the environmental problems of today. Indeed a look at the vast array of technologies employed by the pollution control industries does give credence to this view. The technologies can be divided into techniques which result in the destruction or in phase shift, concentration and immobilization of the pollutants. They can be further characterized as physical, chemical, thermal or biological techniques. The range of technologies can be exemplified by consideration of the range of technologies assessed by the Ecova Corporation for the clean-up of New Bedford Harbor in the United States (Allen and Ikalainen, 1988; Table 3.1).

Table 3.1 Identification and screening of hazardous waste treatment technologies for New Bedford Harbor (from Allen and Ikalainen 1988)

Technology	*Applicable to sediment matrix*	*Applicable to water matrix*	*Applicable for PCB treatment*	*Applicable for metals removal*
Biological				
1. Advanced biological methods	Yes	No	Yes	No
2. Aerobic biological methods	No	Yes	No	No
3. Anaerobic biological methods	Yes	No	No	No
4. Composting	Yes	No	No	No
5. Land spreading	Yes	No	No	No
Physical				
6. Air stripping	No	Yes	No	No
7. Soil aeration	Yes	No	No	No
8. Carbon adsorption	No	Yes	Yes	No
9. Flocculation/ precipitation	No	Yes	Yes	Yes
10. Evaporation	Yes	Yes	No	No
11. Centrifugation	Yes	No	No	No
12. Extraction	Yes	No	Yes	No
13. Filtration	Yes	No	No	No
14. Solidification	Yes	No	Yes	Yes
15. Granular media filtration	No	Yes	No	Yes
16. *In situ* adsorption	Yes	No	Yes	No
17. Ion exchange	No	Yes	No	Yes
18. Molten glass	No	No	Yes	No
19. Steam stripping	No	Yes	No	No
20. Supercritical extraction	Yes	No	Yes	No
21. Vitrification	Yes	No	Yes	Yes
22. Particle reaction	No	No	Yes	No
23. Microwave plasma	No	No	Yes	No
24. Crystallization	No	Yes	No	No
25. Dialysis/electrodialysis	No	Yes	No	No
26. Distillation	No	Yes	No	No
27. Resin adsorption	No	Yes	No	Yes
28. Reverse osmosis	No	Yes	No	Yes
29. Ultrafiltration	No	No	No	No
30. Acid leaching	Yes	No	No	Yes
31. Catalysis	No	No	No	No
Chemical				
32. Alkali metal dechlorination	Yes	No	Yes	No
33. Alkaline chlorination	No	No	No	No

Table 3.1 continued.

Technology	*Applicable to sediment matrix*	*Applicable to water matrix*	*Applicable for PCB treatment*	*Applicable for metals removal*
34. Catalytic dehydrochlorination	No	No	Yes	No
35. Electrolytic oxidation	No	No	No	No
36. Hydrolysis	No	Yes	No	No
37. Chemical immobilization	Yes	No	No	Yes
38. Neutralization	Yes	No	No	No
39. Oxidation/hydrogen peroxide	Yes	Yes	No	No
40. Ozonation	No	No	No	No
41. Polymerization	Yes	No	No	No
42. Ultraviolet photolysis	No	No	Yes	No
Thermal				
43. Electric reactors	Yes	No	Yes	No
44. Fluidized bed reactors	No	No	Yes	No
45. Fuel blending	No	No	Yes	No
46. Industrial boilers	Yes	No	Yes	No
47. Infrared incineration	Yes	No	Yes	No
48. *In situ* thermal destruction	No	No	Yes	No
49. Liquid injection incineration	No	No	Yes	No
50. Molten salt	No	No	Yes	No
51. Multiple hearth incineration	Yes	No	Yes	No
52. Plasma arc incineration	No	Yes	Yes	No
53. Pyrolysis processes	Yes	No	Yes	No
54. Rotary kiln incineration	Yes	No	Yes	No
55. Wet air oxidation	No	Yes	No	No
56. Supercritical water oxidation	Yes	Yes	Yes	No

Evaluation of waste management options

As each pollution problem is addressed, the choice of which technology or technologies is most appropriate to alleviate the particular problem has to be made. The first stage of any assessment must be a pollution audit. In order to alleviate the problem, the problem must first be defined. At sites polluted with toxic chemicals the audit would take the form of an investigation of the site in order to determine the types, concentrations and geographical spread of the pollutants and a geological survey of the site must also be undertaken. For end-of-pipe treatment strategies for the control of pollution, a pollutant audit would identify which components of the waste stream constitute the threat to the environment. It would then go on to ask whether the threat arises as a result of deliberate discharge, a side-effect of the process, mining, or of disposal of dumped-container or landfill run-off leachates.

Once the pollution audit has been completed selection of an approach to preventing or curing the environmental problem using these technologies can only be based on knowledge of a large number of variables for each specific situation. These variables can be grouped as: legislative requirements, environmental impact, and economic and technical feasibility. Meeting all the criteria defined under these headings will often mean that no single technology will suffice to completely alleviate the problem. Hence, mixed technologies are often employed. However, this approach ensures that the combined treatment strategies achieve the required effectiveness in terms of statutory discharge levels and/or the most advantageous cost–benefit ratios, where effectiveness is defined in terms of the extent to which the technology permanently reduces the toxicity, mobility or volume of the toxic components at the site or in the waste stream. The treatment alternatives typically range from those which treat only the principal threats to those which completely destroy, detoxify or immobilize all the hazardous substances. Under CERCLA, assessment of competing technologies for hazardous waste site remediation involves consideration of nine criteria.

1 *Protection of human health and the environment*. Does the technology provide adequate protection at each stage of the process?

2 *Compliance with environmental statutes*. Does the technology meet all applicable statutes?

3 *Long-term effectiveness and permanence*. Does the technology maintain reliable protection of human health and the environment over time?

4 *Reduction of toxicity, mobility and volume*. Does the technology meet the necessary and statutory performance criteria?

5 *Short-term effectiveness*. Does the technology pose any adverse human health and/or environment threats during clean-up period?

6 *Implementability*. Is the technology technically and administratively feasible?

7 *Cost*. Are the estimated capital, operating and maintenance costs realistic in economic terms?

8 *State acceptance*. Does the local state government accept the proposed technology?

9 *Community acceptance*. This criterion defers to the local community's views on the proposals.

In the final analysis the adoption of a particular environmental protection strategy should result in the best practicable environmental option (BPEO) and the best available technology not entailing excessive cost (BATNEEC) (Ch. 1). The implementation of short-term solutions is to be avoided. Indeed it is the legacy of previous 'quick-fix' options that has led to the need for multibillion dollar remediation programmes today. There also needs to be a degree of flexibility in the waste treatment strategies employed such that over longer periods of time the strategies are able to maintain the environmental acceptability of the downstream results against changes in waste contents and volume.

Biotreatment technologies for pollution control

Biotechnology for hazardous waste management can be defined as the development of systems that use biological catalysts to degrade, detoxify or accumulate environmental pollutants. The application of biological agents to this field is far less developed than applications in other fields of biotechnology such as pharmaceuticals or agriculture. Early over-optimistic claims for the wonders of biological systems for site decontamination set back the application of bioremediation strategies by at least a decade. When the 'magic soups' failed, companies involved in clean-up returned to the use of the more familiar proven physico-chemical treatment processes which, although more expensive, could guarantee nine-nines efficiency (see p. 50). The rejection of the biostrategies was further exacerbated by the fact that the major companies involved in environmental clean-up usually have an engineering base and, hence, there is a tendency for there to be a lack of perception of the metabolic potential of microorganisms.

More recently, innovative biotechnologies, bringing together applied microbiology and engineering, have been shown to be a realistic alternative or adjunct to the more conventional physico-chemical technologies. The present constraints on the application of biotechnological approaches arise from the fact that the methods have only been used on a limited number of full-scale demonstration processes. Environmental biotechnologies have been proven at the laboratory scale; this is witnessed by the United States National Institute of Health survey of 20 research groups in the United States, which found that organisms had been isolated which could degrade 41 of 49 pollutants found in priority sites. Consideration of the types

Table 3.2 Predicted treatment effectiveness for soil contaminated with toxic organic compounds (Offut *et al.*, 1988)

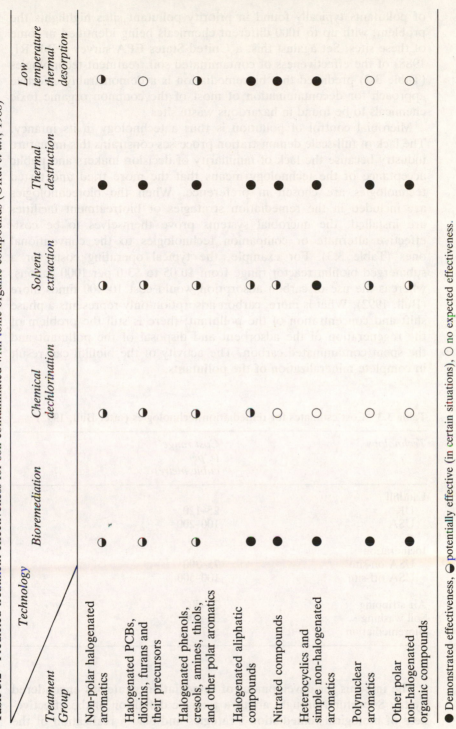

Treatment Group \ *Technology*	*Bioremediation*	*Chemical dechlorination*	*Solvent extraction*	*Thermal destruction*	*Low temperature thermal desorption*
Non-polar halogenated aromatics	◒	◒	◒	●	◒
Halogenated PCBs, dioxins, furans and their precursors	◒	◒	◒	●	○
Halogenated phenols, cresols, amines, thiols, and other polar aromatics	◒	◒	◒	●	◒
Halogenated aliphatic compounds	●	◒	◒	●	●
Nitrated compounds	●	○	◒	●	●
Heterocyclics and simple non-halogenated aromatics	●	○	●	●	●
Polynuclear aromatics	●	○	◒	●	○
Other polar non-halogenated organic compounds	●	○	◒	●	○

● Demonstrated effectiveness, ◒ potentially effective (in certain situations), ○ no expected effectiveness.

of pollutants typically found in priority pollutant sites highlights the problem, with up to 1000 different chemicals being identified in some of these sites. Set against this, a United States EPA survey (HMCRI, 1988) of the effectiveness of contaminated soil treatment technologies (Table 3.2) predicted that bioremediation is a demonstrably effective approach for decontamination of most of the common organic toxic chemicals to be found in hazardous waste sites.

Microbial control of pollution is thus a technology in its infancy. The lack of full-scale demonstration processes constrains this immature industry because the lack of familiarity of decision makers and public acceptance of the technology means that the more tried and tested technologies are chosen in preference. When the biotechnologies are included in the remediation strategies or biotreatment facilities are installed, the microbial systems prove themselves to be cost-effective alternate or companion technologies to the conventional ones (Table 3.3). For example, the typical operating costs for a submerged biofilm reactor range from \$0.05 to \$3.0 per 1000 gallons, whereas the use of carbon adsorption would cost 10–400 times more (Bull, 1992). What is more, carbon adsorption only represents a phase shift and concentration of the pollutant; there is still the problem of the regeneration of the adsorbent and disposal of the pollutant and the spent contaminated carbon. The activity of the biofilm can result in complete mineralization of the pollutants.

Table 3.3 Cost estimates for remediation technologies (after Bull, 1992)

Technology	*Cost range (£ per cubic metre)*
Landfill	
UK	25–120
USA	100–200
Incineration	
USA on-site	75–300
USA off-site	100–500
Air stripping	20–50
Soil washing	35–100
Bioremediation	5–75

The impetus for development of remediation strategies engendered by the Superfund brought about a realistic definition of the effectiveness of biological remediation strategies and also a realization of the scale of the environmental problems faced. Part of this new approach has been to recognize the importance of a preliminary site assessment for treatability – the pollution audit. In terms of the application of biological agents, this means an audit of all the parameters that will affect the activity of the natural or supplemented microbial populations (p. 59) and their effects investigated in the laboratory. These tests can then lead to determination of the most appropriate treatment regimes and provide a meaningful estimation of the rate of degradation and, hence, enable definition of a time scale for the bioremediation of the site. Such considerations then enable cost comparisons to be made with the competing technologies.

As with all the technologies, the environment will dictate the biotreatment strategy employed. For end-of-pipe bioreactor technology, utilizing single microbial species or enzymes might be applicable for well-defined waste streams, especially if the bioreactor is situated in the process stream rather than at the end of the waste pipe. However, given the complexity of most waste streams it is likely that in most instances microbial consortia would have to be employed to achieve the desired or legislated level of treatment. Where whole environment remediation is the objective, the complexity of the problem will dictate a biostrategy that involves a whole ecosystem approach.

In terms of treatment strategies the biotechnologies have been considered and/or applied to pollution of soil, groundwater, leachates, sludges, sediments and air. The techniques employed range from new methodologies to modifications of old technologies. It is apparent that there can be no universal biocatalyst for all toxic wastes, but waste treatment bioreactor configurations are generally based on stirred tank, activated sludge, anaerobic or biofilm systems. The effectiveness of the biotreatment facility for a specific remedial or waste treatment process is thus defined by the biocatalyst(s) used in the bioreactor. It is the biocatalyst that is 'customized' for the particular needs of the process. Attainment of a more detailed understanding of the mode of action of the biocatalysts, largely through the application of modern techniques in molecular biology, has enabled a more rational approach to be taken to bioreactor design and operation rather than the empirical approaches previously employed. This is not to belittle the involvement of chemical and biochemical engineering as the specific requirements of each biocatalyst will have to be met by the detailed consideration of bioreactor configuration. In addition, the engineering elements will also have a major effect on the process economics; and these will, in the final analysis, play a significant role in determining the commercial success of the treatment process.

Environmental biotechnologies have been involved in the progressive development of pollution control strategies. Firstly, in the cure of environmental pollution, bioremediation provides a cost-effective technology for independent application or for use as part of a mixed technology approach. Secondly, with the increase in emphasis on waste minimalization, *bio*adsorption and accumulation have shown their worth in recovery and recycling. For instance, exchange columns with matrices consisting of dead algal cells have proved effective in the removal of heavy metals from waste streams and bacteria have been shown to accumulate a variety of heavy metals (Chs 14 and 15). Thirdly, in waste reduction, bioreactors have been employed as end-of-pipe treatment facilities thereby reducing the environmental impact of human activity.

The advent of the concept of the 'clean technologies' offers a new field for the application of the biotechnologies. It is of note that industrial processes based on modern biotechnology often consume less raw material, water and energy than the traditional production processes. Against this background one can confidently predict that the continued development of the environmental biotechnologies will provide new ways to protect the environment from the ever more demanding technosphere (Nicholson, 1970).

References

ALLEN, D.C. and IKALAINEN, P.E. (1988) Selection and evaluation of treatment technologies for the New Bedford Harbor (MA) Superfund Project. In *Superfund '88 Proc. 9th Nat. Conf. 23–30 November, 1988, Washington, DC*, pp. 329–337. The Hazardous Materials Control Research Institute, Silver Spring, MD, USA.

BULL, A.T. (1992) Degradation of hazardous wastes. In Bradshaw, A.D., Southwood, Sir R., Warner, Sir F. (eds) *The Treatment and Handling of Wastes*. Chapman and Hall for The Royal Society, London.

NICHOLSON E.M. (1970) *The Environment Revolution : A Guide for the Masters of the New World*. Hodder and Stoughton, Sevenoaks, UK.

OFFUT, C.K., KNAPP, J.O'N., CORD-DUTHINH, E., BISSEX, D.A., ORAVETZ, A.W., KENNEY, P.J., GREEN, E.L., BHINGE, D. (1988) *Superfund '88. Proc. 9th Nat. Conf., 23–30 November, 1988, Washington, DC*. The Hazardous Materials Control Research Institute, Silver Spring, MD, USA.

Further reading

BRADSHAW, A.D., SOUTHWOOD, SIR R., WARNER, SIR F. (eds) (1992) *The Treatment and Handling of Wastes*. Chapman and Hall for The Royal Society, London.

FREEMAN, H.M. and SFERRA, P.R. (eds) (1991) *Innovative Hazardous Waste Treatment Technology Series*, Vol. **3**, *Biological Processes*. Technomic, Lancaster, Basel.

Chapter 4

BIOCATALYST SELECTION AND GENETIC MODIFICATION

Enrichment and screening strategies

The investigations of environmental microbiologists, bringing together the study of microbial physiology, biochemistry, genetics and ecology, are directed towards the study of the interaction of microbes with their biotic and abiotic environments. Such work has led to a realization of the degree of biochemical and genetic plasticity that exists within natural microbial communities. This plasticity enables the microbes to respond to their environment, that is, to adapt their physiology in order to maintain their competitiveness in their constantly changing world. The ability of microbes to modulate their physiology through phenotypic rather than the more permanent genotypic adaptation (mutation) is a reversible response to environmental change. Genotypic responses are to be considered more along the lines of Darwinian selection. The genetic make-up of the organisms is determined by genetic rearrangements and selection over longer periods of time. This type of plasticity 'fits' the organism to the state of its environment over longer periods of time, effectively the average conditions in which it lives. The phenotypic responses, defined by the genotype, enable adaptation to transient conditions. A definition of time in microbial evolutionary terms is difficult, especially in view of their rapid generation times and their ability to acquire new genetic material (see below).

Biological and abiological environmental factors exert selective influences on the microbes present in the environment thereby determining which members of the population dominate at any given time. It is these adaptive capabilities that have been harnessed by microbial technologists for the developing environmental biotechnologies.

Such natural selection leads one to the conclusion that if you want an organism with particular characteristics then the first step is to consider which natural environments are most likely to have exerted selective conditions which would favour the evolution of inhabitants with the desired or closely related traits. This leads to the concept of enrichment and screening of natural environments for biocatalysts through environmental selection. If, for example, a desirable characteristic of the new biocatalyst is that it be tolerant of high temperatures then geothermal sites would be appropriate sampling sites. If the biocatalyst is required to operate under conditions of low nutrients say, for example, in the *in situ* biotreatment of some contaminated groundwaters, then environments naturally oligotrophic would be a starting point. Almost all unpolluted fresh and marine waters contain very small amounts of dissolved organic material. Microbes adapted to such conditions are often so well adapted that they are obligately oligotrophic. Such organisms tend to be metabolically versatile, such that they can grow on whatever is there and their enzyme and metabolite transport systems show high affinities and low saturation constants. Major limitations on the potential application of oligotrophs in biotreatment strategies are their intrinsically low metabolic rates and also a tendency to be more sensitive to the toxic effects of environmental chemicals.

The geological diversity of the Earth and the apparent wealth of biological diversity, especially, in this context, the microbial biodiversity, leads one to the belief that if you know how, and where, to look then it will be possible, if not highly probable, that a natural microbe will be found that is capable of the biotransformation and hopefully the biodegradation of all environmental chemicals.

Design of enrichment strategies relating to the environmental source

Five stages in the isolation of microbes can be recognized:

1 selection and sampling of the environment
2 pre-treatment of samples
3 growth in laboratory media
4 incubation
5 isolate selection

The objective for exercising care in the choice of sampling sites is to *increase* the chance of finding the desired microbe. An ecological approach can achieve this in that, for instance, thermophilic microbes are more likely to be isolated from geothermal sites than from environments that are not so heated. That is not to say that there is no chance of finding a sufficiently thermotolerant microbe in a temperate soil, just that by going to a geothermal site you are increasing your chances of finding a thermotolerant organism. Whilst one is *most likely* to isolate a chlorobutane-degrading organism from a site which has been contaminated with chlorobutane or related haloalkanes, it is not to say that if samples were taken from virgin unpolluted soils haloalkane-degrading microbes could not be isolated.

The objective is to manipulate the microbes' environment in such a way as to select and isolate the 'most effective' organisms from within the total microbial population of the selected samples, where the 'most effective' is defined as those which are best at the expression of targeted phenotypic trait(s). Selection is directed towards the gene pool present in the environmental sample. Under such regimes 'the best' is determined at a physiological level by the expression of phenotypic traits already present in members of the microbial population in the sample or as a consequence of genetic modifications through mutation and genetic recombinations. Different selection procedures will favour one or more of the selective mechanisms. For instance, the establishment of a microbial community as a biofilm which promotes cell–cell contact will aid horizontal transfer of genetic material, which is less likely to occur in completely mixed systems such as a stirred tank reactor. However it should be noted that during long-term experiments in stirred systems wall growth on the sides of the reactor can lead to the necessary stable contact.

Given these considerations it is important to define all the characteristics of the organism(s) in terms of the reasons for wishing to isolate them. Hence, if the reason for searching for organisms is simply to isolate those with a specific phenotypic trait, say production of a particular catabolic enzyme, then selective pressures should be designed to isolate only organisms capable of producing that enzyme. However, if a biotechnological application is the ultimate aim, then the enrichment conditions should be based on multiple criteria to incorporate not only selection for the catabolic capability but also key properties, based on the final process envisaged. In this way the selected isolates will provide customized effective biocatalysts for the defined technological application.

Once the selection criteria have been defined and environmental samples have been obtained, the first stage in the selection procedure may be applied to the samples directly. The microbial populations therein may be subjected to dilution or concentration, to selective inhibitory treatments, for example, heat treatment in order to kill all vegetative cells with a view to isolating only spore-producing microbes, ultraviolet irradiation can be used to select for cyanobacteria or enrichment strategies can be employed whilst still in the environmental sample. An example of the latter would be the use of perfusion columns in which selected nutrients are recycled through a soil column to stimulate the growth of competent microbes present in the soil. Whether the first stage is to pre-treat the sample or to go straight to culture in a laboratory medium, the sooner the samples are used after collection, the more environmentally representative they will be.

In order to isolate the microbes as individual species or as consortia it will be necessary to culture the organisms in laboratory growth media. Herein lies an intrinsic problem for all isolation strategies; all laboratory media are to a greater or lesser extent selective, not all the microbes in the environmental samples will be able to grow in the medium chosen. Most isolation media rely on prior knowledge of nutritional and growth requirements of the microbes being sought. This comes from experience, but means that such screening procedures are by their very nature conservative. Truly novel organisms with more esoteric growth requirements will be subjected to a negative selection pressure and those that are selected can often represent rediscoveries of known organisms or phenotypic traits. As a consequence it is generally accepted that less than 10% of fungi and soil bacteria (total numbers determined by microscopic examination and nucleic acid analyses) and less than 0.1–0.001% of marine bacteria have been cultured in the laboratory.

Accepting this limitation, the object of the laboratory cultures is to select for those microbes with the desired traits. This involves incorporating physico-chemical features into defined media and determining growth conditions best suited to select for the target microbes. Defining the required medium components is vitally important and nowhere is this demonstrated better than in attempts to recreate seawater in the laboratory. Numerous recipes exist but nothing is as good as the real thing. Incorporating specific inhibitors into the medium so preventing the growth of undesirable groups of organisms can also be employed. The presence of antibiotics in the medium can select against bacteria in fungal enrichment strategies and the selection can be further enhanced by reducing the pH of the medium. Acidification can be used to selectively isolate lactobacilli and less extreme pHs of say 4.5 can be used to reveal some streptomyces which are not isolated at neutral pH values. The presence of tellurite in the medium selects for corynebacteria, whilst triphenylmethane dyes select for Gram-negative bacteria and mycobacteria. The use of dilute media selects for caulobacters, spirilla and sphaerotilus. The rationale behind some selective media formulation is not always apparent. The 'muck and mystery' approach results in additions of more ill-defined substances; whole chicken egg for nocardiae or sheep dung extract for thermophilic streptomyces.

Defining the selective procedure is critical; not just in terms of the eventual success at isolating suitable organisms, but also in terms of time taken to identify the 'best' isolates. The classical methods tend to have a low success rate and so involve large numbers of repetitive procedures, often consuming piles of agar plates all of which is labour-intensive. By defining precise targets and imposing stringent selective conditions the drudgery can be reduced. The rational approach to designing highly specific screens has become known as 'target-directed' or 'intelligent' screening and this approach has been shown to increase the rate of success and to increase the chance of isolating truly novel organisms.

The period of incubation is of considerable importance, with accepted periods ranging from a few days for thermophilic bacteria to several months for the detection of nitrifying bacteria. Recent experience in our laboratories would also indicate that even when selecting for biodegradative capabilities it is sensible not to be in too much of a rush to discard old and apparently uninteresting plates. Even after 2 months new biodegradative capabilities have been identified.

Bacterial isolation strategies usually involve a period of enrichment in liquid culture followed by spatial separation of organisms in or on solid media where they are allowed to grow as colonies. It is, however, worth noting that failure to obtain colonies on such plates may not necessarily be a sign that no biotransformation potential exists in the enrichment culture. In terms of complex organic substrates concerted metabolic activity deriving from a consortium of microbes may be required in order for biodegradation and concomitant growth of the culture to occur. Spatial separation of components of a consortium may restrict metabolite exchange so preventing sufficient metabolism of the substrate to support growth of the consortium. A second factor for consideration when using solid media in enrichment and selection strategies is that when attached to surfaces, microbes are capable of growth in the presence of lower nutrient levels as a consequence of surface concentration effects, which may translate into an enhanced ability to degrade recalcitrant molecules. Such biofilms are also able to withstand higher levels of growth inhibitors, here considered as the toxic environmental chemicals. All of this goes to explain the reasons behind the advantages of biofilm reactors for waste treatment, but in terms of the isolation of competent microbes there should be a realization that growth on selective agar plates does not necessarily translate into ability to grow on the same substrates in a liquid environment.

Microbiological techniques for enrichment and selection

In situ enrichment prior to isolation

In situ enrichment techniques involve the addition of defined materials to the actual environment so as to enrich the competent section of the microbial population present in that environment prior to isolation. Such baiting experiments have been developed into bioremediation strategies for *in situ* remediation of polluted sites through the addition of nutrients (Ch. 7).

The heat generated by biological activities in compost heaps selects for a population of thermophilic organisms in the centre of the heap. If this is dosed with, say, olive oil, it may subsequently be possible to isolate a subpopulation of thermophilic lipase-producing microbes. 'Baiting' sites provide a defined way for enrichment *in situ*. Microscope slides coated with cellulosic material can be buried in soil to isolate cellulolytic microbes. Practice golf balls containing material soaked with target environmental chemicals can be buried and subsequently retrieved for microbiological isolations.

Laboratory microcosms and soil columns

A step towards laboratory-based enrichment and selection is the use of *ex situ* microcosms – laboratory models of defined environmental situations. Whilst in most instances these are crude approximations to the real environments their use is essential where direct *in situ*

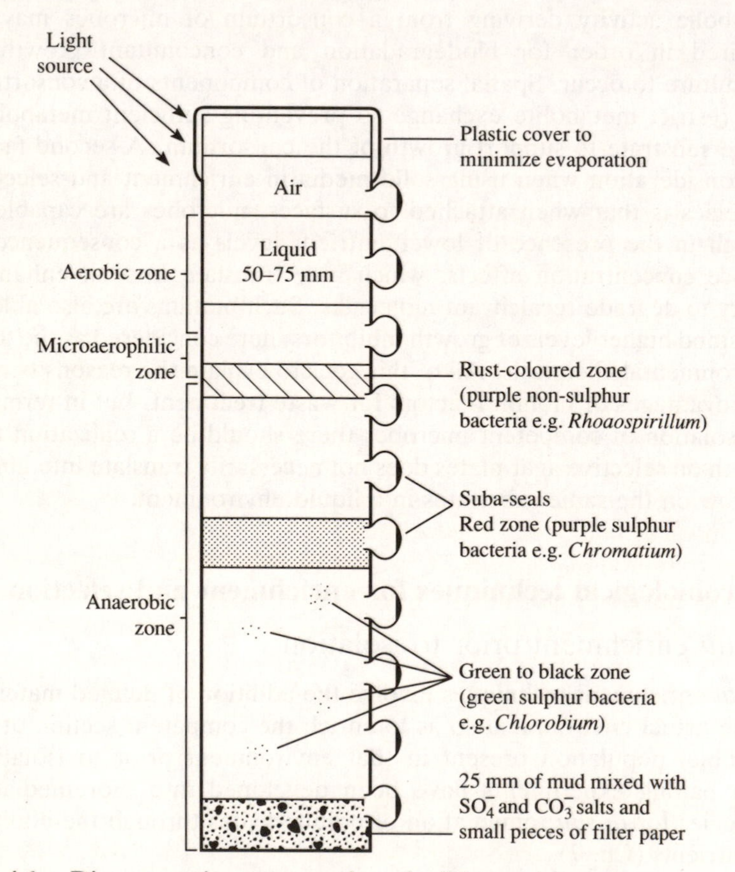

Fig. 4.1. Diagrammatic representation of a Winogradsky column.

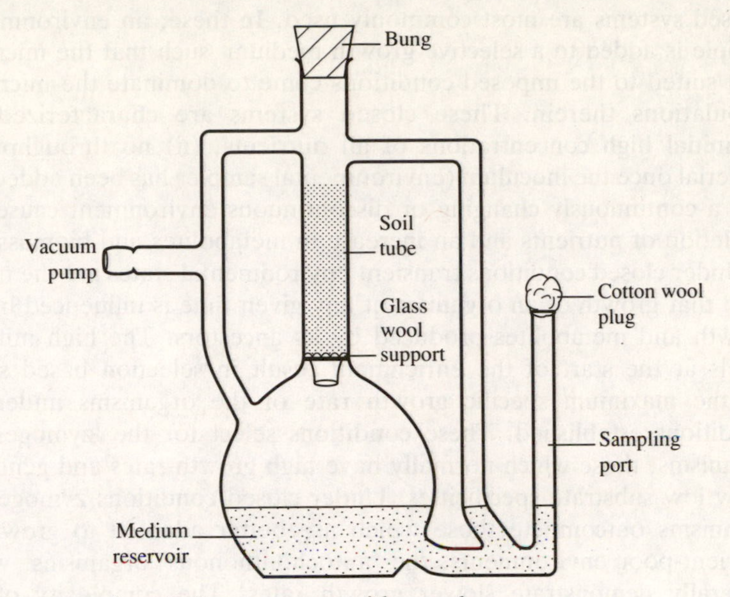

Fig. 4.2. A soil perfusion apparatus with recycle.

techniques cannot be employed because the enrichment strategy involves the use of regulated environmental chemicals. The Winogradsky column provides an extremely simple but effective way of simulating natural sediment environments in the laboratory (Fig. 4.1). The addition of selective agents will enrich for microbes capable of expressing the targeted metabolic activity(ies) from the initial and diverse microbial populations in the sediment sample. The range of organisms is significant as the columns provide a range of microenvironments which will initially select for different metabolic potentials.

Soil perfusion columns (Fig. 4.2) can also be dosed with selective substrates with a recycle loop to maintain the circulation and aeration of the soil. The recycle can be maintained until such time as competent populations of microbes are generated.

More complex microcosms may also be established with a view to achieving the desired isolation. An artificial stream was used to isolate the pentachlorophenol-degrading microbes which subsequently formed the basis for a full-scale bioremediation system for the treatment of sites and groundwaters contaminated with wood preservatives (Ch. 7).

Laboratory-based liquid culture techniques

Two types of liquid enrichment culture techniques are used by microbiologists – batch or closed culture, and open or continuous culture.

Closed systems are most commonly used. In these, an environmental sample is added to a selective growth medium, such that the microbes best suited to the imposed conditions come to dominate the microbial populations therein. These closed systems are characterized by: (i) initial high concentrations of all nutrients, (ii) no throughput of material once the inoculum (environmental sample) has been added and (iii) a continuously changing or discontinuous environment caused by depletion of nutrients and an increase in metabolites and biomass.

Under closed conditions transient environmental states are the norm, such that growth of an organism at any given time is influenced by the growth and metabolites produced by its ancestors. The high nutrient levels at the start of the enrichment result in selection based solely on the maximum specific growth rate of the organisms under the conditions established. These conditions select for the 'zymogenous' organisms, those which normally have high growth rates and generally show low substrate specificities. Under closed conditions zymogenous organisms outcompete those which are better adapted to growth in nutrient-poor environments, the 'autochonthonous' organisms, which generally demonstrate slower growth rates. The complexity of the changes which occur during growth in closed cultures means it is often impossible to predict which organisms will predominate. The high nutrient regime is not a normal natural situation, hence the isolates from closed culture enrichment are likely to represent a minority population in the actua,l environment from which the sample was taken. There the oligotrophic environment will favour the autochonthonous microbes. The attraction of closed enrichment systems based as they generally are on shake-flask cultures, is their simplicity, cheapness and opportunity for multiplicity.

In terms of microbes for use as biocatalysts in pollution control or remediation, both autochonthonous and zymogenous microbes have their place. Where contamination is at the micropollutant level, as is often the case in groundwaters, the oligotrophic environment will favour the former group. However, heavily polluted sites and end-of-pipe treatment facilities would at least initially benefit from the activity of the faster-growing, hence faster-utilizing zymogenous organisms.

In open systems, commonly the chemostat, nutrients are continuously added and spent medium, biomass and metabolites are continuously removed. Under such a regime steady state conditions can be established and maintained indefinitely and so defined selective conditions can also be imposed for prolonged periods. The culture is thus divorced from its history and long-term selection and adaptation experiments can be considered. In terms of selecting competent organisms for the biotreatment of environmental chemicals the chemostat offers several advantages over closed systems. Firstly, the rate of growth of the culture is defined by a single limiting substrate, which, when considering the biodegradation of organic compounds, means that the target pollutant can be used as the limiting carbon source. This provides an ideal selection pressure, not only for the isolation of competent microbes, but also it will tend to select for the most efficient strains in terms of their ability to utilize the carbon source. Secondly, the nature of the culture system does not depend on providing sufficient nutrients to support the complete cycle of biological activity in the culture vessel at the time of inoculation. This means that the reactor can be supplied with extremely low levels of nutrients to simulate oligotrophic conditions and hence, select for the autochonthonous organisms at low population densities. The ability to use low nutrient levels also means that if target pollutants are potentially toxic to microbial activity, the levels supplied as the limiting nutrient can be such as to be below toxic thresholds. Subsequent adaptation of the culture to elevated concentrations can be achieved over time by gradually increasing the amount of the pollutant in the feed stream. Where the pollutant cannot be used as a nutrient and the biocatalyst is to be used to adsorb the pollutant, as in, for example, heavy metal decontamination, continuous bioreactors can be used to acclimatize biocatalysts to the presence of the metals and to provide a continuous supply of biomass as the adsorptive agent.

The steady state nature of such cultures also allows for adaptation to previously non-utilizable substrates by introducing a mixed substrate regime and gradually increasing the concentration of the non-utilizable substrate and decreasing the growth-supporting one. An early example of the success of this approach was in the isolation of a microbe which was capable of degrading 2,4,5-trichlorophenoxyacetate (2,4,5-T) previously considered to be recalcitrant. The initial growth of the culture was supported by supplying the reactor with the biodegradable derivative 2,4-dichlorophenoxyacetate (2,4-D).

The establishment of steady states in these reactors also enables the development of stable mixed communities of microbes as no autogenic succession occurs. This is important for the environmental biotechnologist as in many instances the biotransformation of target chemicals requires the combined activity of microbial consortia.

As will become apparent throughout this consideration of the environmental biotechnologies, the concept of 'horses-for-courses' in terms of the choice of biocatalysts for specific environmental pollution control problems means that even in the early stages of developing screening strategies for the isolation of potential biocatalysts, one should consider the final process requirements. This approach will enable the successful development of customized pollution control biocatalysts. In this sense the planning of the search and discovery programme is no less important than the process design.

Genetic approaches to the isolation of biocatalysts for environmental biotechnologies

The screening strategies considered above all rely on the isolation of natural competent organisms through selection for growth or tolerance, on or in the presence of the target pollutant. The successful exploitation of the natural microbial diversity depends upon the effective detection of novelty; in these terms this generally refers to novel catabolic potential. The question arises, are methods that rely on growth of the microbes for the detection of the desired novelty, the best way to ensure the widest possible range of candidate microbes are isolated, especially given the medium-dependent bias of such isolation techniques? With the developments in modern molecular biology the answer to this question is increasingly becoming 'no'.

The limiting nature of the culture-based techniques is further compounded by the genetic plasticity of microbes which continually undermines the constancy of phenotypic traits. Consideration of the microbial gene pool as being contained in a continuum of host cells which exchange genetic material at micro- and macrolevels, leads one to regard laboratory selection techniques as operating to freeze the genetic flux, by the application of strongly selective conditions. The modular nature of a number of the genetic determinants of catabolic activities further suggests that freezing the genetic flux at one particular point does not necessarily mean that it is at the optimal point. If selection has been on the basis of isolation in selective media the genetic combination identified may not be as good as one that failed to grow under the chosen conditions. This would mean that the selective techniques may miss the more effective biocatalyst.

The use of molecular techniques can get away from the need to culture the isolate before the target capability is detected. Such culture-independent techniques reflect more accurately the genetic diversity present in the natural gene pool and indeed it has already been demonstrated that there is a greater diversity even in isolates derived from the culture-dependent methods than had previously been realized.

The use of nucleic acid probes was one of the first applications for the modern molecular biology techniques in screening microbial populations. Sequences of DNA which encode one or more genes that define a specific phenotype are used to search the genomes of new isolates for the presence of homologous sequences. Once such homology has been identified, these strains are screened for the targeted phenotype or the specific genes are isolated. The advantage of such techniques is that the isolation is not dependent on expressed phenotypes, hence if the conditions normally used to select for the phenotype in question do not support the growth of all isolates, the use of less selective media in conjunction with probe technology will allow a greater range of isolates to be collected. In addition these techniques will also identify microbes with the genetic potential to express the phenotype even if at the time of screening it is cryptic.

The recent developments in the field of probe technology have been directed towards increasing sensitivity and specificity. Techniques have been reported that can detect 10–100 cells per gramme of soil. The application of the polymerase chain reaction (PCR), whereby target DNA sequences are amplified, has enabled sequence analysis of genes from microbes that are difficult to obtain or cultivate. Using these techniques, 2,4,5-T-degrading organisms were detected against a background of non-degrading organisms to a detection limit of 1 cell per gramme of sediment. Such techniques have been used to determine the presence of biological activity in and around the deep sea thermal vents. Indeed the problems associated with culturing marine microbes in general means that the advent of PCR has greatly aided the study of many non-culturable organisms.

In the same way the use of ribosomal RNA sequences has also provided a powerful tool with which to overcome the problem of media-bias. The techniques can use just the *in situ* biomass concentrations and again they have been used very effectively for screening aquatic microbial populations. When used these techniques have often resulted in the identification of additional constituents of microbial communities which had previously been missed using culture-dependent techniques.

The application of the DNA and RNA techniques has also often resulted in a revised estimate of the microbial diversity within culture collections. Collections classified using traditional approaches to define functional characteristics as a means of species categorization are now being seen to underestimate the diversity present there, so much so that in view of the extent of horizontal gene flow which apparently takes place within natural populations, the whole species concept as applied to bacteria is called into question. Whilst this has little effect on the consideration of the use of such microbes in the environmental biotechnologies, it does further emphasize the extent of the resources available for these applications. It also brings one to the realization that culture collections are a major source of microbes. Whilst collections are far from complete they represent a resource in which each isolate has often only been considered for one or two particular phenotypic traits and hence its full potential has yet to be realized. Also where multiple strains of a given species are held the extent of the genetic diversity within the species has not been investigated.

This brings one to consider one of the latest molecular biology techniques and its potential for predictive screening, that of DNA fingerprinting. This technique involves DNA identification analyses of one or more features of the genomes from different individuals of a genus or species. DNA profiling has been applied to human forensic science, diagnostic medicine, family relationship analyses, animal and plant sciences and to conservation studies on endangered species. Although its application to the study of prokaryotic organisms has been more limited, it has been used in determining species identity and strain relatedness and has proven its worth as a tool for taxonomic studies. As databases are developed, the exciting possibility of using DNA profiling for predictive screening becomes achievable. Predictive DNA profiling provides a technique with which to screen culture collections of organisms which have to a greater or lesser extent been characterized using the more standard biochemical and physiological tests and to enable one to categorize individual isolates into groups according to the degree of genetic relatedness based on a number of defined sequences of DNA. Where this has been done it has become apparent that one group may express a phenotypic trait to a greater or lesser extent than another. Simply by assigning an isolate to one or other of these genetic groupings will thus enable prediction of its potential in terms of the defined trait.

In the early days of the environmental biotechnologies, when perhaps more structurally simple xenobiotics were being investigated, the competent microbes were often found to belong to the *Pseudomonas* genus. However, today as we look for microbes to degrade the more complex environmental chemicals it is becoming apparent that the enrichment cultures are not yielding just pseudomonads; other genera are represented, such as *Agrobacterium, Arthrobacter, Corynebacterium* and frequently members of the genus *Rhodococcus*. The taxonomy of many of these genera has been developed in recent years using molecular biological techniques, but now DNA profiling offers the opportunity to provide subgroupings of the various strains of each species present in major culture collections. From these results predictive strategies, based on defined activities of members of the different groups and the use of specific probes for target structural genes, will open culture collections as a source of new biological activities not previously identified.

Genetically engineered microbes

Thus far the application of the techniques of modern molecular biology has been considered as a tool for search, discovery and characterization of natural microbes for use as biocatalysts in biotreatment of environmental chemicals. The techniques also offer the opportunity to design and construct pollution control agents in the laboratory, and to produce genetically engineered microbes (GEMs) tailored as biotreatment agents, by designing new catabolic pathways for the degradation of target pollutants.

The precise nature of *in vitro* manipulation of selected DNA sequences offers the ability to introduce new genetic material which can be expressed in the new host under defined and controlled conditions. Well-characterized genes and DNA sequences can be cloned and selectively combined with pre-existing catabolic genes present in the host creating new or improved catabolic pathways. These techniques bring to the laboratory a method for specifically engineering to order, defined new combinations of genetic information which originated in the natural gene pool. In this regard it is no different from establishing enrichment conditions which promote the natural exchange of genetic material (*in vivo* genetic engineering), where apparently random gene reassortment eventually results in a combination of genes favoured by the selective conditions employed. The controlled nature of the *in vitro* techniques means that such recombinations can be achieved much more efficiently than by *in vivo* methods. The continuous culture enrichment techniques that developed the 2,4,5-T-degrading pseudomonad represented *in vivo* engineering, but the whole process took 18 months to complete.

There are two basic approaches used in the *in vitro* production of a GEM with a new catabolic activity. If the target compound is structurally related to a compound that an organism can already degrade and the pathway for the degradation is known, then the first step is to identify which steps in the biochemical sequence prevent the target compound being degraded. For instance, it may be that there is an additional substituent which, if removed, would allow the pathway to function towards the target compound. This represents an extension of the existing pathway in a vertical sense (Fig. 4.3). Alternatively, perhaps the limited substrate range of key enzymes in the pathway can be overcome such that analogous compounds can also be degraded by the pathway. This can be considered as a horizontal development of the pathway (Fig. 4.3). Once the blocks to the degradation have been identified, alternate enzymes to catalyse the transformation of the troublesome metabolites can be introduced. These methods represent the addition of bolt-on-extras to existing pathways and as such extend the substrate profile of the organism. This procedure apparently mimics what natural bacteria have done previously through the exchange of cassettes of genetic information which build up to form the catabolic plasmids. This is graphically represented by considering the closely related pathways encoded by the salicylate, naphthalene, toluene and octane degradation plasmids found in the pseudomonads.

If it is not possible to identify a pathway which degrades compounds that are closely related to a target compound then it is feasible to construct a complete pathway in a GEM. Once a catabolic sequence has been determined it would then be necessary to identify the enzymes which will be required to catalyse the sequence and then to find the enzymes in natural organisms. Cloning each of the enzymes into a single

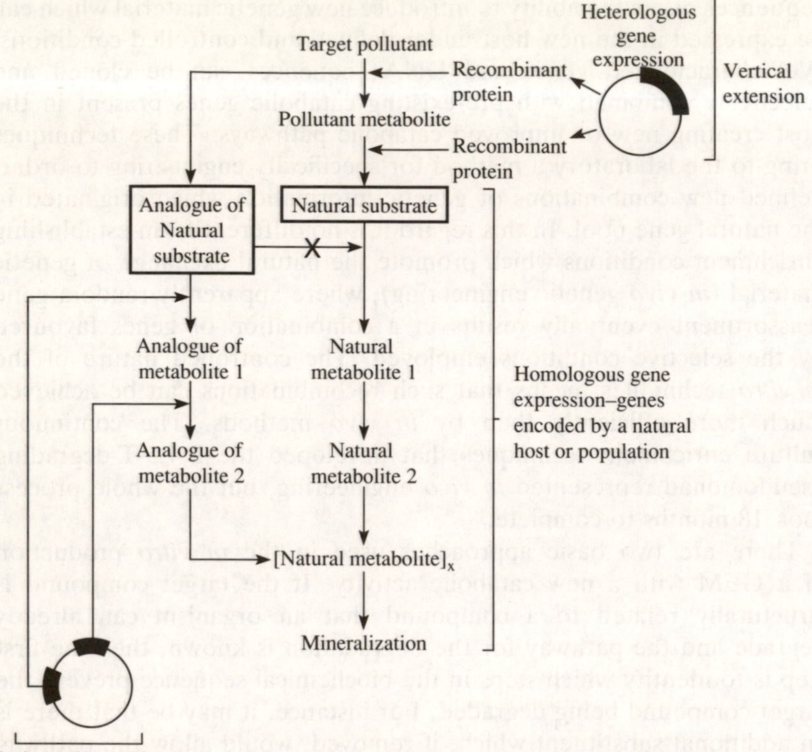

Fig. 4.3. The use of recombinant DNA techniques to augment natural catabolic capabilities.

bacterium, or a number which are later to be used as a consortium, and obtaining sufficient levels of expression in the GEM(s) would then enable a functional pathway to be constructed.

Recombinant DNA technologists look to clone enzymes with relaxed substrate specificities, so that the GEMs have the widest possible substrate range. Once these genes have been cloned they represent cassettes of genetic information which can be recombined in a mix-and-match fashion to generate new catabolic capabilities. As the library of cassettes increases so does the potential for constructing ever more complex degradative pathways to create 'designer' microbes for the biodegradation of toxic environmental chemicals.

GEMs for the treatment of toxic environmental chemicals

It is apparent that GEMs can be created in the laboratory which, in that laboratory, are capable of the degradation of target pollutants. But, as with all the environmental biotechnologies, unless this translates to successful application in the field they are of no commercial interest.

To date GEMs have played no role in the remediation or control of environmental pollution. The major reason for this is the concern over release of such organisms into the environment. However, this has also meant that full-scale trials have not been undertaken in order to test the claims of the GEM protagonists. Hence, it is not possible at this stage to say whether the use of GEMs would represent an environmental panacea or, as many environmental pressure groups believe, be equivalent to opening one of Pandora's boxes, or whether in reality nothing would happen because the GEMs would fail to compete in the natural environment. The supporters of the use of GEMs for environmental pollution control should constantly be aware of the problems caused by the early over-optimistic claims for the 'magic soups' of microbes and the damage to the development of environmental biotechnology caused by the failure of these preparations when applied in the field. Wild claims for GEMs without field trials to support them could again set back the application of the whole technology if they are subsequently shown not to achieve what is claimed.

This is not the place to debate the pros and cons of environmental release but assuming that all parties can be convinced of the safety of GEMs, would organisms engineered in the test tube be effective agents for pollution control outside the laboratory? Firstly, there is the problem of their creation. A look at the recent scientific literature shows that increasingly it is new bacterial species, even isolates belonging to new genera, that are being identified as being capable of degrading the complex organic pollutants. The recombinant DNA technologists are continually developing new techniques for application to an ever wider group of organisms but techniques with which to modify a number of the more significant biocatalysts for pollution control have not yet been developed. This means that the concept of building on natural pathways to further extend the catabolic potential has its limitations.

Complex metabolic pathways have been constructed in the laboratory; both anabolic, as, for example, the construction of plasmids encoding for the synthesis of antibiotics, and catabolic plasmids have been created. However, the maintenance of their integrity is based on selective conditions imposed in the laboratory. For the catabolic recombinant systems the selective pressure is represented by the supply of the target pollutant as the sole carbon source. The question arises as to what degree of stability the recombinant strains will show away from these selective conditions. If the GEMs are to be employed as pollution control agents in the environment they must firstly survive in that environment. To do this they must compete successfully with the microbes already present which are already adapted to that environment. An environment such as a hazardous waste site contains thousands of different chemicals. Any GEMs introduced would have to become established in the pre-existing mixed microbial community which will consist of competitive species, already exploiting most if not all energy and growth substrates, including those for which the GEMs have been designed. Darwinian fitness would thus be a prerequisite for success, but as a consequence of the test tube engineering, GEMs, which may be engineered to degrade a few target pollutants effectively, are to be considered as metabolic cripples in more general terms. This would appear to make it unlikely that they would survive in the environment long enough to achieve the purpose for which they were designed.

If a 'superbug' was developed which could out compete members of the natural microbial populations the chances of it being considered as being safe to release would appear to diminish. The risk factor would increase as the GEM becomes more 'fit', especially as the early release experiments are apparently suggesting that considerable dispersal does occur. It raises the question as to whether there is a balance that can be struck between sufficient fitness to achieve the objective of the release and an in-built instability to ensure public acceptance of the initial release.

GEM protagonists suggest that the first use of GEMs will be in end-of-pipe bioreactors where physical containment will overcome the 'release' problem and the selective conditions of the waste stream will ensure the GEMs have a competitive edge. There are, however, problems with both stances. When considering containment, microbial action on the pollutants will involve mineralization which is usually growth-associated and so results in biomass production. This will result in contamination of the effluent stream with bacterial cells. Whilst these can be removed, the cost of ensuring 100% removal, which would be a prerequisite for total containment, is likely to be prohibitive. The low unit value of waste requires that the treatment process is simple and cost-effective; such processes cannot be run under sterile conditions. The result of this is that natural organisms will continually 'infect' the bioreactor and, given time, as a consequence of the very selective pressures cited as the means of ensuring the stability of the GEMs in the reactor, natural organisms will be selected and evolve to be more effective in the reactor environment than the GEMs and so displace them. Indeed one of the advantages of using microbial systems for pollution control is that the very process tends to select for more efficient biocatalysts.

Natural versus genetically engineered biocatalysts for pollution control

Consideration of the genetic, physiological and biochemical versatility and plasticity of the microbes isolated to date, together with the untapped natural genetic resource represented by those microbes yet to be cultured in the laboratory, raises the question as to the need to even consider the use of GEMs. Why should the pollution control industry use a biocatalyst which after it has affected the biotreatment can then be perceived as a hazard?

Microbial technologists have hardly looked below the surface of the natural pool of microbial diversity. When new organisms have been isolated as organisms capable of degrading pollutants their biochemical versatility has been seen to be immense. Thus, setting aside the other drawbacks of GEMs, the question arises of why we even need to consider them? There will be a natural organism to do the job which will be isolated if a targeted intelligent approach is taken to screening the environment for its presence.

Further reading

BULL, A.T. (1991) Biotechnology and biodiversity. In Hawksworth, D.L. (ed.) *The Biodiversity of Microorganisms and Invertebrates: Its Role in Sustainable Agriculture*, pp. 203–219. CAB International.

BULL, A.T. (1992) Isolation and screening of industrially important organisms. In Vardar-Sukan, F., Sukan, S.S. (eds) *Recent Advances in Biotechnology*, pp. 1–17. Kluwer Academic Publisher, The Netherlands.

BULL, A.T. and HARDMAN, D.J. (1991) Microbial diversity. *Curr. Opin. Biotechnol.*, **2**, 421–428.

GOODFELLOW, M. and O'DONNELL, A.G. (1989) Search and discovery of industrially-significant actinomycetes. In Baumberg, S., Hunter, I.S., Rhodes, P.M. (eds) *Microbial Products: New Approaches*, pp. 343–383. Cambridge University Press.

HAWKSWORTH, D.L. (1991) The fungal dimension of biodiversity: magnitude, significance and conservation. *Mycol. Res.*, **95**, 441–452.

HOPWOOD, D.A. and CHATER, K.E. (1989) *Genetics of Bacterial Diversity*. Academic Press, London.

SAYLER, G.S. and LAYTON, A.C. (1990) Environmental application of nucleic acid hybridization. *Ann. Rev. Microbiol.*, **44**, 625–648.

TIMMIS, K.N., ROJO, F. and RAMOS, J.L. (1990) Design of new pathways for the catabolism of environmental pollutants. In Kamely, D., Chakrabarty, A., Omenn, G.S. (eds) *Biotechnology and Biodegradation*, pp. 61–82. Gulf Publishing Co., Houston.

ORGANIC POLLUTANTS

Chapter 5

THE CARBON CYCLE AND XENOBIOTIC COMPOUNDS

The fate of organic material entering the natural environment is determined by the global elemental cycles. Hence, xenobiotic organic carbon enters the cycles as a result of biological activity. Most elements are subject to some degree of cycling, although the rates vary as a consequence of the role of the element in the life of the planet. Those intimately involved as components of living organisms are subjected to the greatest turnover and from a quantitative point of view the carbon cycle is the most important. Carbon, as with all elements, exists within pools, that is, reservoirs which lock up the element in a given form (Fig. 5.1). Putting tonnage values on the various pools can only ever be an approximation, but on the basis of values presented in Fig. 5.1, 60% of the world's carbon is locked up in geological deposits with 1.8×10^{16} tonnes found in rocks, especially as carbonates in limestone and dolomite, and 2.5×10^{16} tonnes as coal, shale, mineral oil, natural gas and bitumen. Dissolved bicarbonate in oceanic waters accounts for 3.6×10^{13} tonnes whilst living and decaying organic matter represents 2.7×10^{12} tonnes. Atmospheric carbon dioxide (CO_2) accounts for 6.4×10^{11} tonnes of carbon.

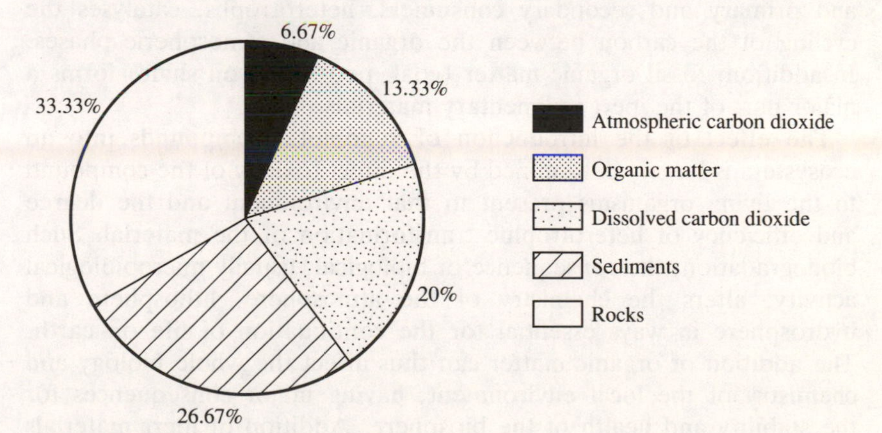

Fig. 5.1. Relative carbon pool sizes.

The comparative size of the reservoirs is important when considering stability in respect of the disruptive influence of influxes of fresh organic material, as a result of abiological and biological activities. Changes in transfer rates between reservoirs alter the chemical and physical environments which in turn affect the biochemistry and living populations in the environment in quantitative and qualitative terms. Ecosystems are in a dynamic equilibrium with populations changing in response to levels of matter and energy in the system. The most active cycling occurs within the organic matter and between the organic matter and the atmosphere, and as a consequence it is these pools that are most noticeably affected by pollution when considered in terms of time scales defined by living organisms. The 'fast' carbon cycle involves the dissolved CO_2 and the biota in the mixed layers of the oceans, the atmosphere and the terrestrial biosphere. This cycle period is measured in years or decades whilst exchange with the deep sea is measured in hundreds of years. This is in contrast to the largest reservoir, the geological deposits, which cycles so slowly that in terms of the same time scales, turnover is insignificant. The 'slow' cycle is measured on a geological clock involving turnover from rocks and sediments by weathering of rocks and dissolution and precipitation of carbonates, considered to be in the order of 100 000 years. In the organic reservoir large quantities of carbon are bound up in the world's plant life as a consequence of photosynthesis. Oceanic algae account for more than 5×10^9 tonnes of organic carbon whilst terrestrial plants represent 4.5×10^{11} tonnes. However, the majority of organic carbon is present in dead organic matter.

Biological activity is the major driving force of the carbon cycle, with composition of, and movement between, the carbon reservoirs largely dictated by the efficacy of the flora and fauna in transformation of the various carbon forms. The activity of primary producers, autotrophs, and primary and secondary consumers, heterotrophs, catalyses the cycling of the carbon between the organic and atmospheric phases. In addition, fossil organic matter (coal, petrol and oil shale) forms a major part of the inert sedimentary materials.

The effect of the introduction of xenobiotic compounds into an ecosystem is largely determined by the direct toxicity of the compound to the living organisms present in that environment and the degree and efficiency of heterotrophic transformation of the material. Such biodegradation, the consequence of biological, usually microbiological activity, alters the chemistry of the atmosphere, lithosphere and hydrosphere in ways essential for the continuation of life on earth. The addition of organic matter can thus affect the whole biology and chemistry of the local environment, having major consequences for the stability and health of the biosphere. Addition of inert materials considered to be persistent in the environment, such as natural compounds like humus, lignin, tannin, melanin and other polyaromatic compounds a consequence of, for example, agricultural practices, may have detrimental effects. However, it is the addition of large quantities of biodegradable organic matter and toxic xenobiotic compounds that is particularly damaging to the environment.

Exogenous organic matter entering the environment

The increased biological activity caused by the addition of biodegradable organic material alters the whole chemistry of the ecosystem. Redox potentials fall as a result of heterotrophic activity leading to a loss of nitrogen from the environment and the reduction of iron and related compounds which have major effects on the future productivity of the affected environment. Addition of degradable organic matter will have a temporal, quantitative and qualitative effect on the heterotrophic microbial populations. As a consequence, xenobiotic compounds have both direct and indirect, beneficial and adverse effects on natural microbial communities. A typical response (Fig. 5.2) is caused by initial death of sensitive organisms. The utilization of released nutrients and selection of competent organisms to degrade the new organic material then leads to an increase in biomass. This is followed by a slow decline as the excess organic material is utilized until such time as the population returns to a level which is supported by the normal organic material in the environment. The overall quantitative effect on microbial biomass in

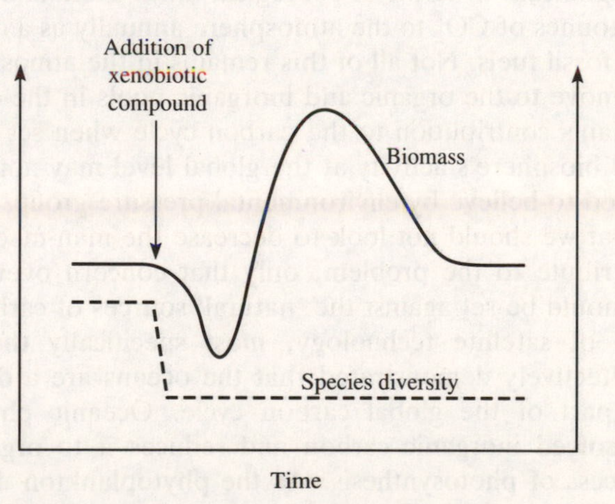

Fig. 5.2. Qualitative and quantitive effects of addition of a xenobiotic compound on natural microbial communities.

the ecosystem is negligible. However, a qualitative examination of the components of the population reveals more significant effects. A consequence of the death of sensitive organisms and competition induced by the presence of the new organic material has an adverse effect on species diversity. For example, in more extreme cases a number of soil experiments have shown that the diversity of fungal species can drop 10-fold as a consequence of such additions, whilst the subsequent microbial populations come to be predominated by *Pseudomonas* species and spore-forming microorganisms. Conversely, the metabolism of microbial communities in the environment has direct and indirect effects on the fate of the xenobiotics.

There is a tendency when considering the biogeochemical cycles, the biological activities affecting the elemental reservoirs, to consider them independently. However, all the cycles are, to a large extent, interdependent; factors influencing one cycle will have a concomitant effect on the others. Hence, exogenous organic material has a major influence on the productivity of an environment by providing additional carbon and also by influencing the equilibrium of other elements present there.

The cycling of carbon between the pools has, over geological periods of time, reached a dynamic equilibrium. However, of major concern today is how human activities are influencing the equilibrium. There is considerable uncertainty about the size of the pools and the turnover times, thus the long-term effect of man's activities leading to the dumping of organic material and burning of fossil fuels and plant residues cannot be gauged accurately. The indications of increased atmospheric CO_2 levels, leading to expressions of concern about global warming, may be the result of human activity, but are the increases observed significant in terms of geological time? Human activities add $5–6 \times 10^9$ tonnes of CO_2 to the atmosphere annually as a consequence of burning fossil fuels. Not all of this remains in the atmosphere; large quantities move to the organic and inorganic pools in the oceans. The effect of man's contribution to the carbon cycle when set against that of the total biosphere's activity at the global level may not be as great as we are led to believe by environmental pressure groups. This is not to argue that we should not look to decrease the man-made emissions which contribute to the problem, only that concern over man-made pollution should be set against the 'natural' sources of carbon.

The use of satellite technology, most specifically the Nimbus-7 satellite, effectively demonstrated that the oceans are a dynamic and important part of the global carbon cycle. Oceanic phytoplankton utilizes dissolved inorganic carbon and reduces it to organic carbon by the process of photosynthesis. As the phytoplankton dies it sinks, so removing carbon from the upper ocean, creating a deficit of carbon in these waters which is in turn replaced from the atmosphere. This 'biological pump' shifts carbon from the 'fast' cycle to the 'slow' cycle defined by the deep oceans and sediments. As such it provides a buffering capacity to cushion the effect of continued addition of carbon dioxide to the atmosphere. The satellite imaging of oceanic phytoplankton pigments has led to a realization of the extent of these blooms and they indicate that the buffering capacity of this photosynthetic process is much greater than had been previously realized. The global importance of phytoplankton as a carbon sink is a contentious issue as indeed are the relative source/sink activities of the other obvious biological reservoir, the tropical rain forests.

The presence of halohydrocarbons in the environment provides a second example for consideration in the debate of natural versus xenobiotic input into our environment in terms of the greenhouse effect and ozone depletion (Symonds *et al.*, 1988). During the period 1982–1984 industrial production of chloride and fluoride as halohydrocarbons is estimated to have led to the release of 2.28×10^{12} and 0.273×10^{12} g of chloride and fluoride, respectively, into the environment, whilst the biological contribution, mainly as chloromethane from the oceans and burning vegetation, was $1.4–3.5 \times 10^{12}$ g (Suida and Debernardis, 1973; Lovelock, 1975).

At a local level, human activity can have a demonstrably devastating effect on the quality of the environment. Point source pollution can destroy an ecosystem, for example, even domestic sewage discharge, unless controlled through a municipal treatment facility, can result in destruction of the local environment and in human terms cause both aesthetic and health risk problems.

The role of natural microbial communities

All natural environments contain a great diversity of microorganisms competing for low levels of unevenly distributed substrates whilst constantly being subjected to fluctuating physico-chemical conditions. In natural environments, no matter how extreme or selective, it is unlikely that axenic (pure) cultures of single species will be found. The existence of such biological diversity can in part be explained by spatial and temporal features of microenvironments. The addition of organic materials to soils greatly increases the number and activity of the microbial populations. These microorganisms are not uniformly distributed in soils when considered in the horizontal plane but are characteristically largely associated with the upper soil levels in a vertical profile. Whilst the availability of carbonaceous substrates plays a major role in defining distribution and species composition, factors such as the soil's gaseous phase will also be a factor. For instance, the availability of oxygen may result in the non-uniform vertical profiles.

The gross distribution of substrates remains relatively stable with time in an undisturbed soil, but, on a microscopic scale, there are major changes occurring all the time establishing microenvironments which exert different selective pressures on the microbial communities. For example, a small soil particle may create a microenvironment in terms of oxygen content; the outside may be aerobic whilst the centre is anaerobic, with an oxygen concentration gradient extending between the two extremes. This implies that microbes with completely different metabolic constitutions can co-exist even in soil particles. In a world where the functional environment is measured in nano- or micrometres, pH, oxygen and nutrient concentration gradients play a major role in determining the extent of microbial activity and on this scale such environments are dynamic and thus account for the vast range of different microorganisms found in one locale.

It is also apparent that microorganisms interact through community structures and that these interactions provide a selective advantage for the constituents of the mixed populations. The form of the interactions extends over a complete spectrum from loose associations based on the sort of interactions which exist in food chains or as a result of population successions, to more complex situations where tight associations are based on specific mutually beneficial relationships between interacting populations.

Simple interactions take the form of autogenic succession, a reflection of the reciprocal relationship between the microbe and its environment. The environmental conditions influence which type of microorganism is present at any given time. In turn the microbial activity determines future environmental conditions. When those conditions no longer represent those best suited to the dominant organism a second species will succeed the first. A simple example of this is one in which an initial aerobic environment containing glucose supports the growth of a bacterium which results in a decrease in the environmental pH and the production of metabolites which become toxic to the first organism. The conditions are then suitable for a second group of bacteria whose metabolism leads to anaerobic conditions such that the second group is succeeded by the third. In any given environment the relative number of each species will be determined by the prevailing conditions; when conditions favour one species it will dominate the total population and as conditions change its fortunes will wane and a second will succeed as the dominant constituent of the population. Table 5.1 lists the types of microbial communities based on more specific interactions.

Under balanced conditions the number of microorganisms found in soils exceeds that in freshwater or marine environments. A diverse range of bacteria, fungi, algae and protozoa are to be found in soils of those groups. The bacterial numbers, typically 10^6–10^9 per gramme of soil, are the greatest, although in view of their size in well-aerated

Table 5.1 Types of microbial community interactions (after Slater and Lovatt, 1982)

1. Structure due to reciprocal provision of specific nutrients
2. Structure due to concerted removal of inhibiting metabolic products
3. Structure due to improvements in individual component growth parameters
4. Structure due to concerted combined metabolic capabilities
5. Structure due to cometabolic stage in metabolism
6. Structure due to transfer of hydrogen ions
7. Structure due to primary and secondary utilizers

soils they usually represent less than half of the soils total microbial mass. However, as the redox potential drops, bacteria tend to account for almost all the viable biomass present in the environment.

The fate of organic xenobiotic compounds in the environment

In terms of quantity produced and the potential for adverse effects on the environment the most important types of xenobiotic compounds are those considered as organic. The compounds listed in Tables 1.1a and b have been identified in soil systems and a number are known or suspected to be carcinogens, teratogens and/or mutagens.

Xenobiotic compounds enter the natural environment either directly through the application of pesticides or accidental spillage, or more indirectly through improper disposal techniques resulting in pollution as a consequence of gaseous emissions, contamination of aqueous run-off or by leachates. The fate of a xenobiotic compound in the environment is largely determined by the chemical structure of the compound and the consequential interactions between the compound and the physical, chemical and biological parameters of the environment. This can be exemplified by considering 'the soil environment'. The type of soil, its mineral and organic content, moisture, oxygen content and temperature will all affect the rate of degradation resulting from abiological processes, such as oxidation, photolysis, hydrolysis or adsorption and determine the availability of exogenous organic materials for biological transformation.

Soils serve as a receptacle for most organic chemicals released by man, either intentionally, as with agrochemicals, or as a consequence of accidents, and thus have a major role in determining the quality of our environment. Hence, pollution in soils tends to be in a more concentrated form than in other environments where the pollutants are less mobile. Their presence in the soil also results in interactions between the pollutants and the higher plants. Plants play a major role in introducing organic material into soils, with 10–50% of fixed carbon entering the soils via this route. Plants accumulate pollutants in their tissues and so start the biomagnification of pollutants in food chains. They then cause subsequent reintroduction of the pollutants into the soils as they die and decompose, and when they act as food for herbivores they can promote the spread of pollutants in the faeces of the animals. The soil environment is represented predominantly by a solid phase consisting of inorganic minerals, organic matter comprising decaying organic material derived from plant, animal and microbial life in early stages of degradation, and soluble components comprising first stage degradation macromolecular products like carbohydrates and proteins, and complex polyaromatic components derived from lignin degradation and microbial biosynthesis collectively termed humic material.

The ratio of minerals to organic matter present in the soil, determines the physico-chemical properties of the soil. The relative levels of these components define the porosity, water-holding capacity, ionic exchange ratios and aggregate stability. In terms of the activity of xenobiotic compounds in the soil, the most influential components appear to be the colloid-sized minerals or organic matter (humus). This is in part due to their high surface-to-volume ratios, ionic properties and affinities for water molecules.

The aqueous phase in soil is important in that it affects the growth of the microbial communities and also the fate of xenobiotics in the soil as a consequence of its solvent properties. It also controls the gaseous content of the soil which in turn affects the biology and chemistry of the local environment and the swelling of clays, thus changing the surface area-to-volume ratios. The present understanding of the interaction of water with clay surfaces suggests water molecules will be highly ordered as a consequence of charge interactions thus decreasing the availability of the water molecules to the constituent microorganisms. These interactions will define the microenvironment in which the microbial populations are active.

The soil gaseous phase is enriched with CO_2 compared to the atmosphere above the soil as a consequence of plant respiration. The extent of the enrichment is determined by the rate of respiration. Oxygen concentrations can fall to zero, creating anaerobic microenvironments where reductive metabolism such as fermentation, denitrification and methane formation predominates.

A consequence of the interactions between the solid, liquid and gaseous phases is that microorganisms, substrates, products, inorganic ions and water molecules interact at soil particle–liquid or colloid–liquid interfaces, where the colloids are predominantly humic substances, viruses, bacteria and clay minerals (hydrous aluminosilicates). Hence, the indications are that microenvironments created by surface interactions are important in influencing the activity and survival of microbes in soil. This phenomenon is also significant in water columns where microbes are seen to concentrate on particulate matter where nutrient levels are higher than in the surrounding waters. Any such concentration in an oligotrophic environment will greatly affect the rate of microbial metabolism, where the rate of bacterial growth may be less than 1% of the maximal rates, as compared with the growth rates in laboratory cultures, because in the natural oligotrophic environments 'famine' determines the rate of growth.

In such nutrient-poor environments as defined by most soils and waters it is believed that as much as 90% of the total bacterial population can be attached to surfaces as defined by interfaces whether the surface represents a solid one as provided by clay particles or the 'particulate' unicellular algae, or air–water interfaces, bubbles or the surface of a water body. Bacteria isolated from oligotrophic environments are those capable of growth on media containing 1–15 mg carbon l^{-1}, levels which are equivalent to those found in open oceans. Physiological adaption enables these bacteria to survive. Important in these adaptive processes are biochemistries with high substrate affinities coupled with low specific growth rates. These microbes are classed as autochonthonous, and their relative importance in natural unpolluted nutrient-limited environments is apparent. Whereas, in polluted environs, and indeed in the laboratory as a consequence of the nutrient-rich isolation procedures usually employed (up to 1000-times greater than the oligotrophy levels), the zymogenous organisms, with higher growth rates and lower substrate affinities, prevail.

The constituents of natural microbial populations present in the soils are determined by the same physico-chemical environment. Characteristics of pH, redox potential, moisture and nutrient concentrations and availability will define the selective conditions which determine the type of microorganisms present in the environment. Different microbial populations will represent communities with different catabolic potentials such that the biological fate of the xenobiotic compound will differ from environment to environment depending on the catabolic potential found in each system.

The availability of the xenobiotic compounds to the microbial populations is also determined by the structure and mineral content of the soil. The clay mineral content also affects the toxicity of inorganic and organic xenobiotics. Whilst colloid–surface interactions provide a concentrating mechanism which is important in oligotrophic environments, the adsorptive capacity of clays also provides a buffering system, reducing high level contamination by reducing the concentration of reactive toxicants through immobilization, so protecting sensitive constituents of the microbial populations in the contaminated soil. Such adsorption also renders xenobiotics inaccessible to microbial action. A prime example of this is the fate of the bipyridilium herbicide Paraquat in soil. In the laboratory, microbes have been isolated which are capable of utilizing Paraquat as a carbon and nitrogen source for growth. However, in an aqueous environment Paraquat becomes strongly cationic and in a soil environment it binds strongly to humic matter and irreversibly to clays (Fig. 5.3). This adsorptive stabilization results in a pesticide, which, whilst chemically defined as biodegradable, is in practice recalcitrant simply because it is inaccessible to biological activity. There are, however, instances where adsorption is vital for metabolism of compounds, for example, physical attachment to cellulose, and for bitumen degradation. Organic matter itself also acts as a buffering system, reducing the soluble concentration of toxic organic compounds through mechanisms such as ion exchange, protonation covalent bonding, H-bonding, van der Waals' forces and hydrophobic interactions.

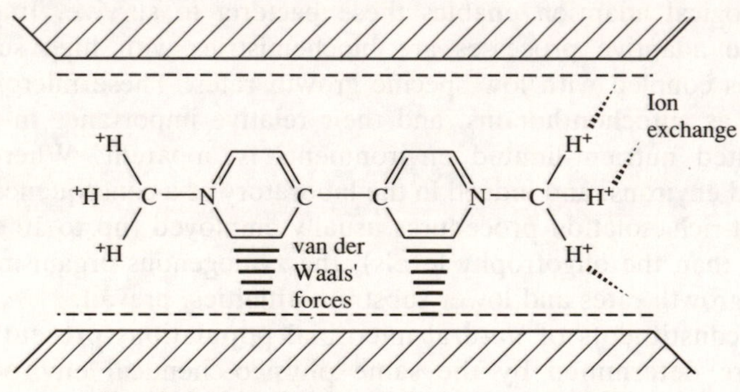

Fig. 5.3. Paraquat adsorption to clays.

Biodegradation

Most significant degradation of organic compounds in the natural environment is mediated by biological agents. Consideration of rates for the degradation of surface compounds reveals that the rates are too great to be fully catalysed by physico-chemical processes. Energy flow through the biosphere is intimately associated with the processing of organic matter, its degradation and ultimately mineralization. The degradation and mineralization processes are largely driven by the chemoheterotrophic microorganisms. Since the activities of these organisms are dependent on substrate availability these microbes live under a 'feast or famine' regime and even in soils apparently rich in organic matter the prevailing substrate levels usually represent famine conditions.

The effectiveness of biodegradation in the natural environment is largely determined by availability, but the degree of recalcitrance is also influenced by the chemical structure of the xenobiotic compound. If the structure is similar to that of a natural substrate present in the contaminated environment, such that existing catabolic pathways, perhaps after minor modifications, can metabolize the compound, then it is more likely to be biodegradable in that environment. If the structure is complex the rate of degradation will be slower and is more likely to be incomplete. This is more likely to be so when the compound is an insoluble polymeric material like lignin or plastic. In these instances the polymeric nature is responsible for the recalcitrance as the corresponding monomers, dimers or trimers are often biodegradable. Where such degradation does occur it is catalysed by extracellular enzymes. The degree of branching of the molecule is also inversely related to the rate of degradation.

The degree of recalcitrance is further enhanced by the presence of exotic substituents, those not usually encountered in the environment as a result of biological processes. This may be in the nature of the complexity of the substituent (Fig. 5.4a) or in the position of the substituent on the molecule (Fig. 5.4b). In these instances enzymes

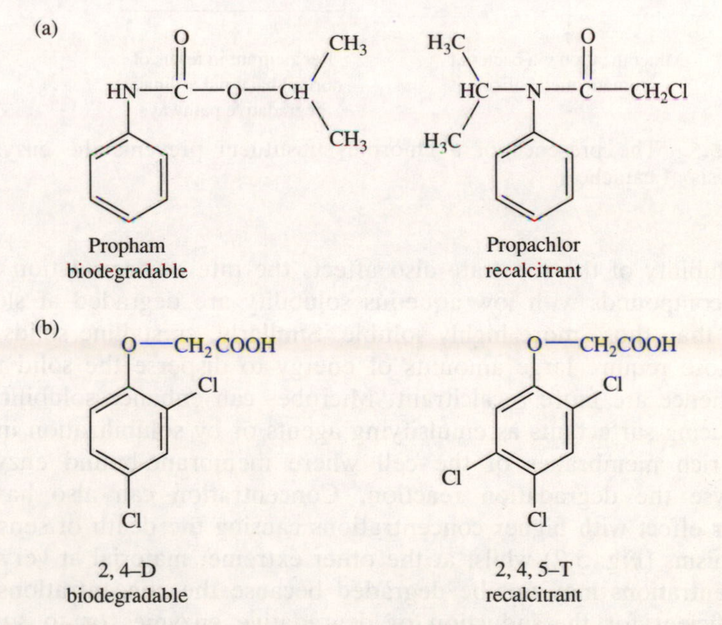

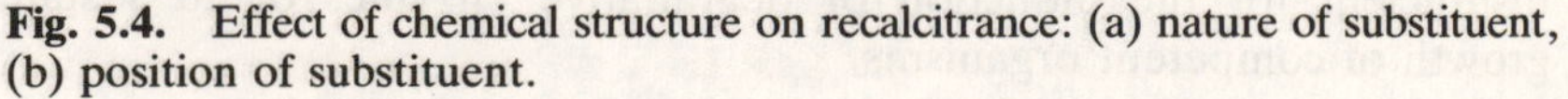
Fig. 5.4. Effect of chemical structure on recalcitrance: (a) nature of substituent, (b) position of substituent.

which would recognize the molecule as a substrate under normal circumstances fail to do so because of the interference of the catalysis by the presence of the novel substituents (Fig. 5.5).

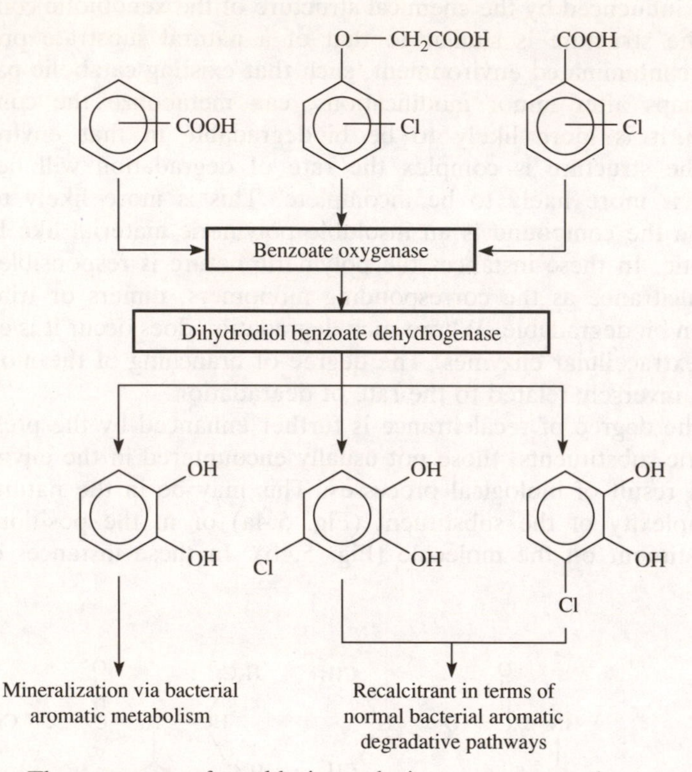

Fig. 5.5. The presence of a chlorine substituent prevents the enzymatic catalysis of catachols.

Solubility of the substrate also affects the rate of degradation such that compounds with low aqueous solubility are degraded at slower rates than those more highly soluble. Similarly, crystalline solids like cellulose require large amounts of energy to disperse the solid form and hence are more recalcitrant. Microbes can enhance solubility by producing surfactants as emulsifying agents or by solubilization in the lipid-rich membranes of the cell where membrane-bound enzymes catalyse the degradation reaction. Concentration can also have a major effect with higher concentrations causing the death of sensitive organisms (Fig. 5.2) whilst at the other extreme, material at very low concentrations may not be degraded because the concentrations are insufficient for the induction of degradative enzymes or to sustain growth of competent organisms.

Recalcitrance tends to imply a complete stability in the environment. There are, however, degrees of recalcitrance defined by the reference time scales employed. The herbicide 2,4-D is readily described as biodegradable such that when used at recommended field application rates it is removed from the environment over a period of 2–4 weeks. Over a similar time scale Monuron could be considered as recalcitrant as it will persist for 22–100 weeks depending on the nature of the environment. Dichlorodiphenyltrichloroethane (DDT) is extremely recalcitrant in the environment, taking 3–15 years to degrade.

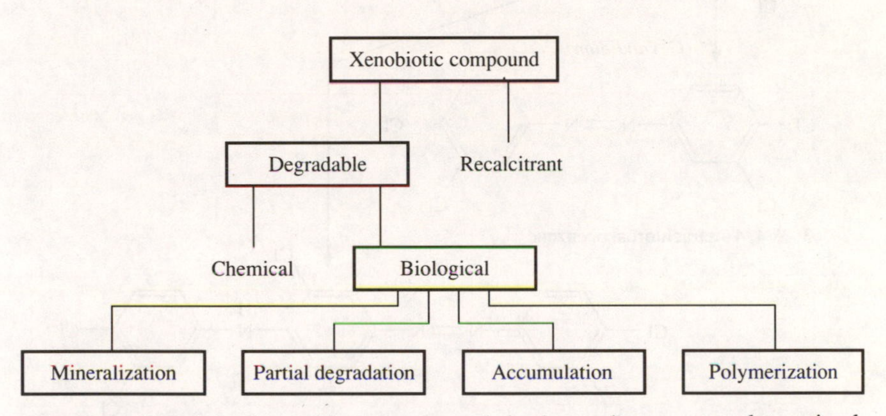

Fig. 5.6. The fate of xenobiotic compounds in the environment as determined by biological activities.

The biological fate of an organic xenobiotic entering the environment can be considered as described in Fig. 5.6. The advent of the study of environmental microbiology was as a consequence of the observed biomagnification or bioaccumulation of xenobiotic compounds through food chains. Such evidence leads microbiologists to question the principle of microbial infallibility as proposed by Bejenick (see Alexander, 1965) which was based on a view of the vast degradative potential of natural microbial communities. Pesticides like DDT were seen to accumulate in the food chain with numerous examples of adverse effects on the ultimate predator in the chain. It should be noted that biotransformation of complex organic molecules does not necessarily lead to structural simplification. Polymerization as a means of detoxification has been observed as in the case of the biotransformation of 3,4-dichlorophenylpropionamide (Fig. 5.7). Biodegradation can take two forms, the most desirous (from an environmental point of view) is the mineralization of the substrate to its constituent elements. This is most likely to occur when the molecule

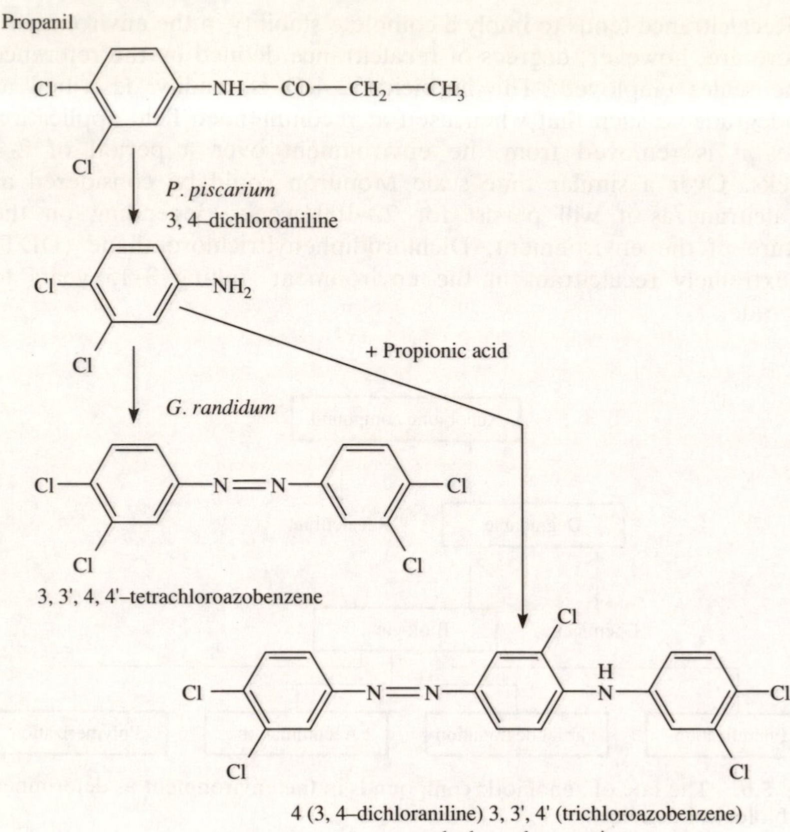

Fig. 5.7. The biopolymerization of 3,4-dichlorophenylpropionamide.

can be utilized as a source of carbon and energy by competent organisms. Where this is not the case, biotransformation may occur as a consequence of co-metabolism. Under these circumstances the microbial populations present there are unable to utilize the xenobiotic compound as a sole source of carbon or energy but whilst metabolizing other substrates they are capable of the transformation of the target xenobiotic. Where the products of co-metabolism are amenable to further degradation they can be mineralized. However, in many instances such processes lead to incomplete degradation and the accumulation of metabolites which in some instances are more toxic than the original xenobiotic compounds. An example of such incomplete degradation is found in the anaerobic degradation of perchloroethylene. Reductive dechlorination, it has been suggested, is linked to methanogenesis (Vogel and McCarty, 1985; Fathepure *et al.*,

1987). The reaction pathway leads to production of vinyl chloride, a compound which is more toxic than the starting material, which accumulates under anaerobic conditions as no further biotransformation occurs.

The nature of the genetic make-up of microorganisms (Ch. 4) is such that the presence of a new metabolizable substrate will lead to the selection of a competent microorganism or a consortium of microorganisms which can utilize the new substrate as a source of nutrients. The time taken for the development of this structured community will depend on the nature and complexity of the new substrate. Hence, on the first challenge with a xenobiotic compound there is invariably a lag phase before the compound is degraded. Once the adapted organisms have proliferated they remain in the environment such that subsequent additions of the same compound result in its more rapid removal from the environment (Fig. 5.8).

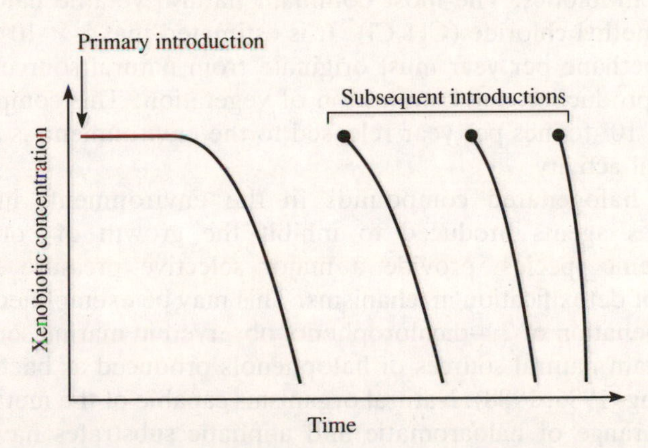

Fig. 5.8. Adaptation of natural microbial communities to the repeated addition of a xenobiotic compound.

Haloorganic compounds in the natural environment

Halogenated compounds are widely distributed in the biosphere as a result of natural production. The human contribution to the pool of haloorganics should thus be seen against a background of such natural compounds and degradative processes.

Suida and Debernardis (1973) listed more than 200 organic compounds containing covalently-bound halogens which they considered to have been produced naturally. It would appear that most natural organohalogens have been produced by biological activity. Geothermal activity, through passive degassing and actively erupting volcanoes, contributes significantly to global levels of hydrogen fluoride and hydrogen chloride but the only organohalogens released by volcanic activity, for instance, the methyl halides, mostly methyl chloride, are released as a result of burning vegetation commonly associated with lava flows.

It might be expected that the number of chlorine-containing compounds far exceeds those of the other halogens since the concentration of chlorine in the environment is much greater than the other three halides. Of the organohalogens previously cited, 75% were chlorinated substances. It is, however, apparent that almost all chlorinated molecules have been isolated from terrestrial, primarily fungal organisms, whilst a large majority of the brominated substances have been obtained from the marine environment. A large number of the halogenated compounds possess significant biological activities. Generally chlorinated metabolites of bacteria and fungi demonstrate antimicrobial activities and several of these substances such as chlortetracycline, chloramphenicol and grisofulvin have been used as clinical antibiotics. The most dominant natural volatile halohydrocarbon is methyl chloride (CH_3Cl). It is estimated that 5×10^6 tonnes of chloromethane per year must originate from natural sources, from biological production and combustion of vegetation. This compares to only 2.6×10^4 tonnes per year released to the environment as a result of industrial activity.

Natural halogenated compounds in the environment, in many instances as agents produced to inhibit the growth of competing or pathogenic species, provide a major selective pressure for the evolution of detoxification mechanisms. This may be exemplified by the biodehalogenation of 2,4-dichlorophenol observed in marine sediments which contain natural sources of halophenols produced as bactericidal agents (King, 1986, 1988). Natural organisms capable of the metabolism of a wide range of haloaromatic and aliphatic substrates have been isolated and there is little doubt that these are important factors in determining the fate of halogenated substances in the environment.

Consideration of the constitutions of the EPA organic pollutants list shows a large proportion of the chemicals contain one or more halogen substituents (Ch. 1). The importance of halogenated xenobiotic compounds in the environment is beyond doubt. Consideration of the ecological implications, biodegradation routes and technological approaches to preventing further biological pollution by these compounds provides the basis for further discussion of the ecology and biotreatment of toxic organic chemicals.

References

ALEXANDER, M. (1965) Persistence and biological reactions of pesticides in soils. *Soil Sci. Soc. Am. Proc.*, **29**, 1–17.

FATHEPURE, B.Z., NENGA, J.P. and BOYD, S.A. (1987) Anaerobic bacteria that dechlorinate perchloroethylene. *Appl. Environ. Microbiol.*, **53**, 2671–2674.

KING, G.M. (1986) Inhibition of microbial activity in marine sediments by a bromophenol from a hemichordate. *Nature*, **323**, 257–259.

KING, G.M. (1988) Dehalogenation in marine sediments containing natural sources of halophenols. *Appl. Environ. Microbiol.*, **54**, 3079–3085.

LOVELOCK, J.E. (1975) Natural halocarbons in the air and in the sea. *Nature*, **256**, 193–194.

SLATER, J.H. and LOVATT, D. (1982) The significance of microbial communities in biodegradation. In Gibson, D. T. (ed) *Biochemistry of Microbial Degradation*, Marcel-Dekker.

SUIDA, J.F. and DEBERNARDIS, J.F. (1973) Naturally occurring halogenated organic compounds. *Lloydia*, **36**, 107–143.

SYMONDS, R.B., ROSE, R.I. and REED, M.H. (1988) Contamination of Cl- and F-bearing gases to the atmosphere by volcanoes. *Nature*, **334**, 415–418.

VOGEL, T.M. and MCCARTY, P.L. (1985) Biotransformation of tetrachloroethylene, dichloroethylene, vinyl chloride and carbon dioxide under methanogenic conditions. *Appl. Environ. Microbiol.*, **49**, 1080–1085.

Further reading

BULL, A.T. (1980) Biodegradation: some attitudes and strategies of microorganisms and microbiologists. In Elwood, D.C., Hedger, J.N., Latham, M.J., Lynch, J.M., Slater, J.H. (eds) *Contemporary Microbial Ecology*, pp. 107–136. Academic Press.

BULL, A.T. and SLATER, J.H. (1982) Microbial interactions and community structure. In Bull, A.T. and Slater, J.H. (eds) *Microbial Interactions and Communities*, pp. 13–44. Academic Press.

Chapter 6

BIODEGRADATION OF ORGANIC COMPOUNDS

The most dynamic part of the carbon cycle is that involving the transfer of carbon dioxide to and from the atmosphere. This is driven by the two naturally opposing processes of photosynthesis and respiration. Photosynthesis is the only significant way new organic carbon is synthesized and hence, forms the foundation of the cycling of carbon between the inorganic and the organic state. Whilst there are organisms capable of photosynthetic processes under anaerobic conditions the major activity is carried out by aerobic phototropic organisms and hence occurs in habitats where light is readily available. The fixation of carbon in this way results in an accumulation of organic material, initially polysaccharides, which can then be subjected to metabolic processes whether anabolic or catabolic. In the first instance, anabolic processes result in the transformation of simple organic molecules into more complex molecules, so producing the building blocks of life. These biopolymers – proteins, polysaccharides, fats and oligonucleotides – are subsequently recycled within the biosphere by a combination of catabolic and anabolic metabolic pathways. Turnover of more complex molecules produces simpler ones which can then serve as precursors for the synthesis of new complex molecules, or biodegradation leads to mineralization, yielding CO_2. Catabolism is thus the degradative phase of metabolism and is active towards complex nutrient molecules produced endogenously, in the cell, or exogenously and imported into the cell. Catabolism results in the release of the energy inherent in the structure of the nutrient and this is conserved in the form of adenosine triphosphate (ATP) which represents the energy-transferring molecule in cells. By this route the energy to maintain the life forces of the cell is produced. The energy so produced is then used to fuel the anabolic processes of the cell.

Carbon dioxide is released into the environment as a result of respiration and by fermentation with the single most important source being that of microbial degradation of dead organic matter. Where organic materials are not mineralized as a consequence of biological or physico-chemical activities, they can be transformed under the influence of geophysical and geochemical processes into new compounds. The compounds so produced add to the limited number of different organic molecules produced by biosynthetic processes and hence have played an important role in generating the spectrum of organic materials found in the environment. It is this spectrum of organic material that has provided the selective pressure for the evolution of the catabolic diversity of microorganisms. The microbes have evolved to occupy new riches by developing abilities to utilize previously unused carbon sources.

The converse natures of anabolism and catabolism are reflected in the end-products of each process. In anabolism the biosynthetic pathways diverge as complex molecules are synthesized and used to generate new and more complex structures. In catabolism the biodegradative pathways converge as structurally complex molecules are converted to a limited number of structurally simple molecules which can enter the central metabolic pathways of the cell.

The catabolism of complex aliphatic molecules involves β-oxidation steps whereby a sequence of reactions results in the removal of two carbon units as acetyl-Co A, with each reaction a consequence of the oxidation of the β-carbon (second from the carboxyl carbon). Repetition of the sequences results in a decrease in the length of the carbon skeleton of the molecule until such time as the products of the reaction can enter the central metabolic pathways represented by the tricarboxylic acid (TCA) cycle. For aromatic molecules ring cleavage is followed by β-oxidation to generate succinate and acetyl-CoA, the TCA cycle intermediates (Harayama and Timmis, 1989).

The importance of ring cleavage in the maintenance of the carbon cycle should not be overlooked; after glucosyl residues the benzene ring is the most widely distributed structure in the biosphere. The first

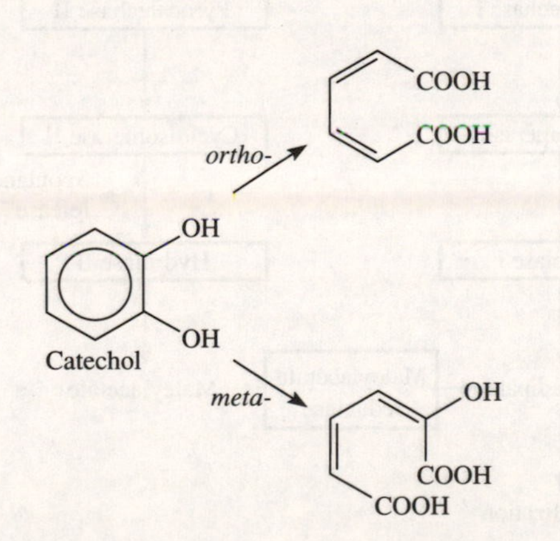

Fig. 6.1. Aromatic ring cleavage pathways: between two hydroxyl groups (*ortho*-cleavage) or proximal to one group (*meta*-cleavage).

step in the pathway for the degradation of aromatic compounds is the introduction of two hydroxyl groups on adjacent (*ortho*) or opposing (*para*) carbons in the aromatic ring. If the structure already contains one hydroxyl group the second is introduced by a monooxygenase or in its absence a dioxygenase inserts both groups. The presence of the two hydroxyl groups destabilizes the aromatic ring structure, an important

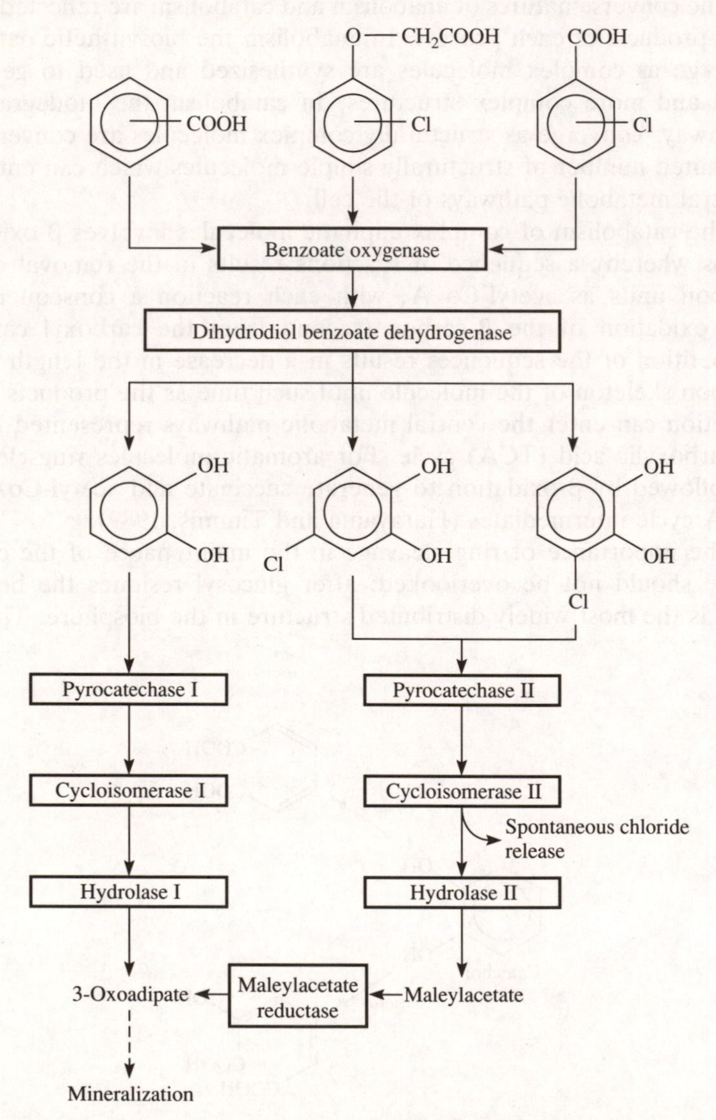

Fig. 6.2. Catabolism of aromatic compounds with or without chlorine substituents.

prerequisite for ring cleavage. Cleavage either occurs between the two hydroxyl groups (*ortho*-cleavage) or proximal to one group (*meta*-cleavage). Where the hydroxyl groups are in the *para* position, as in the case of gentisate or homogentisate, dioxygenase activities result in a ring cleavage mechanism resembling *meta*-cleavage (Fig. 6.1).

After ring cleavage, catabolism is directed towards producing metabolites which can enter the TCA cycle. Haloaromatic compounds are usually degraded by the *ortho*-cleavage (β-ketoadipate) pathway by which protocatechuate derived from hydroxybenzoates and catechol derived from benzoate are cleaved by protocatechuate 3,4-dioxygenase

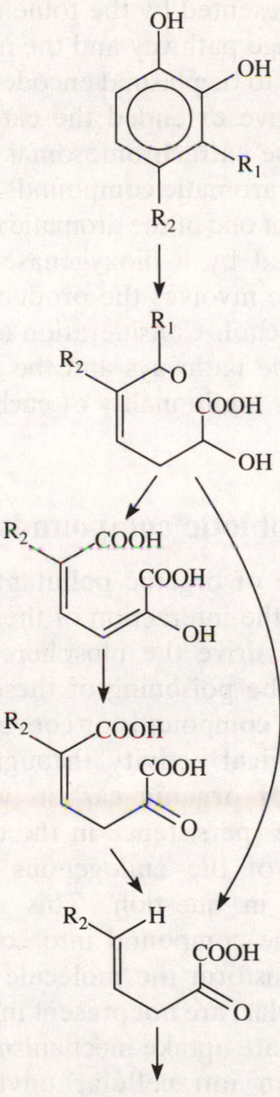

Fig. 6.3. Alkylaromatic metabolism via the *meta*-cleavage pathway.

and catechol 1,2-dioxygenase, respectively. Both branches converge at a common metabolite β-ketoadipate enollactone which is then metabolized to succinate and acetyl-CoA. The *ortho*-cleavage pathway is not effective towards alkylaromatic compounds; however, a modified *ortho*-pathway has evolved at least in some bacteria to overcome this problem (Fig. 6.2). In these organisms, represented in the laboratory by *Pseudomonas* sp. B13 (Dorn *et al.*, 1974), there are two parallel *ortho*-pathways. These are catalysed by isomeric proteins which demonstrate different substrate specificities. This isolate was identified as being able to utilize 3-chlorobenzoate and as such is considered further later in this chapter.

However, in general alkylaromatics are metabolized via the *meta*-cleavage pathways, as represented by the toluene degradative pathway (Fig. 6.3). The *meta*-cleavage pathway and the modified *ortho*-cleavage pathways have been shown to be plasmid encoded and as such represent bolt-on pathways which have extended the catabolic potential of the host strains possessing these extrachromosomal genetic elements.

In the case of polycyclic aromatic compounds, catabolism is initiated by an initial destabilization of one of the aromatic rings by dihydroxylation and ring cleavage mediated by a dioxygenase. As an example, the degradation of naphthalene involves the production of salicilate which is then metabolized to catechol. Consideration of these pathways again shows the modularity of the pathways and the extension of metabolic capabilities afforded by the additionality of each module.

Biodegradation of xenobiotic compounds

The ecological significance of organic pollutants entering the natural environment is defined by the interaction of these compounds with the metabolic processes which drive the biosphere. The most significant effects will be caused by the poisoning of these processes as a result of the introduction of toxic compounds or conversely as a consequence of vastly increased biological activity through the introduction of a localized new source of organic carbon which is subsequently metabolized. Recalcitrance (persistence in the environment) is a consequence of an inability of the endogenous metabolic systems to transform the compound in question. This may be the result of (i) an inability to bring the compound into contact with a sequence of biocatalysts able to transform the molecule because such enzyme systems, extra- or intracellular, are not present in the local environment, (ii) the absence of appropriate uptake mechanisms preventing transport of the compound into an intracellular environment or (iii) the compound failing to interact with the biocatalytic control mechanisms which determine the production of the required enzyme(s). The offending substituent may have its effect as a consequence of its complexity or its novelty in terms of constitution or position on the molecule. The presence of the substituent may cause the usual catabolic enzymes not to recognize the substituted molecule as a substrate or the substituent(s) may interfere with the activity of an enzyme.

As degradation of organic compounds is mediated by β-oxidation for aliphatic compounds and by ring cleavage followed by β-oxidation for aromatic compounds, novel substituents which interfere or prevent the action of the ring cleavage or β-oxidation enzyme systems, cause the molecule to demonstrate a greater degree of recalcitrance. If such compounds are to be degraded by biological activities then either alternative enzymes or complete pathways must be activated or evolved. Ones which are able to recognize the substituted molecules as substrates or the novel substituents must be removed so as to allow the normal catabolic pathways to function on the unsubstituted substrates. Despite the array of natural aromatic compounds, the turnover of these compounds involves only a limited number of catabolic pathways. The substituted molecules undergo transformations which give rise to a limited number of dihydric phenols (catechols) which then are subjected to ring cleavage. There are fewer than 12 pathways which direct the catabolism of such compounds to the catechol intermediates.

The biochemistry of carbon–halogen bond cleavage

The type of halogen, the number of halogens associated with the molecule and the position of the substitution define the biodegradability of haloorganic molecules. The mechanisms involved in the cleavage of carbon–halogen bonds in order to make the substituted compounds amenable to further metabolism provide an example of the ways in which microorganisms can adapt their metabolic processes to utilize xenobiotic compounds as growth substrates and hence catalyse the mineralization of a group of organic pollutants that are potentially toxic, carcinogenic and teratogenic in the natural environment.

As described above the catabolism of aromatic compounds like benzoate, benzene, anilines and phenoxyacetates proceeds via a two-step reaction to catechol (Fig. 6.2) which then undergoes *ortho*-ring cleavage catalysed by a catechol 1,2-dioxygenase. However, the dioxygenase involved in the ring cleavage does not recognize chlorocatechols as substrates, hence whilst the first two enzyme-catalysed reactions do occur, that is, the benzoate oxygenase and the dihydrodiol benzoate dehydrogenase do recognize the chlorinated aromatic compounds as substrates, the 1,2-dioxygenase and the subsequent pyrocatechase, cycloisomerase and hydrolase do not, so chlorocatechols accumulate as partial degradation metabolites. Hence, the recalcitrance of these metabolites is caused by the inability of microorganisms possessing the catabolic *ortho*-cleavage pathway to degrade chlorocatechol.

The ability of microbes to adapt to this situation is graphically illustrated by the isolation of a pseudomonad which was capable of the utilization of 3-chlorobenzoate for growth. *Pseudomonas* sp. B13 (Dorn *et al.*, 1974), as described earlier, possesses a duplicate set of enzymes providing an alternative *ortho*-cleavage pathway. It possesses two catechol 1,2-dioxygenases; one effective towards catechol, the other efficient at transforming 3-chlorocatechol. In addition, this organism possesses a second pyrocatechase, cycloisomerase and hydrolase (Fig. 6.2) and it is these four isofunctional enzymes which are critically responsible for the organism's ability to utilize the chlorinated substrate by complementing the activity of the highly specific β-ketoadipate pathway enzymes. The dioxygenase, pyrocatechase and cycloisomerase recognize the halogenated metabolites as substrates and the halogen substituent is spontaneously eliminated as a result of labilization, a consequence of the isomerase activity.

The effect of substrate specificity is also seen when considering the degradation of 4-chlorobenzoate (4-cba). Moving the halogen substituent further around the ring has the effect of rendering *Pseudomonas* sp. B13 incapable of utilizing the new compound as a growth substrate. This is because the catabolic pathway of B13 cannot recognize the new substrate. This new metabolic ability was conferred on B13 by a plasmid mediated genetic exchange which resulted in B13 acquiring part of the toluene degradative plasmid (Reineke and Knackmuss, 1978). The new strain could utilize 4-cba because the toluene 1,2-dioxygenase acted as an isofunctional enzyme to the endogenous enzyme in B13. The exogenous enzyme could catalyse the conversion of 4-cba to 4-chlorodihydro-1,2-dihydroxybenzoate, which could not be achieved by the endogenous enzyme. Once the first catabolic step had been achieved the modified *ortho*-cleavage pathway (Fig. 6.2) was capable of catalysing the mineralization of this substituted benzoate.

Microbial dehalogenases

The biochemical pathways of *Pseudomonas* sp. B13 lead to the spontaneous dehalogenation of the haloaromatic compounds. There are also enzymes, 'dehalogenases' (Jensen, 1960), which specifically catalyse the cleavage of the carbon–halogen bond. The presence of these enzymes in natural microbes and the observation that in many degradative pathways the first catabolic step leading to the mineralization of organohalogens is dehalogenation, means that these enzymes are of great scientific and practical interest when considering the prevention or remediation of pollution caused by toxic haloxenobiotics.

Microbial dehalogenases cleave carbon–halogen bonds through catalysis of oxygenolytic, reductive and hydrolytic reactions and the enzymes are active towards substituted aliphatic and aromatic compounds although individual enzymes show specificities for either haloaliphatic acids, alkanes, alcohols or aromatic substrates. Thus, enzymes active towards the haloacids are not active towards haloalkanes or haloaromatic compounds and *vice versa.* A fourth enzymic mechanism is represented by an epoxidase reaction leading to the dehalogenation of haloalcohols. These enzymes are specific for the haloalcohols, showing no activity towards haloalkanes whereas the hydrolytic haloalkane dehalogenases show some activity towards haloalcohols.

Oxygenolytic dehalogenation As seen above, oxygenase activities can lead to spontaneous dehalogenation as a consequence of ring cleavage. These enzymes have also been seen to act as dehalogenases, by actively catalysing the dehalogenation step. Under anaerobic conditions the metabolism of halogenated hydrocarbons is mediated by the strictly anaerobic methanogenic bacteria (Bouwer and McCarty, 1983; Belay and Daniels, 1987). These bacteria, whilst utilizing H_2, CO_2, or formate, acetate or methanol for growth, are able to co-metabolize haloalkanes leading to complete mineralization. Reductive and dehydrohalogenation have been observed to act alone or in concert to convert bromoethane to ethylene or ethane, respectively (Vogel and McCarty, 1985). Whilst perchloroethylene and trichloroethylene are resistant to aerobic metabolism, except in the presence of methanotrophs with a supply of natural gas, under anaerobic conditions both are degraded by sequential reductive dechlorination yielding toxic vinyl chloride as the final metabolite. In general, the dechlorinated products from the more highly substituted compounds are subsequently more susceptible to aerobic biodegradation.

Isolating aerobic microorganisms capable of utilizing haloalkanes as carbon and energy sources was considerably more difficult and this was initially taken as evidence that these substrates were transformed by co-metabolic processes as is the case in anaerobic environments. However, as more effort was directed towards the isolation of competent aerobes, such microbes were found. Early studies on the enzymology of carbon–halogen bond cleavage of haloalkanes indicated that prokaryotes and eukaryotes possessed two catalytic mechanisms, either oxidative or reductive reactions, catalysed by cytochrome P_{450} and other monooxygenases, or, by a glutathione-dependent nucleophilic replacement catalysed by glutathione-*S*-transferase (Jakoby and Habig, 1980; Stucki *et al.*, 1981).

Oxygenases are broad substrate multifunctional enzymes, induced by the presence of the *n*-alkane moiety, which catalyse a range of reactions including oxidations, condensations and dehalogenations. For example, the methane monooxygenases usually catalyse the oxidation of methane to methanol via an NADH, oxygen dependent reaction. However, halomethanes also act as substrates for these enzymes and alkanes with a carbon skeleton of up to five carbon units (pentane) are oxidized at rates comparable to that for methane, although increasing the size or number of halogen substituents has an adverse effect on oxidization rates.

Methane monooxygenases from the various methanogens catalyse different reactions when haloalkanes are the enzymes' substrates. For instance, the methane-oxidizing bacterium strain 46-1 (Little *et al.*, 1988) oxidized trichloroethanes to the corresponding acids and halo-acids, which in the natural environment may then become substrates for 2-haloacid dehalogenases. Some monooxygenases act as terminal alkane hydroxylases, whilst others act to produce a mixture of 1- and 2-alcohols from *n*-alkanes. *Methylococcus capsulatus* converted bromomethane to formaldehyde via the putative intermediate bromoethanol (Stirling and Dalton, 1980). Dibromomethane has also been shown to be converted to ethylene via bromomethanol by a monooxygenase-mediated reductive dehalogenation by a methanogenic consortium. These enzymes have also been implicated in the dehalogenation of chloroethylenes and a methane-oxidizing methanotrophic bacterium represented the first isolate capable of axenic biodegradation of trichloroethylene.

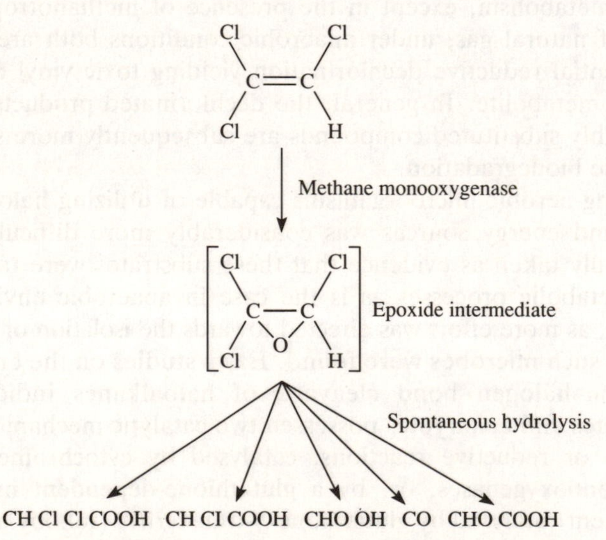

Fig. 6.4. Methanotrophic pathway for the degradation of trichloroethylene.

The degradation was the result of co-metabolic activity, supported by growth on methane or methanol. The concomitant accumulation of metabolites suggested that the axenic culture could not metabolize the halosubstrate completely. This would explain the observed role of microbial communities in the mineralization of the chloroethylenes. The mechanism outlined in Fig. 6.4 has been suggested as the degradative route of trichloroethylene metabolism, based on the accumulation of glyoxylate and dichloroacetate (Little *et al.*, 1988).

The involvement of cytochrome monooxygenases has been suspected since mammalian systems were shown to mediate detoxification of halogenated hydrocarbons. Growth of a strain of *Pseudomonas putida* on camphor has been seen to induce high levels of cytochrome P_{450} (Lam and Vilker, 1987). Resting-cell assays using such cells to examine the dehalogenation products derived from bromotrichloromethane suggested that the cleavage of the C–Br bond was the result of a reductive dehalogenation mechanism:

$$2PFe'' + R_2CX_2 + H^+ \text{ to } 2PFe' + R_2CHX + X^-$$

whilst the nitrifying bacterium *Nitrosomonas europaea* has been shown to be capable of the degradation of a range of aliphatic compounds with the degradation apparently catalysed by an ammonia monooxygenase, in that the degradative activity was stimulated by the presence of ammonia (Vanelli *et al.*, 1990).

The degradation of trichloroethylene has also been reported to involve enzymes from the aromatic degradative pathway. For example, the degradation of trichloroethylene (TCE) by a strain of *Pseudomonas cepacia* required induction of enzymes from the aromatic *meta*-cleavage pathway (Fig. 2.9) (Nelson *et al.*, 1987). Studies on the enzymes involved demonstrated that toluene dioxygenase catalysed TCE degradation. *Alcaligenes eutrophicus* JMP134 possessed two metabolic pathways for the degradation of TCE (Harker and Kim, 1990). The first, and most rapid, was a phenol-dependent chromosomally-encoded pathway whilst the other was a plasmid-encoded pathway dependent on the presence of 2,4-dichlorophenoxyacetic acid. This again represents a co-metabolic process and may explain the observation that TCE in soils occurs faster in the rhizosphere, the plant root region of the soil environment, where additional nutrients are supplied by plant exudates, than in the edaphosphere.

Dioxygenase activities have also been implicated in the microbial metabolism of polychlorinated biphenyls (PCBs). The principal metabolic pathways catalyse the hydroxylation of unsubstituted carbons at positions 2 and 3 or 3 and 4. In either instance the mechanism appears to require adjacent unchlorinated carbons. The relationship between PCB structure and degree of recalcitrance is defined by the degree of substitution; degradation decreases as chlorine substitution is increased. The presence of two chlorines at the *ortho*-position of a single or on both rings greatly increases recalcitrance. When chlorine substitution is confined to one ring the PCB is less recalcitrant and ring cleavage occurs preferentially on the non- or lesser-chlorinated ring.

Degradation of PCBs by many bacterial genera has been reported. Most of the aerobic PCB-degrading bacteria are capable of metabolizing mono-, di- and trichlorobiphenyls although some can metabolize congeners containing up to five chlorine substituents (Clark *et al.*, 1979; Furukawa *et al.*, 1979). The white rot fungus *Phanerochaete chrysosporium* can go one step further and degrade hexachlorobiphenyls as a consequence, it is believed, of catalysis by several extracellular lignin peroxidases (see below).

Reductive dehalogenation Reductive dehalogenation results in the direct substitution of the halogen substituent by hydrogen and has been reported in both aerobic and anaerobic organisms. Anaerobic degradation of *meta*-chlorobenzoates was first reported for an anaerobic microbial community isolated from sewage sludge and subsequently reductive dechlorination has been shown to be the first step in the anaerobic degradation of many aromatic compounds (Quenson *et al.*, 1988). Methanogenic communities taken from lake sediments and sewage sludge have been shown to possess the ability to mineralize halobenzoates after they were isolated by enrichment for growth on 3-chlorobenzoate.These communities are able to dehalogenate all mono-, iodo- and bromobenzoates and then mineralize the unsubstituted benzoate. However, only chlorobenzoates substituted in the *meta*-position, were susceptible to dehalogenation; 3-chlorobenzoate was mineralized but 2- or 4-chlorobenzoates were not. With these microbial communities *meta*-halogens were more susceptible than the *ortho*- or *para*-isomers. However, under anaerobic and aerobic conditions the order of recalcitrance is seen to change for the different isomeric series of halobenzoates. For instance, the degradation of monochlorophenols in unacclimated sludge samples has also been seen to demonstrate an order of recalcitrance: *para*- > *meta*- > *ortho*-substituted phenols. For further reading on reductive dehalogenation see the recent review of microbial degradation of haloaromatics by Reineke and Knackmuss (1988).

Hydrolytic dehalogenation There are two groups of enzyme which catalyse the hydrolytic dehalogenation of haloorganic compounds, namely the glutathione-dependent dehalogenases and the halidohydrolases (Hardman, 1991).

A glutathione-dependent conversion of dichloromethane to formaldehyde was catalysed by cell-free extracts of a *Hyphomicrobium* sp. The dehalogenase activity was strongly inducible and was strictly dependent on GSH, without leading to its consumption. The mechanism proposed for the dehalogenation of dichloromethane is analogous to that of the rat liver system (Fig. 6.5). This mechanism proposed a non-enzymic nucleophilic substitution of the carbon–halogen bond after the formation of an *S*-glutathione conjugate, hence these dehalogenases are a type of glutathione-*S*-transferase.

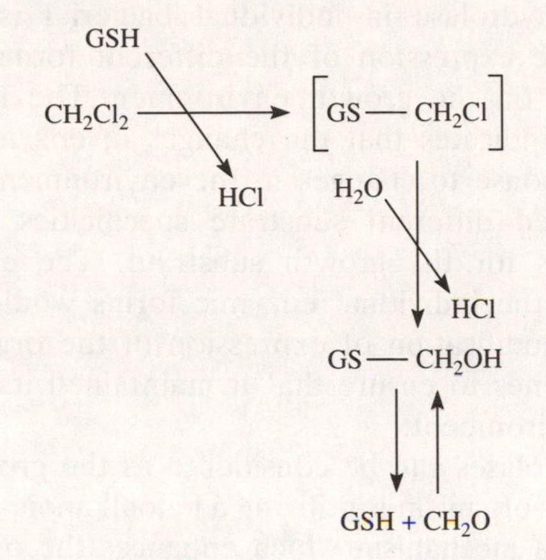

Fig. 6.5. Glutathione-dependent conversion of dichloromethane to formaldehyde by a *Hyphomicrobium* species.

Halidohydrolase-type dehalogenases have been isolated as being active towards both aliphatic and aromatic compounds, although it is the aliphatic, most specifically the haloaliphatic acid halidohydrolases, that have been studied in the most detail. Whilst natural haloalkanoic acids have been identified it is likely that many are also found as intermediates in the degradation of more complex natural halogenated organic compounds.

Dehalogenation is the key step in the catabolism of 2-haloacids and it is normally the first step in a degradative pathway. Once the halogen substituent has been removed, the alkanoic acid can be assimilated by the organisms' central metabolism. The halidohydrolases catalyse the cleavage of the carbon–halogen bond, resulting in the formation of hydroxyalkanoic acids from monosubstituted compounds and oxoalkanoic acids from disubstituted compounds. The haloaliphatic acid halidohydrolases fall into two groups depending on their substrate range. The haloacetate halidohydrolases are characterized by their relative inactivity towards halopropionates and can be further classified into two types depending on their activity towards the carbon–fluorine bond of fluoroacetate. The 2-haloacid halidohydrolases are active towards halo-acetates and propionates. The physiology of the microbes producing the 2-haloacid halidohydrolases and the biochemistry and genetics of these enzyme systems have been studied in considerable detail (Hardman, 1991) and this work has highlighted the range of halidohydrolase isofunctional enzymes present in natural organisms.

Whilst these enzymes may not be of great significance for the environmental biotechnologist, they do provide models with which to study catabolic dehalogenating systems. The presence of more than one halidohydrolase in individual bacterial isolates has been described and the expression of the different forms has been seen to be dependent on the growth environment.The reversible nature of this response indicates that the changes in enzyme profiles are a physiological response to changes in the environment. Each enzymic form demonstrated different substrate specificities and presumably different affinities for the growth substrate. The changes in levels of expression of the individual enzymic forms would thus represent a physiological modification of expression of the organisms portfolio of catabolic enzymes to ensure that it maintained its competitiveness in a changing environment.

The halidohydrolases can be considered as the growth-rate-limiting enzymes when the organism is utilizing a haloalkanoic acid as its growth substrate. Thus, a mechanism which enhances the organism's overall halidohydrolase specific activity should impart a selective advantage during enrichment culture, especially in closed culture, which has been the favoured selection procedure. Such an increase in specific activity could arise as a result of selection of constitutive mutants, the duplication of existing halidohydrolase genes, the expression of previously cryptic genes or the aggregation, within one organism, of heterologous halidohydrolase genes. These mechanisms could explain the selection of organisms capable of utilizing haloalkanoic acids and also the selection of strains containing more than one enzymic form.

The importance of the haloacid halidohydrolases as enzymes forming part of a catabolic pathway for the degradation of environmental chemicals is demonstrated by the degradation of 1,2-dichloroethane by *Xanthobacter autotrophicus* (Janssen *et al.*, 1985). This ability was conferred by the presence of two halidohydrolases. The first cleaved a chlorine substituent from 1,2-dichloroethane and the second removed the chlorine from monochloroacetate which, it was suggested, was a metabolic intermediate of the catabolic pathway (Fig. 6.6). The haloalkane halidohydrolase was seen to be of a similar size to the previously studied 2-haloacid halidohydrolases but there were no common substrates and no immunological similarity between the two enzymes. A second example is again provided by the degradation products of TCE by the methane monooxygenases present in a methanotrophic community (Little *et al.*, 1988) (Fig. 6.4). In this instance mono- and di-substituted haloacids are generated via an epoxide intermediate, and these can be subsequently mineralized through the activity of a haloacid halidohydrolase.

The halidohydrolases active towards haloalkanes are not as abundant as for the haloacids and the degradation pathways for the haloalkanes are dependent on the carbon chain length. Where substrates have carbon skeletons of greater than five or six the dehalogenation tends to be mediated by an oxygenase activity, whereas, with C_1 to C_4 compounds halidohydrolase enzymes are active. The *Xanthobacter* enzyme was capable of direct hydrolytic dehalogenation of C_1 to C_4 α-halo- or α,ω-dihalo-*n*-alkanes although in comparison with other haloalkane halidohydrolases isolated from organisms capable of growth on chlorobutane this substrate range is limited (Hardman, 1991).

The catabolic pathway proposed for the degradation of 1,2-dichloroethane provides an example of an important physiological or biochemical procedure adopted by bacteria to enable them to degrade novel organic compounds, namely enzyme recruitment. In the catabolic pathway (Fig. 6.6) two dehydrogenases were involved in the catalysis of the chlorinated alcohol and aldehyde metabolites to produce monochloroacetate. These enzymes act fortuitously in the metabolism of 1,2-dichloroethane so enabling the complete degradation, even though the enzymes' usual substrates were non-halogenated compounds. Such recruitment of enzymes in order to achieve mineralization of a substrate was also seen in the degradation of dichloromethane as a primary

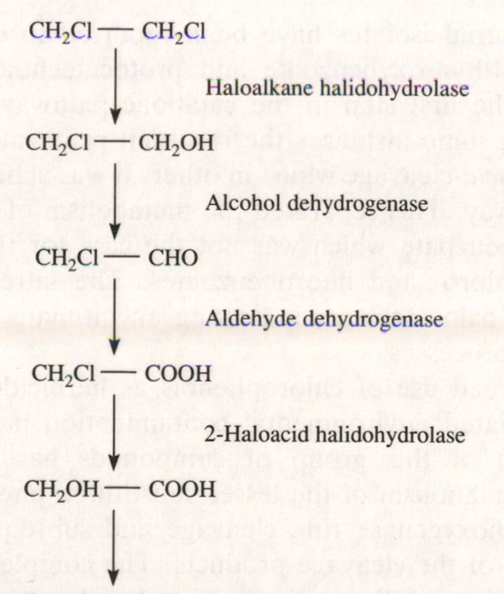

Fig. 6.6. Hydrolytic dehalogenation of 1,2-dichloroethane by *Xanthobacter autotrophicus*.

growth substrate following glutathione-dependent dehalogenation. The organism concerned appeared to recruit enzymes usually involved in methanol catabolism and hydroxypryruvate reductase and serine glyoxylate aminotransferase in order to utilize the novel substrate. The utilization of broad substrate range enzymes in order to complete metabolic pathways is essential if metabolic intermediates derived from the xenobiotic starting materials are not to accumulate.

An alternative reaction for haloaromatic compounds resulting in the hydroxylation of the aromatic ring with concomitant release of the halogen substituent is that catalysed by a halidohydrolase-type dehalogenase (Klages and Lingens, 1979). In this hydrolytic reaction the hydroxyl group is derived from water rather than molecular oxygen as is the case in the oxygenase system (Fig. 6.7).

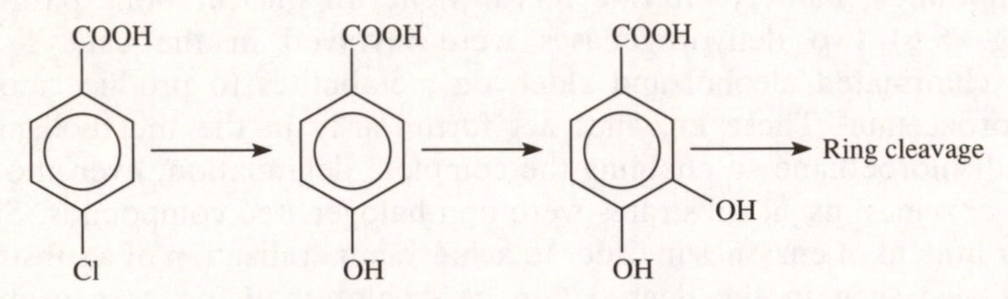

Fig. 6.7. Hydrolytic dehalogenation of 4-chlorobenzoate.

Several bacterial isolates have been reported to degrade 4-chlorobenzoate via 4-hydroxybenzoate and protocatechuate with dechlorination being the first step in the catabolic pathway. It is interesting to note that in some instances the resultant protocatechate was metabolized via a *meta*-cleavage whilst in others it was achieved by an *ortho*-cleavage pathway. This separated the metabolism of 4-chlorobenzoate from that of benzoate which was not the case for the β-oxygenolytic cleavage of chloro- and fluorobenzoates. The latter resulted in the production of halocatechols, indicating an intimate link between the two catabolic processes.

The widespread use of chlorophenols as herbicides and fungicides and the associated environmental contamination has meant that the biodegradation of this group of compounds has been extensively studied. The catabolism of the lesser substituted phenols, (mono- and di-) involves dioxygenase ring cleavage and subsequent spontaneous dechlorination of the cleavage products. The complete catabolic pathway of a number of these compounds has been determined. With more highly substituted phenols (pentachlorophenol, PCP) some of the chlorosubstituents must be removed before ring cleavage since

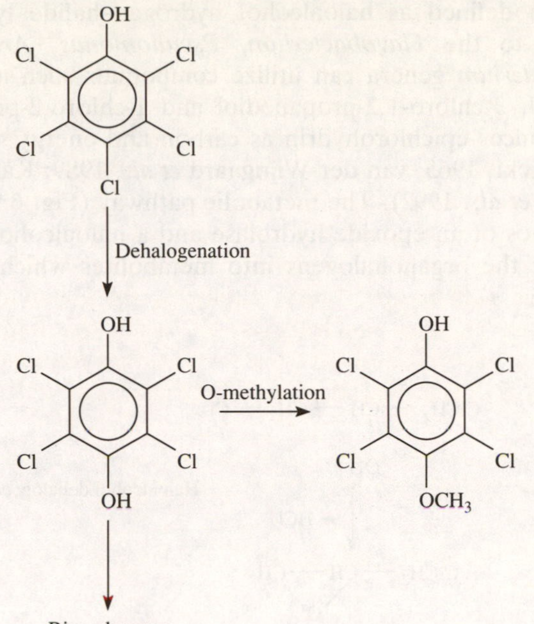

Fig. 6.8. *Para*-hydroxylation of chlorophenols by *Rhodococcus chlorophenolicus*.

the halogens deactivate the aromatic nucleus in respect of electrophilic attack by dioxygenases.

A number of bacteria and bacterial communities have been shown to be able to mineralize PCP. The actinomycete, *Rhodococcus chlorophenolicus* PCP-11 (Apajalahti *et al.*, 1986) isolated as an organism capable of mineralizing 11 different chlorophenols has been studied in detail. It was shown to hydroxylate the *para*-position, with respect to the existing hydroxyl group, whether the 4-position was chlorinated or not (Fig. 6.8). This *para*-hydroxylation is a hydrolytic reaction and it would appear to be an example of a specific hydroxylase which also acts as a *para*-specific dehalogenase and as such may be a first evolutionary step towards the production of an aromatic halidohydrolase. As such this may demonstrate the role of natural selection in the evolution of new catabolic activities.

Haloalcohol degradation Haloalkane halidohydrolases are not only active towards haloalkanes, they also catalyse the dehalogenation of haloalcohols. However, recently a new class of dehalogenases has been described which are only active towards haloalcohols. These have been defined as haloalcohol hydrogen–halide lyases. Bacteria belonging to the *Flavobacterium, Pseudomonas*, *Arthrobacter* and *Corynebacterium* genera can utilize compounds such as 1,3-dichloro-2-propanol, 3-chloro-1,2-propanediol and 1-chloro-2-propanol and in some instances epichlorohydrin as carbon and energy sources (Castro and Bartnicki, 1965; van der Wijngaard *et al.*, 1989; Kasai *et al.*, 1990; Nagasawa *et al.*, 1992). The metabolic pathway (Fig. 6.9) incorporates the activities of an epoxide hydrolase and a haloalcohol dehalogenase to convert the organohalogens into metabolites which can enter the TCA cycle.

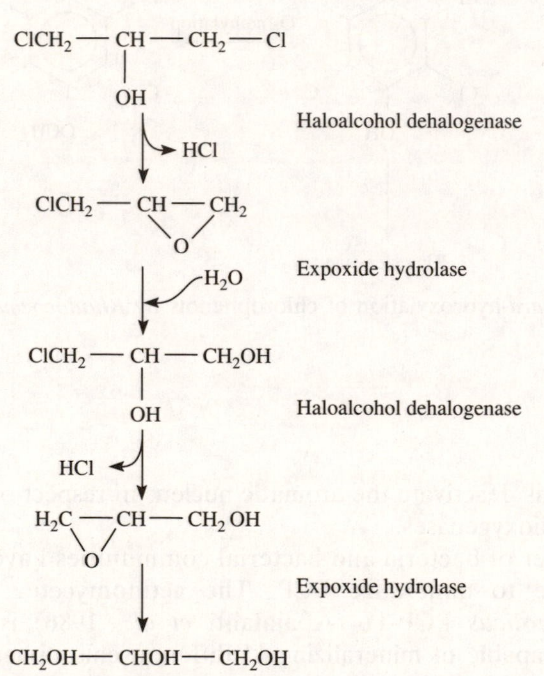

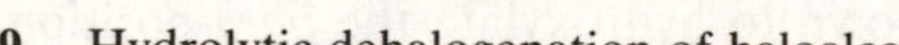

Fig. 6.9. Hydrolytic dehalogenation of haloalcohols.

Ligninases The dehalogenases described above are in general enzymes which demonstrate a restricted substrate range, although the oxygenases function as dehalogenases because of a limited relaxation of substrate specificity. Ligninases by comparison demonstrate an extremely broad substrate profile. The importance of ligninases in the carbon cycle is self-apparent; however, they are also of interest in terms of environmental control and remediation of pollution.

Lignin is a cell wall polymer of wood fibre and vascular tissue in higher plants and ferns forming 20–30% of these materials. It is formed by enzyme-catalysed free radical co-polymerization of coumaryl, coniferyl and sinapyl alcohols resulting in the production of a polymer of phenylpropane units. The structural complexity of lignin means that the initial stages of degradation must be extracellular, non-specific and not hydrolytic. Lignin is not degraded anaerobically, indeed even under highly aerobic conditions the rate of degradation is low. Bacteria belonging to the Actinomycetes and fungi belonging to the Ascomycetes and Fungi Imperfecti can degrade lignin to some extent. However, the white rot Basidiomycetes are the only organisms which can effectively degrade this natural polymer. These organisms bring to bear their complement of ligninases, cellulases and hemicellulases and are thus the only microbes able to mineralize lignin. It should, however, be noted that the extracellular depolymerization of lignin is considered to be a secondary metabolic process. Lignin degradation is supported by the primary utilization of celluloses or other carbohydrates.

Ligninases are a group of extracellular enzymes which catalyse the degradation of lignin and include lignin peroxidases, aryl methoxy demethylase and phenol oxidase, indeed lignin peroxidase is one of perhaps 15 extracellular enzymes produced by the most studied white rot fungi *Phanerochaete chrysosporium*. The peroxidase-dependent enzyme is the best characterized of all the ligninases, although its existence was only first reported in 1983. The enzyme requires hydrogen peroxide as an oxidant. Details of the activity are still not well understood, but the use of low molecular weight model compounds show that the reaction involves the production of unstable aryl cation radicals as a consequence of direct oxidation which then undergo a variety of non-enzymatic secondary reactions. These reactions lead to the depolymerization of lignin with the production of smaller water-soluble products which are then further modified by the ligninase system. These reactions also involve cleavage of aromatic ring structures; however, it is not known whether this occurs before or after depolymerization. The products of these reactions then enter the cell and the intracellular components of the TCA cycle complete the mineralization process.

As well as lignin and lignin-related aromatic compounds, ligninases also oxidize a variety of structurally diverse organic pollutants which are generally considered to be resistant to microbial attack. Ligninases have been shown to oxidize polyaromatic hydrocarbons such as (i) biphenyls like pyrene, anthracene and dibenzo(*p*)dioxin, (ii) chlorinated aromatics like 4-chlorobenzoate, pentachlorophenol and 2,4,5-trichlorophenoxyacetic acid, (iii) polycyclic chlorinated aromatics as, for example, in the oxidation of DDT (1,1,1-trichloro-2,-2 bi (4-chlorophenyl) ethane) and the Arochlors 1254 and 1242 and (iv) chlorinated alkylhalides, such as Lindane and Chlordane and a range of triphenyl methane dyes.

The nature of the ligninase systems indicates that the action of ligninases away from the producing organism would be of only limited value given the multiplicity of enzymic forms in the ligninase 'complex' and the combined extra- and intracellular activities required to complete the mineralization process. As an organism's success in degrading a xenobiotic is considered in terms of whether or not the product of the catabolism is amenable to further degradation, the white rot fungi may also find use as part of a consortium of microbial biocatalysts for the degradation of complex xenobiotic compounds. The nature of the ligninase activities enables these enzymes to catalyse the most difficult first oxidative step, such that the metabolites so derived would subsequently be more amenable to the catabolic processes of other microbes present in the consortium.

The biochemistry of organomercurial detoxification

Whilst enzymatic activity towards organomercurials is largely directed towards detoxification of these compounds, the organic moiety, once separated from the metal, can be used as a source of carbon and energy.

In Chapters 13 and 14 the effects of ecological effects of biomethylation are described in detail. The divalent form of mercury, Hg^{2+}, is converted to much more toxic organomercurial compounds:

$$Hg^{2+} + CH_3Co \longrightarrow Hg^{+}CH_3 \xrightarrow{CH_3Co} Hg(CH_3)_2$$

The methyl donor involved in the methylation of Hg^{2+} is methyl cobalamin, which is usually associated with the conversion of homocysteine to methionine. The resultant organomercurial compounds are volatile and toxic.

There are two types of enzyme-catalysed detoxification reactions. The first is considered as a broad-spectrum system. This involves a two-step procedure with the hydrolytic cleavage of the mercury from the organic moiety yielding benzene, ethane and methane from phenyl mercury, ethyl mercury and ethyl mercury, respectively. The second step is the reduction to elemental mercury, which is essentially non-toxic and volatile. This is then lost from the immediate environment as a result of volatilization. The first reaction is catalysed by organomercurial lyase and the second by mercuric reductase (Fig. 6.10). The biochemistry of the narrow-spectrum mercury resistance determinants is not known. The resistance does not appear to involve degradation of the organomercurials, but instead may represent an effective permeability barrier system.

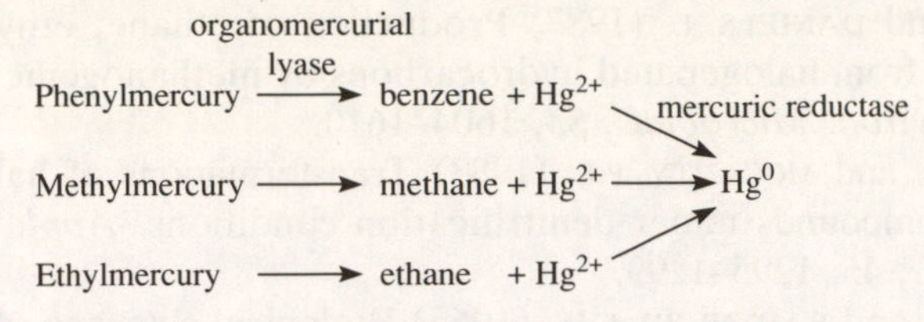

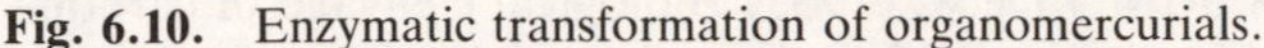

Fig. 6.10. Enzymatic transformation of organomercurials.

Exploitation of the biochemical versatility of microorganisms

The biochemical versatility of microorganisms, a consequence of their genetic plasticity (Ch. 4), and their ability to modify their physiology so as to ensure maximum competitiveness in an ever-changing environment, is one of the, if not *the*, major driving forces of the organic and atmospheric reservoirs of the carbon cycle. The description of the mechanisms by which microbes utilize organohalogens serves to indicate this versatility.

Microbial technologists have hardly dipped below the surface of the natural pool of microbial diversity. When new microorganisms have been isolated as organisms capable of degrading specific pollutants their biochemical versatility has been seen to be immense. The extent of this untapped diversity is at present not known. Attempts to determine microbial diversity indices in natural environments are limited by the inability of microbiologists to culture all microbes present in a particular environmental sample. It is suggested that less than 10% of soil bacteria and less than 0.1–0.001% of marine microbes have as yet been cultured in the laboratory. Whilst the extent of this diversity can only be a matter of conjecture at this time, our present understanding of the biochemical versatility of microbes leads one to suggest that for any given pollutant there will be a natural organism capable of metabolizing it. However, the isolation of that microbe will often require a targeted intelligent approach to screening the biosphere for its presence. It is from this point that the final chapter in this section will consider the technological application of biocatalysts for the control and remediation of organic pollutants.

References

APAJALAHTI, J.H.A., KARPANOJA, P. and SALKINOJA-SALONEN, M.S. (1986) *Rhodococcus chlorophenolicus* sp. nov., a chlorophenol-mineralizing Actinomycete. *Int. J. System. Bact.*, **36(2)**, 246–251.

BELAY, N. and DANIELS, L. (1987) Production of ethane, ethylene and acetylene from halogenated hydrocarbons by methanogenic bacteria. *Appl. Environ. Microbiol.*, **53**, 1604–1610.

BOUWER, E.J. and MCCARTY, P.L. (1983) Transformations of halogenated organic compounds under denitrification conditions. *Appl. Environ. Microbiol.*, **45**, 1295–1299.

CASTRO, C.E. and BARTNICKI, E.W. (1965) Biological cleavage of carbon–halogen bonds. Metabolism of 3-bromopropanol by *Pseudomonas* sp. *Biochim. Biophys. Acta*, **100**, 384–392.

CLARK, R.R., CHIAN, E.S.K. and GRIFFEN, R.A. (1979) Degradation of polychlorinated biphenyls by mixed microbial cultures. *Appl. Environ. Microbiol.*, **37(4)**, 680–685.

DORN, E., HELLWIG, M., REINEKE, W. and KNACKMUSS, H.J. (1974) Isolation and charaterization of a 3-chlorobenzoate degrading pseudomonad. *Arch. Microbiol.*, **99**, 61–70.

FURUKAWA, K., TOMIZUKA, N. and KAMIBAYASHI, A. (1979) Effect of chlorine substitution on the bacterial metabolism of various polychlorinated biphenyls. *Appl. Environ. Microbiol.*, **38(2)**, 301–310.

HARAYAMA, S. and TIMMIS, K.N. (1989) Catabolism of aromatic hydrocarbons by *Pseudomonas*. In Hopwood, D.A., Chater, K.E. (eds) *Genetics of Bacterial Diversity*, pp. 152–175, Academic Press.

HARDMAN, D.J. (1991) Biotransformation of halogenated compounds. *Crit. Rev. Biotechnol.*, **11(1)**, 1–40.

HARKER, A.R. and KIM, Y. (1990) Trichloroethylene degradation by two independent aromatic-degrading pathways in *Alcaligenes eutrophus* JMP104. *Appl. Environ. Microbiol.*, **56**, 1179–1181.

JAKOBY, W.B. and HABIG, W.H. (1980) Glutathione transferase. *Enzymatic Basis of Detoxification*, **11**, 63–4.

JANSSEN, D.D., SCHEPER, A., DIJHUIZEN, L. and WITHOLT, B. (1985) Degradation of halogenated aliphatic compounds by *Xanthobacter autotrophicus* GJ10. *Appl. Environ. Microbiol.*, **49**, 673–677.

JENSEN, H.L. (1960) Decomposition of chloroacetates and chloropropionates by bacteria. *Acta Agricult. Scand.*, **10**, 83–103.

KASAI, N., TSUJIMUA, K., UNOURA, K. and SUZUKI, T. (1990) Degradation of 2,3-dichloro-1-propanol by a *Pseudomonas* sp. *Agricult. Biol. Chem.*, **54(12)**, 3185–3190.

KLAGES, U. and LINGENS, F. (1979). Degradation of 4-chlorobenzoic acid by a *Nocardia* species. *FEMS Micro. Letts.*, **6**, 201–203.

LAM, T. and VILKER, V.L. (1987) Biodehalogenation of bromotrichloromethane and 1,2-dibromo-3-chloropropane by *Pseudomonas putida* PpG-786. *Biotech Bioeng*, **24**, 151–159.

LITTLE, C.D., PALUMBO, A.V., HERBES, S.E., LIDSTROM, M.E., TYNDALL, R.L. and GILMER, P.J. (1988) Trichloroethylene biodegradation by a methane-oxidizing bacterium. *Appl. Environ. Microbiol.*, **54**, 951–956.

NAGASAWA, T., NAKAMURA, T., YU, F., WATANABE, I. and YAMADA, H. (1992) Purification and characterization of halohydrin hydrogen halide lyase from a recombinant *Escherichia coli* containing the gene from a *Corynebacterium* sp. *Appl. Microbiol. Biotechnol.*, **36**, 478–482.

NELSON, M.J.K., MONTGOMERY, S.O., MAHAFFEY, W.R. and PRITCHARD, P.H. (1987) Biodegradation of trichloroethylene and involvement of an aromatic biodegradative pathway. *Appl. Environ. Microbiol.*, **53**, 949–954.

QUENSEN, J.F.I., TIEDJE, J.M. and BOYD, S.A. (1988) Reductive dechlorination of polychlorinated biphenyls by anaerobic microorganisms from sediments. *Science*, **242**, 752–754.

REINEKE, W. and KNACKMUSS, H.J. (1978) Chemical structure and biodegradability of halogenated aromatic compounds: Substituent effects on 1,2-dioxygenation of benzoic acid. *Bochimica Biophysica Acta.*, **542**, 412–423.

REINEKE, W. and KNACKMUSS (1988) Microbial degradation of haloaromatics. *Ann. Rev. Microbiol.*, **42**, 263–287.

STIRLING, D.I. and DALTON, H. (1980) Oxidation of dimethyl ether, methyl formate and bromomethane by *Methyloccus capsulatus* (Bath). *J. Gen. Microbiol.*, **116**, 277–283.

STUCKI, G., GALLI, R., EBERSOLD H-R, and LEISINGER, T. (1981) Dehalogenation of dichloromethane by cell extracts of *Hyphomicrobium* DM2. *Arch. Microbiol.*, **130**, 366–371.

VAN DEN WIJNGAARD, A.J., JANSSEN, D.B. and WITHOLT, B. (1989) Degradation of epichlorohydrin and halohydrins by bacterial cultures isolated from freshwater sediment. *J. Gen. Microbiol.*, **135**, 2199–2208.

VANNELLI, T., LOGAN, M., ARCIERO, D.M. and HOOPER, A.B. (1990) Degradation of halogenated aliphatic compounds by the ammonia oxidizing bacterium *Nitrosomonas europaea*. *Appl. Environ. Microbiol.*, **56**, 1169–1171.

VOGEL, T.M. and MCCARTY, P.L. (1985) Biotransformation of tetrachloroethylene to trichloroethylene, dichloroethylene, vinyl chloride and carbon dioxide under methanogenic conditions. *Appl. Environ. Microbiol.*, **49,** 1080–1083.

Chapter 7

APPLICATION OF BIOTECHNOLOGIES TO THE TREATMENT OF ORGANIC POLLUTANTS

Introduction

Toxic organic compounds, products of modern society, are increasingly being recognized as extreme threats to the self-regulatory capacity of the biosphere. The complex cycles of organic matter described in Chapter 5, which have created and maintained the habitable environments as we know them, are apparently becoming overloaded. Increased public and political awareness of the problems has led to the realization that there is a great need for increased efforts towards the prevention and remediation of environmental pollution.

The problem is two-fold. Firstly, there is the legacy of hazardous waste sites, landfills and dumps, the products of decades of a throw-it-away mentality; as such this is a problem that needs a cure. Secondly, there is a need to prevent the problems getting worse. End-of-pipe or in-process treatment of toxic organic compounds in waste streams would obviate deliberate discharge or costly transport of these toxic substances to landfill sites. When considering environmental pollution, prevention is considerably less expensive than cure. For example, the United States Office of Technology Assessment estimated the remedial cost for 600 hazardous waste sites in the United States could be approaching $100 billion.

Biotechnologies for the prevention of environmental pollution by toxic organic compounds

The application of biological agents in the removal of organic compounds in wastewaters is an established and proven technology. Municipal water treatment plants have been established in Europe for more than 100 years. The history of sewage treatment dates back to 1850 when it was common to treat sewage by land spreading. These methods were superseded by trickle filters, then, as the amounts of sewage increased, more intensive methods were considered and the theory of activated sludge was first described in 1914 in a report to the Society of Chemical Industry. Such considerations mean that aerobic sewage treatment is nearly 100 years old. Anaerobic strategies for the treatment of suspended solids were first developed in France at about the same time and employed on an industrial scale in the United Kingdom at the turn of the century.

These biological systems are effective in the degradation of structurally simple carbonaceous and ammoniacal compounds so protecting receiving water courses by reducing the organic content and hence the biological and chemical oxygen demand (BOD/COD) of the effluent entering the natural environment. Both types of treatment system have their advantages. Anaerobic systems score in terms of volumetric loading rates and energy requirements, and in that sludge yields are lower. However, aerobic systems are applicable to a greater range of waters, achieve higher degrees of BOD, nitrogen and phosphorus removal, and are less sensitive to the poisoning effects of toxic organic compounds entering the system.

In Britain today sewage treatment involves at least primary and usually secondary treatment systems which purify the sewage to 95% in respect of the BOD before the effluent is discharged to the receiving water course. Strict requirements are supposed to be met to ensure that the BOD does not exceed 20 mg l^{-1}, that the suspended solids in the effluent do not exceed 30 mg l^{-1} and that the receiving waters dilute the effluent 8-fold.

Municipal wastewater treatment

Municipal wastewater comprises domestic wastewater or sewage and storm water. Whilst it may also contain industrial discharges this is strictly controlled such that where excessive amounts are produced by individual factories, or when toxic components are included in the waste stream, the effluent must first be treated on-site before it is discharged into the municipal system. There are two types of collecting strategies: the combined system in which domestic sewage is combined with storm water, which represented the first systems installed; and later systems which separated the two sources of water. This overcame a major drawback of the combined system whereby during periods of excessive rain part of the large volume of water can be discharged directly into the receiving water courses.

A typical sewage entering the treatment facility would have a BOD of 275–300 mg l^{-1} and a suspended solids content of 300–350 mg l^{-1}. Municipal treatment facilities are designed to reduce both of these whilst also reducing the number of pathogenic organisms and the level of inorganic nutrients of the type which cause eutrophication, namely phosphorus and nitrogen. In times of drought the conservation of water as a natural resource becomes of great concern, but also with increasing industrialization and consumer demand the recycling of water from municipal works is becoming increasingly important.

The treatment process can involve three stages — primary, secondary and tertiary treatments — but raw sewage is treated to varying levels depending upon circumstances and geographical location. In primary treatment, raw sewage is passed through screens to remove large particulates, then through degritting chambers before entering primary sedimentation tanks, where part of the biomass and flocculated organic material settles. In a number of coastal towns this is the extent of the treatment. The liquid is then discharged to the sea, whilst the raw sludge is subjected to anaerobic digestion to yield methane. The dried sludge can then be used as a fertilizer or until recently dumped at sea. Increased legislation means that disposing of sewage that has only been through primary treatment into the marine environment is becoming increasingly rare. The secondary treatment involves taking the settled sewage and passing it through trickle or percolating filters, where the sewage undergoes oxidation mediated by biofilms supported on solid supports in the filters, or is treated as flocs in suspension in activated sludge systems. The biomass created as the treatment progresses is again separated from the liquid waste in settlement tanks and is removed to anaerobic digesters or is dumped. The much-improved effluent can then be discharged to the environment or passed to a tertiary facility where the inorganic nutrients are removed either chemically or biologically. This final stage reduces the possibility of contributing to eutrophication of the receiving waters (Ch. 8). In certain instances where water is to be recycled for human use, a final sterilization step is incorporated whereby chlorine is added before it is passed to a water reclamation plant.

With time the municipal facilities have been developed and the technology has advanced such that many plants have become energy self-sufficient by utilizing the methane produced by anaerobic digesters. However, during this period the microbiology has not advanced significantly. The biological processes involved are those which would be found in any aquatic environment that has become overloaded with degradable organic compounds. The heterotrophic bacteria metabolize the organic material liberating carbon dioxide, ammonia and water, and this is associated with a decrease in the oxygen concentration without which mechanical aeration would lead to anaerobis. With time, ciliates come to feed on the bacteria, and nitrification, mediated by other microbes, converts the ammonia first to nitrites then to nitrates. Algae and diatoms then utilize the nitrates as sources of nitrogen, so further diversifying the fauna in the system.

A consequence of the nutrient basis of this ecosystem, and hence the microorganisms that inhabit it, is that municipal plants cannot handle many of the toxic environmental chemicals that can be present in trade effluents. If such chemicals enter the municipal system they can lead to poisoning of the biological systems in the treatment facility or they can become bound to the biomass or are subjected to partial degradation. This means that the degree of removal is much less controllable. Trade effluent control can limit the concentrations, but they still pose an environmental threat as a result of breakthrough into receiving watercourses. To date, a large part of the answer has been in the form of dilution, that is, the controlled release of toxic effluents at previously agreed consent levels, dumping, transportation to landfill sites or incineration. However, these are increasingly coming to be regarded as unacceptable solutions to the problem of disposal. Harnessing the novel degradative capabilities of selected microbes in effluent treatment facilities based on designs developed for municipal waste treatment represents the simplest form of end-of-pipe biotreatment facility.Even with the modern developments, it is not economically feasible to develop effective treatment facilities that do not result in the concentration of these toxic components into the sludges from the treatment facility. Hence sewage sludge will always require some form of disposal (see Ch. 3).

The intrinsic large volumes and extremely low unit values of waste require that new treatment processes are simple and cost-effective. The evidence to date leads one to confidently forecast that with continued development the application of new or improved biocatalysts will have major applications in industrial waste management. Superbugs, universal biocatalysts, for the degradation of all toxic substances, cannot be developed and different classes of pollutants will require different approaches. Thus, the treatment of specific environmental chemicals will require the development and application of customized biocatalysts and, where required, novel bioreactor conformations.

As scientists and engineers began to extend the municipal treatment strategies to new applications for the treatment of trade wastes and environmental decontamination, they began to realize the complexities involved in moving from the laboratory to full scale.

The activated sludge systems with their high rates of aeration were shown to remove toxicants from the wastewaters by biotransformation, adsorption to the biomass and by air stripping. The latter, together with the metabolic CO_2 and CH_4, contributed to a problem that was only recently perceived; that of release to the atmosphere of greenhouse gases and toxic volatile compounds. Adsorption of toxic inorganic substances, like heavy metals, or structurally complex or highly substituted organics onto the biomass became a problem as the biomass was removed from the treatment facility. The presence of the adsorbed chemicals, together with those resulting from transformations catalysed by the dewatering phases of the process, achieved through the application of heat or further chemicals, led to the presence of significant concentrations of environmental chemicals in the sludges. Sludge disposal then leads to undesirable environmental pollution.

Partial mineralization of the toxic organic compounds was also observed, that is, rather than the desired end product of the biological activity, CO_2, intermediary metabolites accumulated in the treated waters, for instance, the production of the more biologically toxic vinyl chloride from trichloroethylene under anaerobic conditions (Ch. 6). Partial metabolism under aerobic conditions leads to oxidized intermediates which tend to be less toxic than the parental molecules but are more mobile in the environment. Such compounds may be missed by standard analytical techniques and lead to false claims for mineralization of the toxicants. This highlights the need for confirmatory loss and mass balance studies.

Biotechnologies for the cure of environmental pollution by toxic organic compounds

Soil remediation

Soil contamination has to date represented the largest field of application. Clean-up strategies can be divided into two types of approach: (i) those that are used for reclamation of land where contamination is confined to the upper regions of the soil and (ii) solid phase techniques which involve such approaches as soil tilling, using conventional agricultural equipment, that is, landfarming.

When the pollution has permeated to subterranean levels, to below depths where cost-effective excavation can be considered, *in situ* recirculation systems involving the addition and circulation of nutrients, or nutrients and adapted competent microbes can be used. Such bioaugmentation can take three forms: (i) stimulation of existing soil or aquifer organisms by addition of nutrients, (ii) removal of organisms from contaminated sites for enrichment and selection in the laboratory before returning cocktails of organisms and nutrients to the contaminated site and (iii) bioaugmentation with genetically engineered organisms (GEMs). The third option is still largely at the laboratory stage with work in microcosms because of the concerns about environmental release of GEMs. Stimulation of the existing microbial populations or augmentation with adapted strains has been used successfully in the United States and Europe to decontaminate toxic and hazardous waste sites. Neither involve recombinant DNA techniques, whilst the former does not even induce the mutational changes arising from *in vitro* culture. The addition of nutrients only, promotes the growth of the microbes *in situ* and works on the principle that during the time that the indigenous microbial consortia have been exposed to the pollutants a subpopulation will have developed which can utilize the compounds. Hence the supplementation with growth-limiting nutrients, which may include oxygen and methane, promotes general microbial activity and stimulates the competent subpopulations. Methane is included because of its role in stimulating methanogenic bacterial degradation of haloorganic compounds (see Ch. 6). Another limiting nutrient can be oxygen. This can be added by simply sparging air into the water in a well in the contaminated region. The levels of dissolved oxygen can be further enhanced by sparging with pure oxygen or a third means is through the addition of hydrogen peroxide. The danger with the latter is the possibility of achieving concentrations of oxygen that are toxic to the microbial populations present in the system.

When the *Exxon Valdez* ran aground in the Prince William Sound in 1989, approximately 11 million gallons of crude oil polluted the Alaskan environment. It was estimated that the oil eventually polluted 350 miles of coast line, where the oil settled into the gravel beaches and onto rock and cliff surfaces. As a result of natural weathering 15–20% of the oil was lost through volatilization leaving the residue which remained on the beaches. The first approach was to use physical cleaning techniques – periodic flooding, application of high pressure heated water followed by vacuum extraction and skimming of the oil released. This approach removed the surface contamination but did not remove oil trapped below the gravel surface. This pollution incident provided a much-needed chance to assess the comparative effectiveness of bioremediation in oil-spill clean-up. It was regarded as a supplementary technology; however, it still represented the largest field bioremediation effort ever undertaken. The United States EPA recommended two fertilizers for addition to the beaches to enhance the natural biodegradation of the oil. Within weeks of addition of the fertilizers the beaches were visibly cleaner than the control beaches. Although scientific verification was difficult it was agreed that the bioremediation techniques were to one extent or another successful and represented a technology that was significantly less disruptive to the environment than the physical and chemical alternative technologies.

Bioaugmentation through the addition of laboratory-selected competent microbes effectively accelerates what would eventually happen as a result of the indigenous microbial population (Fig. 7.1). The laboratory strains have passed through enrichment and selection strategies which promote physiological and genetic changes and result in the development of strains which are more effective or efficient at degrading the pollutants in question. Such strains have been termed 'vanguard' microbes (Bewley *et al.*, 1989). Once isolated, the cocktail of microbes usually, if not invariably, a mixed population rather than a single species, can be returned to the polluted site and added back

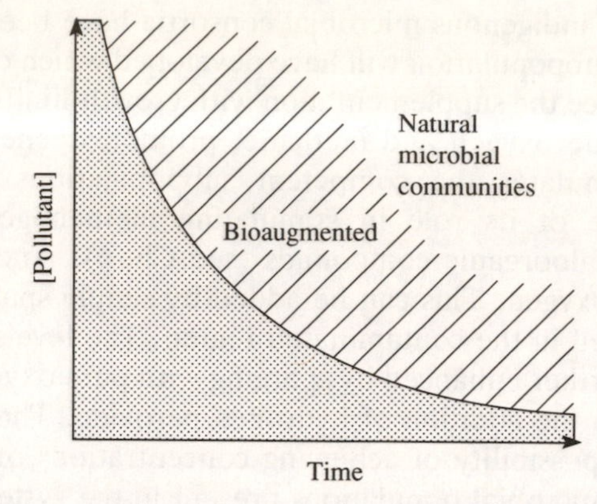

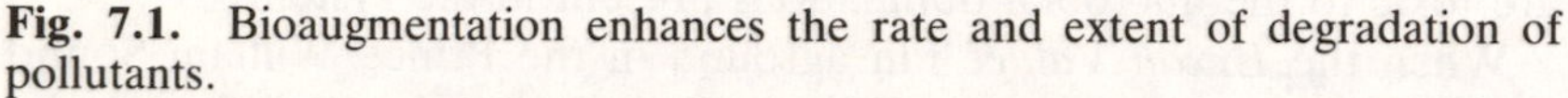

Fig. 7.1. Bioaugmentation enhances the rate and extent of degradation of pollutants.

with growth-promoting nutrients, so stimulating the removal of the pollutants. There are drawbacks to the application of biological remediation agents; they are not capable of reaching the nine-nines efficiency obtained with other technologies and some biodegradative reactions are so inefficient that they are not economically viable. However, in general, bioremediation, especially if used in association with other technologies, is a demonstrably effective approach to remediation.

There are a large number of examples of the remediation of polluted soil sites using *in situ* techniques. These have included remedial action towards gasoline and hydrocarbon fuels after pipe-line fractures, leaking storage tanks and spills, cocktails of organic chemicals and solvent–fuel mixtures. The bibliography provides further reading on details of these applications. These techniques vary from true *in situ* techniques to on-site techniques of enhanced landfarming soil banking or composting (Fig. 7.2) to the use of on-site bioreactors for treatment of washing waters, leachates and contaminated aquifer and ground-waters. In order to provide an outline of approaches taken, two soil remediation projects which have been reported extensively will be considered in detail. The first represents a soil washing system used for the removal of the wood preservative pentachlorophenol (PCP) and creosote from contaminated soils, the second the use of a mixed technology approach for the remediation of a gas works site.

The soil washing system was developed by Biotrol®. The process is based on a series of scrubbing and physical separation steps using water as the carrier. The principle that most of the contaminants are associated with the silt and clay content of the soil is utilized and the scrubbing techniques break up the soil aggregates allowing access to

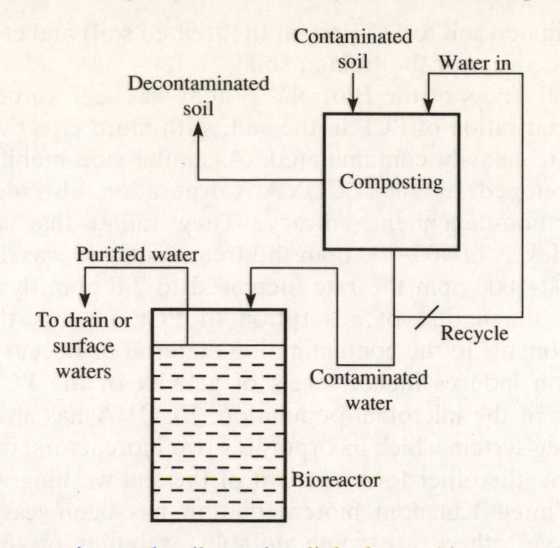

Fig. 7.2. Composting and soil wasting linked to a bioreactor for treatment of contaminated soils.

the smaller particles. The soil is then slurried and put through several screening and scrubbing operations. The fine soil particles suspended in the process water are then combined with froth resulting from the screening phase, which contains hydrophobic compounds, and then allowed to flocculate and settle. The highly contaminated fine solids are then dewatered and can then be subjected to biotreatment, incineration or off-site disposal. The process water is then passed into the water treatment unit, a fixed film bioreactor containing a competent (vanguard) microbial population. The COD and lower molecular weight polyaromatic hydrocarbons (PAHs) are degraded by microbes indigenous to the contaminated site. These microbes are added to the bioreactor from soil leachate samples taken from the site. Once a biofilm of these microbes has been formed, a *Flavobacterium* (Pflug and Burton, 1988) capable of degrading PCP is added and subsequently becomes established in the reactor. The bioreactor is a multicell submerged, packed-bed reactor using plastic tower packing material. As part of the pre-treatment strategy the process water is heated to 25°C, the pH adjusted to 7.2 and inorganic nutrients added to supply nitrogen and phosphorus. The water treatment can be operated continuously with the soil washing process or used to treat stored washing waters. A mobile pilot-scale operation was operated in 1987–88 at a 'Superfund' site when the two stages of the process were mounted on trailers. It had a working capacity of 500 lb soil h^{-1} and was shown to remove 85–99% of the pentachlorophenol (from 1997 ppm in the contaminated soil to 115 ppm in the treated soil) and creosote, with a 73–83% recovery of the treated soil.

The effectiveness of the Biotrol® system was seen to be dependent on the concentration of PCP in the soil, with more effective treatment of soils more heavily contaminated. A similar non-mobile treatment facility developed by the ECOVA Corporation also demonstrated a concentration-dependent efficacy. They found that at soil contamination levels of 400–700 ppm the treatment rate was 3.0 ppm dy^{-1} whilst at 700–1000 ppm the rate increased to 7.8 ppm dy^{-1}. This may result from the nature of adsorption of PCP in soil, the effect of other components in the contaminating material or because the higher concentration induces higher levels of activity of the PCP-degrading components of the microbial population. ECOVA has also developed a soil washing system which incorporates two bioreactors; one for slurry biotreatment, the other for treatment of the soil washing waters.

In the United Kingdom bioremediation has been less extensively used; however, there are some notable examples of the economic success of this approach to site clean-up. One of the most extensively documented is that of the clean-up of the Greenbank gasworks site in Blackburn, Lancashire (Bewley *et al.*, 1989, 1991). The site housed a gas works and tar distillery for nearly 70 years and when the industries were closed the site could not be developed because of the toxic compounds present there. A scheme was developed utilizing mixed technologies, conventional and biological. This approach avoided the need for removal of large quantities of toxic soil to off-site landfill. Following a site audit, 30 500 m^3 soil polluted with coal tar and phenols was treated with microbes, whilst the remaining 12 000 m^3 contaminated with metals and complex cyanides was encapsulated on-site. Organisms competent at degrading the coal tar, PAHs or phenols were selected by batch enrichment techniques in the laboratory, using soil samples taken from the Greenbank site. Laboratory and field trials determined the optimum mix of microbes, nutrients and surfactants to be added to the site to optimize the treatment strategy. The surfactants were added to enhance the availability of the pollutants for microbial action. In the full-scale process the soil was treated in beds and pre-processed to attain maximum possible homogeneity before the microbe–nutrient–surfactant cocktails were added. The cocktails were added to layers of soil using an agricultural boom sprayer and the soil rotovated to ensure sufficient dispersal of the additives. During the treatment period the soil was turned over and additional nutrients, water and fresh microbial inoculants added. The treatment process reduced the PAHs and phenols to below the target concentrations in most parts of the sites so enabling utilization of what was previously derelict land. Whilst cost benefits derived from the application of the different technologies will be site specific, in this instance there was an

11% saving in the total cost of treatment as compared with options which did not involve bioaugmentation techniques.

The technologies of bioremediation and bioreactor processing of contaminated soils and waters can also be applied to the treatment of sludges, using systems adapted from the municipal activated sludge systems. Remediation of river, harbour, bay or lake sediments using biological agents *in situ* is also a promising area for development. Such on-site strategies avoid disturbing the sediments, which in itself causes spread of the pollution problem. Applications such as these are less well-developed but General Electric are developing anaerobic microbial systems for the degradation of PCBs in the Hudson River.

Groundwater remediation

A major developing area is in the biotreatment of waters contaminated with chlorinated solvents. This pollution is causing increasing concern as the levels in potable waters is now being shown to be reaching unacceptable levels. In surveys in the United Kingdom, levels of trichloroethane (TCE) as high as 5.5 ppm, nearly 200-fold higher than that recommended as safe for potable water by the World Health Organisation, have been recorded.

There are, however, considerable problems to be overcome in the treatment of these more highly substituted organic compounds, in part, because of the low concentrations of the contaminants but also because the degradation of some of these compounds is mediated by microbes whilst they are growing on a second carbon source. A significant proportion of industrial effluents contain tetra- and trichloroethylene and these solvents are amongst the most pervasive of the groundwater pollutants. Investigation of the rates of reductive dehalogenation of these substances in anaerobic aquifer sites under methanogenic or sulphate-reducing conditions showed that under the different environmental conditions the degree and form of biotransformation varied for each pollutant. TCE was converted to dichloroethylene faster under methanogenic conditions, but further dehalogenation was severely restricted. *In situ*, where the predominant flow of carbon and energy is through methanogenesis or sulphate reduction, utilization of anaerobic microorganisms capable of reductive dehalogenation could be advantageous, as treatment of large volumes of micropollutant containing groundwater causes severe engineering problems. The oligotrophic conditions will not support large populations, thus decontamination would not be a quick process although this could be accelerated by additions of additional simple carbon sources such as acetate or methanol. The addition of easily degradable naturally occurring carbon substrates not only stimulates degradation of the pollutants by competent microbial communities but it also has the effect of increasing the toxicity threshold of many toxic organic compounds. However, a major problem with anaerobic catabolism of the chloroethylenes is the incomplete degradation which leads to production of toxic metabolites such as vinyl chloride.

Other major problems for *in situ* bioremediation in these situations is that the increase in biomass which results from the biodegradation causes plugging of the soil, so reducing the flow of water; the addition of nutrients can also affect the surface waters associated with the site and the treatment can lead to odour and taste problems. *In situ* techniques have been employed but these have often been combined with pump-and-treat strategies involving extraction of the contaminated waters through wells which are then subjected to filtration, air stripping or treatment in above-ground bioreactors before being returned to the aquifer. Such techniques have been used to treat waters contaminated with petroleum hydrocarbons and alcohols, ketones and organic acids. However, toxicity at high levels of contamination or process instability at low or fluctuating concentrations of contaminants lead to reactor design problems.

Bioremediation of groundwaters contaminated with a wide variety of pollutants has been demonstrably effective as a means of environmental clean-up. Such an approach has been taken by BioTrol® for the remediation of groundwaters contaminated by phenols and PAHs. The BioTrol Aqueous Treatment System (BATS) operates as a continuous bioreactor containing microbial consortia selected for their ability to degrade target pollutants. As such, systems such as this are capable of being used to treat groundwaters contaminated with a variety of chemicals. For each type of chemical a different competent consortium can be used. A two-stage biological process involving an activated sludge system and two BATS was used successfully to treat 2.7 million gallons of water contaminated with wood preservatives and subsequently these systems have been applied to a variety of remediation problems. It has been seen to be a cost-effective method for treating waters contaminated with phenolic compounds, operating at a rate of 20 gallons per minute and achieving effluent phenolic concentrations of less than 0.1 ppm at an operating cost of US$0.75–2.62 per 1000 gallons.

The BATS is in effect the wastewater treatment facility and a soil washing system is a submerged fixed-film plug flow reactor. The nature of this reactor is such that there is a long residence time in the reactor as compared to trickling filter systems. The plug flow means that different parts of the reactor are exposed to different levels of nutrients and pollutants, that is, at the front end of the reactor the biofilm will be exposed to higher concentrations of nutrients and pollutants than at the back end of the reactor (Fig. 7.4). This means that there will be a selective pressure for variation in the microbial consortia present in the biofilm in the different parts of the reactor, which in turn means that as the groundwater passes through the reactor it is contacted with a changing biofilm which consists of microbes that are best adapted to scavenging the pollutants present in the waters, even though the concentration will be dropping as it progresses through the reactor. This, in effect, results in the development of a multistage treatment system within the one reactor. The concentrations of many of the pollutants will fall at a rate directly proportional to the concentration of the pollutant. Such first-order kinetics mean that in effect plug flow systems operate at higher overall removal rates than the mixed tank systems.

Biofilm reactors also offer several advantages over suspended biomass reactors for treatment of wastewaters. Firstly, the fixed film system reduces the amount of biomass produced, as compared with an activated sludge system, and hence reduces the need for post-treatment biomass separation procedures. Biofilms are also characterized by their ability to withstand shock loading such that whilst transient higher concentrations may be toxic to the biomass and so kill the outer surface of the film, the underlayers will have been protected from the toxic effects and as the dead layers slough off, the remaining film will maintain the activity of the reactor. Such reactors are also effective in the removal of suspended solids. A further advantage is that once formed, biofilms can be used to treat waters contaminated at the micropollutant level, at concentrations which could not actually support cell growth. The integrity of the biofilm could be maintained by periodic addition of a carbon source. This approach is well suited to the secondary utilization of halogenated substances by methanogenic bacteria, where the biofilm is grown and supported by the addition of acetate. Biofilm development with different heterogenous inocula was studied in a laboratory-scale methanogenic fluidized bed reactor. The removal of low molecular weight haloaliphatic compounds (10–30 mg l^{-1}), and formation of a biofilm during growth on acetate, has been demonstrated under methanogenic conditions, whilst trace concentrations (10 mg l^{-1}) of chlorinated benzenes could be utilized as secondary substrates under aerobic conditions.

Treatment of halogenated substances using immobilized cells has been investigated at the laboratory scale on model supports such as alginate and as in the case of the immobilization of an *Alcaligenes* sp. capable of degrading 4-chlorophenol, on Lecaton-particles (light expanded clay aggregate) in a packed-bed fermentor used to treat sterile and non-sterile municipal wastewater (Westmeier and Rehm, 1987). Whereas the BATS utilized selected competent microbial consortia, this laboratory study highlighted a problem which can occur when using specialized laboratory strains for waste treatment. When the fermentor, seeded with *Alcaligenes* sp. A7-2, was used to treat an influent of sterile municipal water contaminated with 4-chlorophenol, the water was decontaminated at a rate of 300 m mol l^{-1} h^{-1}. However, when the naturally occurring population was used to seed the fermentor and the reactor was fed with non-sterile municipal water, the 4-chlorophenol was only partially metabolized and it did not prove possible to establish the A7-2 strain in the packed-bed as a stable member of the mixed community.

Activated carbon is used as an adsorptive material for water purification. However, for water polluted with halo-compounds, it is an expensive technology. The cost of carbon adsorption is around £2–3 per 1000 litres, 5–10 times more expensive than biological alternatives, and adsorption technology only results in a phase shift, not the destruction of the pollutant; the spent carbon must either be regenerated or incinerated. Two bacteria, *Pseudomonas putida* P8 and *Cryptococcus elinovic* Hl immobilized onto activated carbon, demonstrated the potential of this combination of support and biocatalyst (Mosen and Rehm, 1987). The carbon-immobilized mixed culture was able to degrade phenol concentrations up to 17 g l^{-1}. The activated carbon acted as a buffer and concentrator, such that the bacteria remained active at phenol concentrations 10-times higher than would be toxic to free cells. In such a combination, the carbon could be used to concentrate organohalogens out of large volumes of groundwater contaminated on the micropollutant level and the immobilized bacteria could mineralize the adsorbed organohalogen, so increasing the adsorptive life of the charcoal.

The same type of bioreactor can be used for the biotreatment of leachates from dump sites, landfills and mine tailings, which are a major source of surface and groundwater pollution. If leachates can be controlled and collected they are often amenable to bioremediation. The contaminants can span the complete range of chemicals and concentrations. Bioreactors can effectively work as end-of-pipe reactors and have been operated under aerobic, anaerobic or combined systems to affect the treatment.

Air remediation

Biofiltration has previously been applied to the control of odours from waste water treatment plants, composting works and the like. Such systems use filter beds consisting of organic materials such as compost or peat. The developments in the regulation of process conditions and new high-porosity packing materials has opened the way to wider applications of this technology. The use of xenobiotic-degrading competent microbes in such filter systems enables the development of biofilters for removal of volatile organics from air. The technology can be applied either as part of a manufacturing process operation or as a means of filtering polluted air contaminated as part of an air-stripping process for water decontamination. So-called bioscrubbers can be developed to remove the volatiles from the air. As a modification of the trickle filter or submerged film reactor, the principle is to culture a biofilm of the competent microbes on high surface area support materials. The biofilm is supplied with nutrients and humidified by wetting the inlet gas stream so as to maintain a surface film of water over the biomass. The contaminated air is passed over the biofilm and the volatile materials are taken up into the water film and then degraded by the biofilm.

The ability of biofilters to scrub a range of volatile organic compounds including dichloromethane and 1,2-dichloroethane has been investigated. For easily degradable compounds, such as alcohols, etc., samples of activated sludge provided suitable organisms for inoculation of the reactors. However, for more recalcitrant substances such as the haloorganics, the support materials were inoculated with specialized organisms previously isolated by enrichment culture.

As the emissions standards are lowered for treatment facilities, the use of direct air stripping of volatile organics is being prohibited. The use of standard absorption techniques, such as passing the air through activated carbon filters, is an expensive alternative and has a severe limitation in that it is only a phase shift, not a destructive process, hence once the carbon is exhausted there is still a disposal problem. Combining a bioscrubber with an air stripping plant (Fig. 7.3)

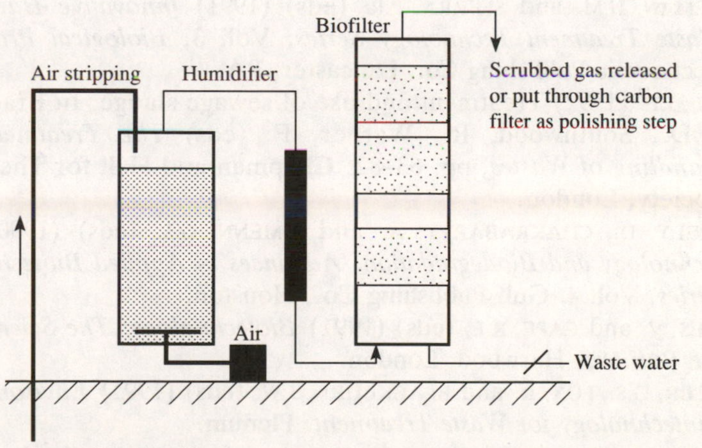

Fig. 7.3. Combined technologies for the treatment of groundwater contaminated with volatile organic compounds.

represents a combination of technologies which may provide a viable strategy for treatment of waters polluted with volatile organic compounds.

References

BEWLEY, R., ELLIS, B., THEILE, P., VINEG, I. and REES, J. (1989) Microbial clean up of contaminated soil. *Chem. Ind.*, **23**, 778–783.

BEWLEY, R.J.K., SLEAT, R. and REES, J.F. (1991) Waste treatment and pollution clean-up. In Moses, V., Cape, R.E. (eds) *Biotechnology: The Science and the Business*, pp. 507–520. Harwood Acad. Publishers, London.

culture of *Pseudomonas putida* and *Cryptococcus elinivii* adsorbed on activated carbon. *Appl. Microbiol. Biotechnol.*, **26**, 283.

PFLUG, A.D. and BURTON, M.B. (1988) Remediation of multimedia contamination from the wood preserving industry. In Ommen, G.S. (ed) *Environmental Biotechnology*. Plenum.

WESTMEIER, F. and REHM, H.J. (1987) Degradation of 4-chlorophenol in municipal wastewater by absorptive immobilized *Alcaligenes* sp. A 7-2. *Appl. Microbiol. Biotechnol.*, **26**, 78.

Further reading

ANDERSON, G.K. and PESCOD, M.B. (1992) Degradation of organic wastes. In Bradshaw, A.D., Southwood, R., Warner, F. (eds) *The Treatment and Handling of Wastes*, pp. 167–189. Chapman and Hall for The Royal Society, London.

FREEMAN, H.M. and SFERRA, P.R. (eds) (1991) *Innovative Hazardous Waste Treatment Technology Series*, Vol. **3**, *Biological Processes*. Technomic Publishing Co., Lancaster, PA.

HALL, J.E. (1992) Treatment and use of sewage sludge. In Bradshaw, A.D., Southwood, R., Warner, F. (eds) *The Treatment and Handling of Wastes*, pp. 63–82. Chapman and Hall for The Royal Society, London.

KAMELY, D., CHAKRABARTY, A. and OMENN, G.S. (eds) (1990) *Biotechnology and Biodegradation: Advances in Applied Biotechnology Series*, Vol. 4. Gulf Publishing Co., Houston.

MOSES, V. and CAPE, R.E. (eds) (1991) *Biotechnology: The Science and the Business*. Harwood, London.

SAYLER, G.S., FOX, R. and BLACKBURN, J.W. (eds) (1991) *Environmental Biotechnology for Waste Treatment*. Plenum.

NITRATE AND PHOSPHATE POLLUTION

Chapter 8

NITROGEN AND PHOSPHORUS IN THE ENVIRONMENT

The environmental behaviour of nitrogen and phosphate

The majority of global nitrogen and phosphorus exists in forms not readily available to the biota. Nitrogen occurs predominately as molecular nitrogen (N_2) in the atmosphere while the majority of phosphorus is immobilized within the rocks and soils of the earth. The supply and environmental cycling of available forms of these nutrient elements is largely dependent on the biological decomposition of nitrogen- and phosphorus-containing compounds accumulated within the biota. Because of the importance of decomposition, the cycling of elements accumulated by living biomass does not occur independently. Turnover and decomposition of biomass varies considerably between habitats, depending on the size and activity of microbial and fungal communities. In warm wet oxic environments, decomposition and nutrient release occurs rapidly. In tropical forests the residence time for leaf litter carbon is approximately 3 months, for temperate forest 4–16 years, while in boreal systems it may exceed 100 years (Recklefs, 1990). Bacterial decomposition is frequently limited by available nitrogen. The average ratio of C : N in bacteria biomass is approximately 10 : 1 (i.e. 1 g of N required per 11 g microbial biomass). This value may be viewed as a measure of the relative microbial requirements for the two elements. Typical plant material has a C : N ratio of 40–80 : 1, i.e. it is deficient in nitrogen, thus rapid decomposition of plant material will depend on the availability of external sources of nitrogen. Animal remains, having a C : N ratio close to that of the decomposers, break down rapidly (Swift *et al.*, 1979; Begon *et al.*, 1990). Decomposition of animal biomass is also aided by the absence of complex polymers not readily broken down, e.g. cellulose and lignin, and the high proportion of biomass initially present in a liquid state. The C : N ratio of soils is remarkably constant at around 10, although in waterlogged or acidic soils, where decomposition is inhibited, the ratio may rise to 17. Begon *et al.* (1990) suggest that the soil decomposer system is remarkably stable, stating that in general, when material with a nitrogen content less than 1.2–1.3% is added to soil, any available ammonium ions are absorbed, while when materials with nitrogen contents higher than 1.8% are added, ammonium ions tend to be released.

The entry of nitrogen into the biological nitrogen cycle occurs predominately through the nitrogen-fixing activity of certain free-living bacteria, blue–green algae and symbiotic bacteria associated with the roots of certain plants (e.g. *Rhizobium*, present in root nodules of some legumes). These organisms are able to reduce N_2 to NH_4. Although representing a small proportion of the global annual flux of nitrogen, nitrogen fixation is the ultimate source of nitrogen in both aquatic and terrestrial habitats. Nitrogen in this form (NH_4^+), is only utilized to a limited extent by plants, nitrogen is most readily accumulated as nitrate. As a result the productivity of both terrestrial and aquatic systems is frequently limited by the availability of nitrate. Organic nitrogen, represents another source of ammonium nitrogen.

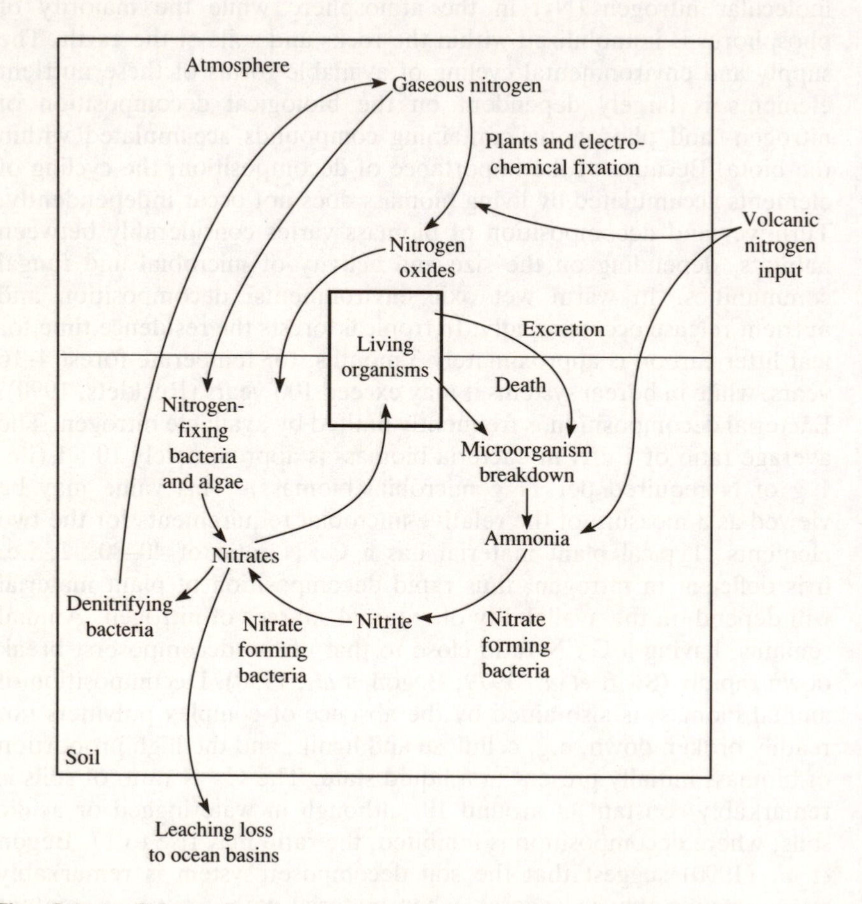

Fig. 8.1(a) Schematic diagram of transformations of compounds in the nitrogen cycle (after Etherington, 1975).

The transformation of organic nitrogen involves ammonification, i.e. the hydrolysis of protein and oxidation of amino acids, a process readily accomplished by all organisms. Free ammonium ions may be oxidized by free-living bacteria in the soil (*Nitrosomonas*) and the sea (*Nitrosococcus*) from the N3− state to N3+ to give nitrite (NO_2^-), which is further oxidized by *Nitobacter* in the soil and *Nitricoccus* in the sea to give nitrate (NO_3^-), the N5+ state. In this form, nitrogen is readily taken up by both terrestrial and aquatic plants. Under anoxic conditions, e.g. waterlogged soils and sediments, denitrification can occur, in which nitrate and nitrite are utilized by bacteria as electron acceptors (oxidizers) (Fig. 8.1). Nitrogen fixed near the soil surface may be lost by denitrification if it leaches into deeper anaerobic layers. Denitrification in soils may be achieved by bacteria such as *Pseudomonas denitrificans*, at redox potentials less than 0.2 V. On a local habitat scale, nitrogen fluxes associated with fixation and denitrification frequently do not balance; however, on a global scale they do, accounting for approximately 2% of total cycling nitrogen (Ch. 9) (Hardy and Havelka, 1975).

Except for very limited microbial transformations, phosphorus exists in the environment as orthophosphate (PO_4^{3-}), valency +5. In this form phosphorus is readily taken up by both aquatic and terrestrial plants. Animals excrete excess dietary phosphorus as phosphate salts

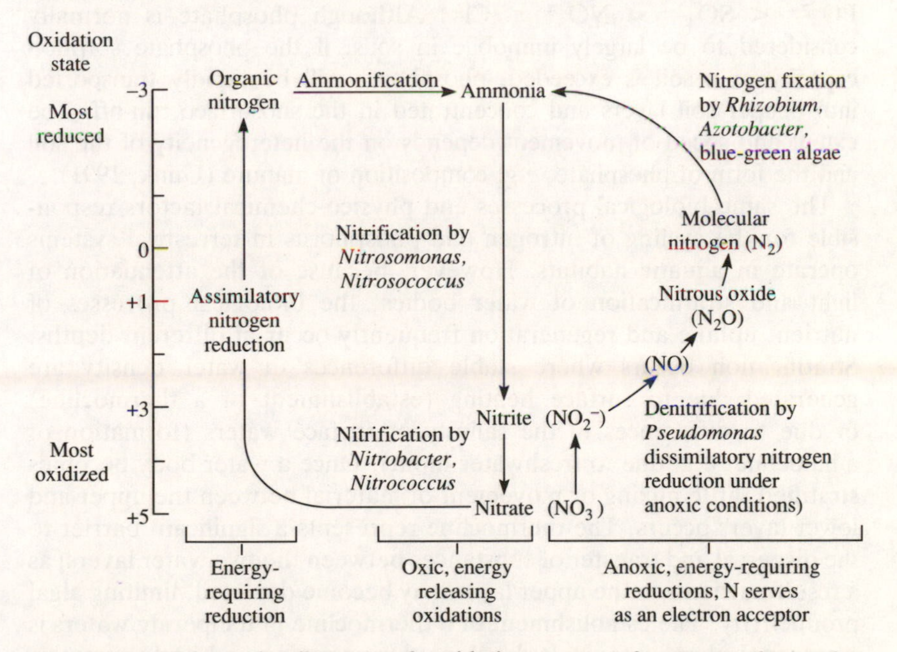

Fig. 8.1(b) Schematic diagram of oxidation states of compounds in the nitrogen cycle (Reklefs, 1990).

in their urine. Phosphorus immobilized within the biota is released during decomposition by the action of phosphatizing bacteria which break down organic phosphate compounds, releasing the phosphate ion. Volatile compounds are not involved in the geochemical cycling of phosphorus; cycling occurs predominately within the soil and aquatic compartments of the biosphere, phosphorus present in the atmosphere is associated with particulate material (Recklefs, 1990).

The biological availability of phosphate is largely pH-dependent. At low acidic pHs, phosphorus is strongly bound to clay particles and readily forms insoluble complexes with ferric iron (e.g. strengite, $Fe(OH)_2H_2PO_4$) and aluminium (e.g. variscite $Al(OH)_2H_2PO_4$). Because of the ubiquitous presence of both ferric iron (Fe^{3+}) and aluminium in soils, sediments and water, very low levels of dissolved phosphate occur under acidic conditions. Where anoxic conditions occur, phosphorus immobilized as insoluble ferric iron complexes may be released as Fe^{3+} is reduced to Fe^{2+} and forms iron sulphide (Edzward, 1977). In alkaline conditions phosphate forms other insoluble compounds, particularly with calcium (e.g. hydroxyapatite $Ca_{10}(PO_4)_6(OH)_2$). Under aerobic conditions in the presence of calcium, aluminium and ferric iron, the concentration of dissolved phosphate is highest at pH 6–7. Due to the reactivity of phosphate, phosphorus has low mobility in the soils. This contrasts strongly with nitrate which is highly mobile and rapidly leached from soils. The mobility of the major soil anions increases in the order $PO_4^{2-} < SO_4^{2-} < NO_3^- \leq Cl^-$. Although phosphate is normally considered to be largely immobile in soils, if the phosphate sorption capacity of a soil is exceeded, phosphorus will be rapidly transported into deeper soil layers and concentrated in the subsurface run-off. The extent and speed of movement depends on the heterogeneity of the soil and the form of phosphate, e.g. composition of manure (Uunk, 1991).

The same biological processes and physico-chemical factors responsible for the cycling of nitrogen and phosphorus in terrestrial systems operate in aquatic habitats. However, because of the attenuation of light and stratification of water bodies, the biological processes of nutrient uptake and regeneration frequently occur at different depths. Stratification occurs where stable differences in water density are generated due to surface heating (establishment of a thermocline) or due to differences in the salinity of surface waters (formation of a halocline, e.g. due to freshwater input). Once a water body becomes stratified, little mixing or movement of material between the upper and lower layers occurs. The thermocline represents a significant barrier to the dispersal and transfer of substances between the two water layers; as a result nutrients in the upper layer may become depleted, limiting algal productivity. The establishment of a thermocline in temperate waters is a seasonal phenomenon. It develops during spring and early summer, as surface waters warm, and is disrupted in winter with the cooling of surface water and storm mixing of the water body. The winter mixing and resuspension of bottom sediments replenish the nutrient content of surface waters. In deep stable oceanic systems and tropical waters the thermocline may be a well developed and stable feature. In coastal waters and lakes the thermocline is a less persistent feature, prone to temporary disruption by tidal movements and persistent high winds. In polar regions little stratification may be evident.

Light intensity decreases approximately exponentially with depth. Photosynthesis is restricted to the photic zone, the depth to which sufficient light penetrates for photosynthesis to exceed respiration. Light attenuation is dependent on the turbidity of the water. In oceans, the photic zone may extend down to 100 m, but typically in coastal waters, due to their high silt and particulate loads, it is much shallower, 6–48 m (Dring, 1982). Phytoplankton sinking below this depth, enter the detritus pathway. In shallow waters, phytoplankton debris reaches bottom sediments, where decomposition and release of nutrients occurs, largely unchanged. In deeper waters, a significant proportion, approximately 18%, of the biotic material cycling at the surface is exported to deeper depths as sinking particulate material, principally zooplankton faecal pellets. Zooplankton grazing serves to effectively repackage phytoplankton into rapid sinking faecal pellets. Nutrient release occurs at depth. The bacterial degradation of sinking detritus frequently produces an oxygen minimum at depths of 500–1000 m (Collier and Edmonds, 1984; Whitfield and Turner, 1987). Due to the active incorporation into the biota, ocean concentration profiles of nitrogen and phosphorus show a pronounced surface depletion and a progressive increase in concentration with depth as sinking biological debris is oxidized (Whitfield and Turner, 1987).

Environmental concerns

Environmental concerns over the concentrations of nitrogen and phosphates in aquatic systems are centred on two issues; the eutrophication of water bodies and the possible health risks associated with the consumption of drinking water containing high concentrations of nitrate.

Eutrophication

The impact of organic effluent on rivers has been well documented. Organic pollutants stimulate and are gradually removed by microbial activity. The increased microbial activity rapidly deoxygenates the water downstream from the discharge, resulting in an oxygen sag curve. The extent of deoxygenation depends on a number of factors including the dilution of the effluent on entering the river and amount of biologically oxidizable material present in the effluent. The latter is assessed by determining the biochemical oxygen demand (BOD) of the effluent, i.e. the oxygen utilized by a sample of water (effluent) incubated at 20°C for 5 days. The determination of chemical oxygen demand (COD),

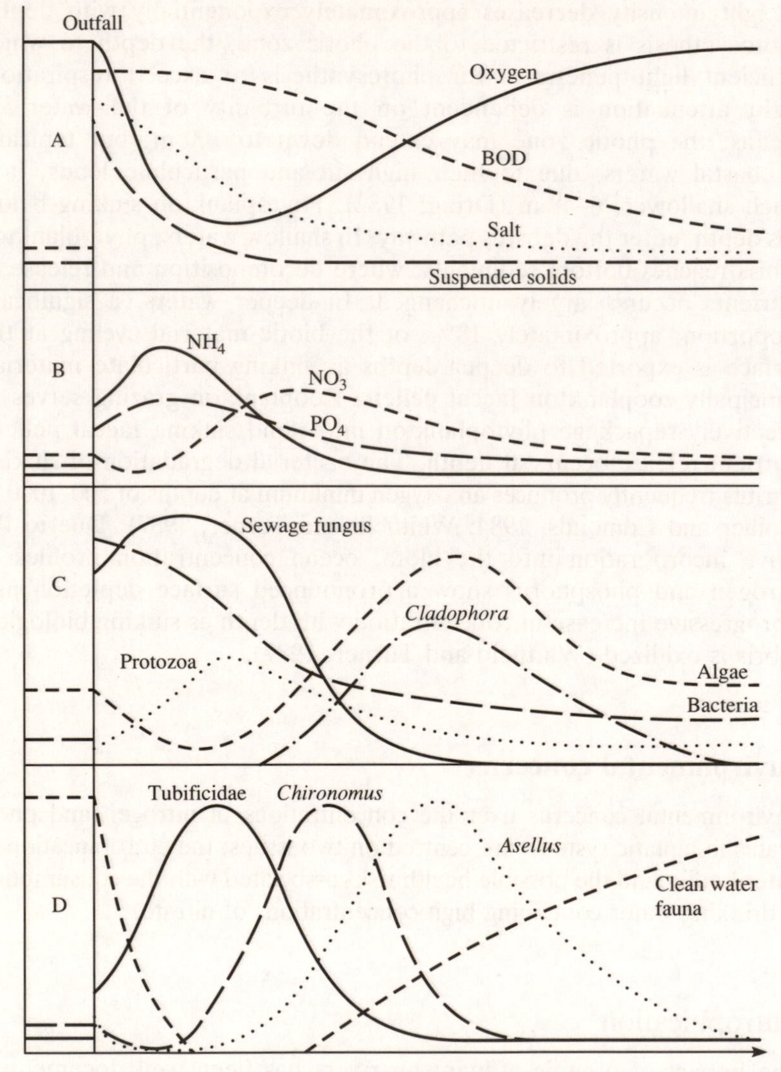

Fig. 8.2. Schematic representation of the changes in water quality and the populations of organisms in a river below a discharge of an organic effluent. A. Physical changes. B. Chemical changes. C. Changes in microorganisms. D. Changes in macroinvertebrates (Hynes, 1960).

which involves measuring the O_2 used when organic compounds present are oxidized by $K_2Cr_2O_7$ or $KMnO_4$, provides another measure of the concentration of organic pollutants in the effluent. High BOD or COD effluent can cause substantial deoxygenation of the receiving water body. For this reason statutory standards of sewage treatment and effluent discharge consents are often framed in terms of permissible effluent BODs. The impact of organic effluent on rivers is frequently transitory, water quality and oxygen content increasing downstream of the discharge. Changes in water quality are reflected by the biota. Near the discharge the biota is impoverished, dominated by species able to tolerate low oxygen concentrations; normal river biota and species diversity re-establishes some distance downstream (Fig. 8.2).

Large discharges or input from numerous sources, may result in the permanent eutrophication of a water system. Water bodies may be classified on the basis of their nutrient content (Table 8.1). The 'natural' nutrient status is largely determined by the fertility of the catchment area and the 'ecological age' of the water body. Oligotrophic conditions are associated with relatively newly-formed water bodies receiving water from nutrient-poor and frequently acidic substrata. As such a system develops, its nutrient status slowly increases as biomass, detritus and sediment gradually accumulate. Primary production in aquatic ecosystems is frequently nutrient limited. Phosphate is normally the limiting nutrient in freshwater habitats, while nitrogen appears to be the major limiting factor in marine waters, although in both cases other elements, e.g. Si, may also be limiting (Barnes and Mann, 1991). Anthropogenic inputs of nutrient result in the cultural eutrophication of water bodies. Initially the effects may be minor, e.g. a small increase in biomass production; however, as the process continues the whole ecology of the system is disrupted (Table 8.2). Changes occur in phytoplankton species composition, which becomes increasingly dominated by rapid growing 'bloom' species, including toxic blue–green algae. With increased production, water turbidity increases, reducing light penetration and causing the loss of submerged macrophytes. These plants are important components of freshwater lake systems, providing microhabitats for invertebrates and fish. With their loss, the invertebrate fauna becomes impoverished and the composition of the fish community changes. In shallow European lakes subject to eutrophication (submerged macrophytes virtually absent, chlorophyll-*a* >100 mg l^{-1}), the normal top predator, the pike (*Esox lucius*) is lost, and species associated with turbid conditions, such as bream (*Abramis brama* L.) and roach (*Rutilus rutilus* L.), dominate fish populations. The benthivorous activities of bream and carp (*Cyprinus carpio*) help, along with wind-induced resuspension of algal detritus and bottom sediments, to maintain water turbidity (Uunk, 1991).

Eutrophic water bodies, because of their nutrient status, are prone to

Table 8.1 Proposed boundary values for trophic categories of inland lakes and reservoirs[a]

Trophic category	*Mean annual total P ($mg\ m^{-3}$)*	*Mean annual chlorophyll ($mg\ m^{-3}$)*	*Maximum chlorophyll ($mg\ m^{-3}$)*	*Mean annual Secchi transparency (m)*	*Minimum annual transparency (m)*
Ultraoligotrophic	<4.0	<1.0	<2.5	>12	>6
Oligotrophic	<10	<2.5	<8	>6	>3
Mesotrophic	10–35	2.5–8	8–25	3–6	1.5–3
Eutrophic	35–100	8–25	25–75	1.5–3	0.7–1.5
Hypertrophic	>100	>25	>75	<1.5	<0.7

[a]Note that the concentration of chlorophyll and Secchi transparency are indices of the standing crop of phytoplankton, and that maximum values correspond to the summer 'bloom' condition. Modified from Vollenweider and Kerekes (1982)

Table 8.2 Characteristics and effects of cultural eutrophication

Biological factors

1. Primary productivity: usually much higher than in unpolluted water, algal blooms, initially increased plant and animal biomass
2. Diversity of primary producers: initially green algae diversity may increase, but blue–green algae rapidly become dominant and diversity declines. Similar trends may be shown by aquatic macrophytes
3. Diversity of heterotrophic organisms (macro- and microinvertebrates and fish) declines, community becomes dominated by species able to tolerate adverse water quality

Chemical factors

1. Anoxic conditions may develop, especially during the night when algae are not photosythesizing or following algal blooms
2. Chemical composition and pH of water may change

Physical factors

1. Mean depth of water decreases as rates of sedimentation increase, shortening the lifespan of the lake
2. Turbidity increases, limiting light penetration and, hence, primary plant productivity

Problems

1. Treatment of potable water may prove difficult and the supply may have an unacceptable taste or odour
2. Water may be injurious to humans and agricultural animals
3. Amenity value of water body may be decreased
4. Increased vegetation may impede water flow and navigation
5. Commercially important fish species (e.g. salmonids and coregonids (whitefish)) may be lost

massive phytoplankton (usually blue–green and dinoflagellate) blooms, particularly in the spring as temperature and light intensity increase and the water body becomes stratified. Algal biomass increases rapidly until it is limited by the available nutrients, normally phosphorus or nitrogen. Rapid growth, is followed by mortality, often causing a green scum of decomposing algae to form on the surface of the water. The rapid decomposition of algal biomass reduces the oxygen content of the water severely, frequently causing substantial fish kills. Since bloom species are frequently toxic to both humans and farm animals, care must be taken to ensure livestock do not gain access and that drinking water supplies do not become contaminated with water from affected lakes or reservoirs. In The Netherlands, the cost to society resulting from (i) the loss of amenity value, (ii) the extra costs of water purification, (iii) the additional cost associated with management of affected water bodies and (iv) the damage to commercial fishing, was estimated in 1979 to be 760 million Dutch guilders (Uunk, 1991).

The impact of the eutrophication, during the late 1960s and early 1970s, on the great lakes of North America and Canada is well-known and documented. In many cases recovery has been achieved by reducing sewage and nutrient inputs; particularly important is the control of phosphate input (Lesht *et al.*, 1991). Management often centres on maintaining or achieving 'permissible' levels of phosphorus and nitrogen. These levels are either derived empirically or from models (Table 8.3). Vollenwieder (1975) has shown that lake nutrient levels may be modelled assuming a steady state with constant inputs and outputs:

$$P = \frac{L}{z(r_s + r_f)}$$

where: P is the total phosphorus concentration in g m^{-3}, often measured in spring at the time of lake turnover, L is the phosphorus loading in g m^{-2} of lake per year, z the mean depth of lake in metres, r_s the sedimentation coefficient, i.e. the fraction of P lost per year to the sediments and r_f is the hydraulic flushing rate, i.e. the number of times the water in a lake is replaced each year.

Table 8.3 Vollenweider's permissible loading levels for total nitrogen and total phosphorus (biochemically active) (g m^{-2} yr^{-1})

Mean depth (m)	*Permissible loading*		*Dangerous loading*	
	N	*P*	*N*	*P*
5	1.0	0.07	2.0	0.13
10	1.5	0.10	3.0	0.20
50	4.0	0.25	8.0	0.50
100	6.0	0.40	12.0	0.80
150	7.5	0.50	15.0	1.00
200	9.0	0.60	18.0	1.20

Based on this conceptual model permissible loading levels for lakes of different depths have been produced. Empirical relationships have been established between chlorophyll-*a* concentrations, water transparency as measured by a simple Secchi disk and phosphate loading (Organisation for Economic Cooperation and Development, 1982). (*A Secchi disk is a round disk with quarters painted alternately black and white. It is lowered into the water body; the depth at which white sections are no longer visible is recorded as the Secchi depth.*) The logarithm of mean

Secchi depth and surface chlorophyll-*a* concentrations are generally linearly related to log phosphorus concentrations. Such relationships may be used to predict the reductions in nutrient loading required to obtain desired water quality objectives. However, in practice, the reductions of phosphate and nitrogen inputs required are often greater than those predicted by these simple relationships and models. The situation is complicated by the mobilization of phosphates and nutrients from sediments, the efficiency with which blue–green algae are able to exploit available phosphorus and the resuspension of algal detritus and sediments by wind and/or bioturbation (Lijklemma, 1991; Uunk, 1991).

Until recently, cultural eutrophication was largely considered to be a problem of inland waters. However, the increasing frequency of algal blooms in coastal waters suggests that this is no longer the case (Paerl, 1988; GESAMP, 1990; Barnes and Mann, 1991). Systematic monitoring of the Baltic Sea since 1980 has shown a progressive increase in nutrient concentrations and production, and a decrease in water oxygen content. Between 1930 and 1980 the nitrogen and phosphorus content of seawater off the Dutch coast increased by factors of four and two respectively. In the inshore waters of the German Bight between 1964 and 1984 winter nitrate levels have increased by a factor of nearly three from 5.6 $\mu g\ l^{-1}$ to 16.5 $\mu g\ l^{-1}$, largely as a result of inputs from the Elbe and other rivers draining into the North Sea. It has been suggested that as a result of anthropogenic nutrient input into the North Sea, phytoplankton are no longer limited by nutrients (Barnes and Mann, 1991). The occurrence of algal blooms has increased along coasts bordering the North Sea and in other coastal regions subject to nutrient enrichment. In 1988, a bloom of the small flagellate algae, *Chrysochromulina polylepis*, damaged seaweeds, invertebrates and fish along a 200 km stretch of the coasts of Denmark, Norway and Sweden. Toxins produced by the algae caused over US$ 10 million of losses to the Norwegian salmon farming industry. The toxin, as yet unidentified, was also found to accumulate in shellfish, presenting a potential human health hazard and preventing their economic exploitation (GESAMP, 1990). While anthropogenic enrichment is an important contributory factor to the occurrence of algal blooms, it is not the only cause. Blooms are most likely to occur where the water body is stratified, horizontal movement/mixing is limited, irradiance intensity is high and day length is long (Paerl, 1988).

Nitrate and drinking water supply

Over the last 30 years nitrate levels in potable water sources have increased steadily. This has popularly been attributed to the increased use of inorganic nitrogen fertilizers and the subsequent leaching of nitrate into groundwaters. There now exists considerable public concern over the possible health hazards associated with elevated levels of nitrate in drinking waters (Addiscot *et al.*, 1991). European Commission directives set a maximum limit of 50 ppm nitrate for drinking water. This corresponds to the recommended level set by the World Health Organisation, which sets a maximum limit of 100 ppm. In the United States the statutory limit is 45 ppm. Nitrate itself does not pose a health threat; however, nitrate is readily reduced to nitrite (NO_2^-) by the enzyme nitrate reductase, which is widely distributed and abundant in both plants and microorganisms (Glidewell, 1990). Nitrite poses two distinct health risks, being potentially carcinogenic and causing methaemoglobinaemia (blue-baby syndrome).

Methaemoglobinaemia (blue-baby syndrome) Bottle-fed babies, in areas where drinking water nitrate content is high, are particularly at risk from blue-baby syndrome. Young babies during the first year of life are susceptible due to the persistence of foetal haemoglobin and because their stomachs are not sufficiently acidic to inhibit the microbial conversion of nitrate to nitrite. Foetal haemoglobin has a much higher affinity for NO ligands than normal haemoglobin. Nitrite, formed in the stomach, readily passes into the bloodstream where it reacts with oxyhaemoglobin, oxidizing the ferrous iron, to form the methaemoglobin in which iron occurs in the ferric form. The conversion of oxyhaemoglobin to methaemoglobin reduces the oxygen-carrying capacity of the blood. Fatality results from 'chemical suffocation'. Fortunately the condition and fatalities are rare. The last recorded case in the United Kingdom was in 1972, and the last recorded mortality in 1951. The majority of the several thousands of cases recorded world-wide are associated with the use of well water, often from privately-dug poorly-sited wells subject to contamination by domestic or animal excreta (Addiscott *et al.*, 1991). The safe level of nitrates in drinking water is put at 100 ppm. In areas where drinking water supplies contain high levels of nitrates, the normal practice is to supply bottled water, low in nitrate, to mothers bottle-feeding infants.

Cancer In the acidic conditions of the stomach (pH *c.* 1.0), nitrite is converted to nitrous acid, HONO, which is in equilibrium with the nitrous acidium ion H_2ONO^+. The acidium ion is a powerful nitrosation agent, readily reacting with a range of food components including the amino acids and iron–porphyrin complexes derived from the protein myoglobin, which serves as the major oxygen store in muscle. Several N-nitroso products formed from amino acids, e.g. nitrosodimethylamine, $(CH_3)_2NNO$, are carcinogenic to a wide range of animal species. Paramagnetic dinitrosyl–iron complexes of the type $[Fe(NO)_2X_2]^{x+}$ (where X, for example is cysteinate), which result from the reaction of nitrite with the iron–sulphur centres of enzymes and proteins, are also implicated as possible carcinogens. Compounds of this type have been isolated from the organs of rats fed on a high sodium nitrite diet and from the livers of rats after the administration of known chemical carcinogens (Glidewell, 1990). However, there appears to be no clear epidemiological evidence to link the nitrate content of drinking water with the incidence of stomach cancers. Studies are complicated by the fact that drinking water forms only part of the dietary intake of nitrate and nitrite. The major source of dietary nitrate is from vegetables. The nitrate content of lettuce, spinach, celery and beetroot is typically around 1000 ppm while peas, broad beans, onions and potatoes normally contain less than 200 ppm. These levels vary seasonally and with growing conditions, particularly with the amount of nitrogen supplied to the crop. Nitrate is also present in cured meats as an additive; traditional curing processes often involve the soaking of the raw meat in salt solutions containing, among other ingredients, sodium or potassium nitrate. The current United Kingdom regulatory upper limit for ham and bacon products is set at 500 ppm sodium nitrate or the equivalent of 595 ppm of potassium nitrate. Average Northern European and American dietary input of nitrate is around 95 mg (Knight *et al.*, 1987). For some people beer may represent the major dietary source of nitrate – four pints of beer of average nitrate content is sufficient to double the normal dietary intake of nitrate (Ministry of Agriculture, Food and Fisheries, 1987). Since the adult stomach is considerably more acidic than that of an infant, conversion of nitrate to nitrite is limited. It is estimated that about only 5% of dietary nitrate is reduced to the more toxic nitrite, adding about 4.75 mg to the average dietary intake of nitrite which is in the order of a few milligrammes a day (Knight *et al.*, 1987). As with nitrate the major dietary sources of nitrite are also vegetables and cured meats; negligible quantities occur in drinking water. The concentration of nitrite in vegetables is low, typically around 1 ppm in fresh produce; however, because of the quantities eaten and their high concentration of nitrate, they represent the most important overall source of nitrite, accounting for 75% of the total exposure. In cured meats, nitrite may be added to a maximum limit of 200 ppm, or formed from the addition of nitrate. Typical pork products contain between 14 and 40 ppm. Although sodium and potassium nitrite give cured pork produce its characteristic pink colour and enhance its flavour, their main function is to prevent spoiling. The addition of nitrites to meat inhibits the growth of potentially harmful bacteria, including the anaerobic bacterium *Clostridium botulinum*, which causes botulism and would readily multiply in sealed food containers. The principal biocide(s) is not thought to be nitrite but a product(s) formed during heat treatment of the product from a reaction between nitrite and components of the meat (Glidewell, 1990).

It is clear that nitrate in drinking water is, in most circumstances, only a minor contributory source of dietary nitrite. However, it may be significant where dietary intake is already high. In an epidemiological study of the geographical distribution of the incidence of cancers in China, the occurrence of oesophageal cancer was found to be very high in one area in the Henan province. This finding has been attributed to dietary factors, in particular the local practice of pickling vegetables by storage in water for several weeks and the consumption of substantial quantities of cornbread. Local water is high in both nitrates and nitrite, thus vegetables stored in the water have high nitrate and nitrite contents. Vegetables pickled in this way have also been found to contain the iron–nitrosyl complex ($Fe_2(SCH_3)_2(NO)_4$). This compound is non-carcinogenic and only weakly mutagenic, but it can increase the carcinogenic action of other compounds. In animal studies it has been shown to substantially increase the tumorigenic properties of N-nitrosamines and aromatic hydrocarbons. Cornbread is often contaminated with the mould *Fusarium moniliforme* which can produce a range of nitrosamines, further increasing exposure to potential carcinogens (Glidewell, 1990).

It is clear from the above, that the possible relationship between the incidence of cancer, drinking water nitrate levels and dietary intakes of nitrate/nitrate is complex. Sufficient evidence exists for the health risks associated with nitrate to be taken seriously. However, it is difficult to balance the cost of reducing drinking water nitrate levels with this perceived health risk. Reducing nitrate levels where drinking water supplies exceed statutory limits is both complex and costly. Often the only viable option is to blend high nitrate supplies with alternative supplies low in nitrates. While efforts are being made to reduce the use of nitrite and nitrate in the food industry, it must be appreciated that the beneficial effects of nitrite use (prevention of spoilage and incidence of human food poisonings) have been substantial. Demands for substantial reductions in the use of nitrogenous fertilizers to reduce high nitrate levels in groundwater sources are probably misplaced. Increased nitrate concentration in groundwater aquifers is linked to the disturbance of established stable vegetation–soil systems and not simply to fertilizer use (cf. p.151).

Sources and loads

The nutrient enrichment of freshwater and marine coastal systems is anthropogenic in origin. Nitrogen and phosphorus are discharged into water courses as raw sewage, effluent from sewage treatment works, as a constituent of trade effluent, urban and rural run-off, and as a consequence of agricultural practice. Undoubtedly, the major point source of organic pollution is discharges from sewage treatment works.

Sewage

A detailed description and discussion of sewage treatment may be found in Gray (1989) and Chapter 7. Essentially the treatment of sewage involves four stages (Fig. 8.3).

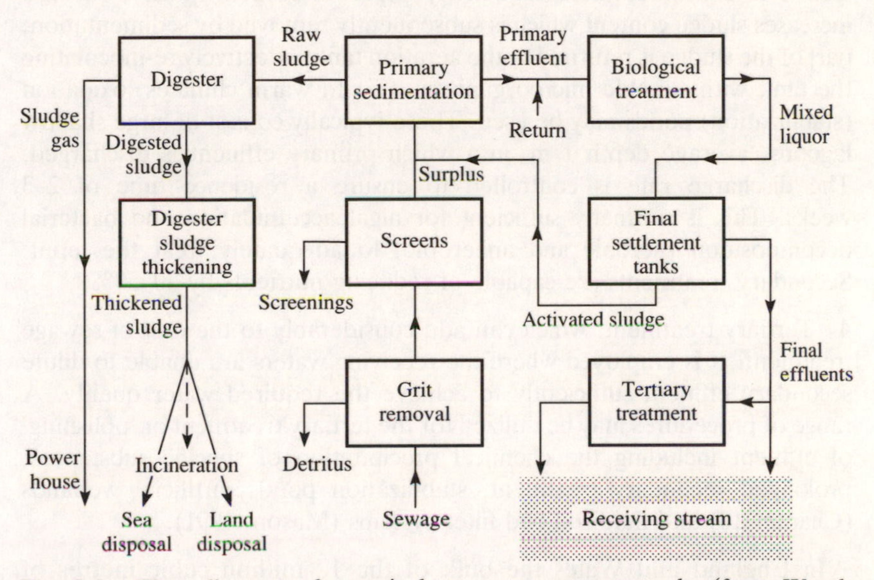

Fig. 8.3. Flow diagram of a typical sewage treatment works (from Wood, 1982).

1 Preliminary treatment: raw sewage is screened to remove large debris. Stones and grit are also removed by passing the sewage along constant velocity channels or by passing it through a grit chamber.

2 Primary treatment: the suspended solids load is reduced by up to 50% in large sedimentation tanks. The raw sludge produced is passed on to the sludge digestion tank where, combined with sludge produced during the secondary treatment process, it is subject to anaerobic digestion and decomposition, a by-product of which, methane, frequently provides the power source for the treatment works. The supernatant liquid (primary effluent) passes on to the secondary treatment stage. Primary treatment typically has a nutrient removal efficiency of between 5 and 15%.

3 Secondary treatment: organic compounds are subjected to biologically mediated oxidation and decomposition. Three methods are commonly utilized. (i) Trickle filters: these consist of large circular or rectangular tanks, 1–3 m deep, packed with graded mineral or plastic 'gravel'. This provides good oxygen diffusion and a large surface area (80–110 m^2 m^{-3}) on which, once primed with settled sewage, a complex and diverse biological community is established which is capable of cleaning top-loaded primary effluent as it percolates through the tank. (ii) In the activated sludge process, primary effluent is mixed with a flocculent suspension of microorganisms in an aerated tank. During aeration, which may last from 1 to 20 hours, microbial activity improves effluent quality, but also increases sludge content which is subsequently removed by sedimentation; part of the sludge is returned to the aeration tank, effectively re-inoculating the tank with suitable microorganisms. (iii) In warm climates, oxidation (stabilization) ponds may be used. These typically consist of large shallow lagoons, average depth 1 m, into which primary effluent is discharged. The discharge rate is controlled to ensure a residence time of 2–3 weeks. This is normally sufficient for algal accumulation and bacterial decomposition (aerobic and anaerobic) to adequately treat the input. Secondary treatments are capable of reducing nutrients by 30–50%.

4 Tertiary treatment, which can add considerably to the cost of sewage treatment. It is employed where the receiving waters are unable to dilute secondary effluent sufficiently to achieve the required water quality. A range of procedures may be utilized for the tertiary treatment or 'polishing' of effluent including the chemical precipitation of specific substances, prolonged secondary treatment, stabilization ponds, artificial wetlands (Chapter 10) and fluidized bed filter systems (Mason, 1991).

In England and Wales the bulk of the 12 million cubic metres of sewage produced each day is of domestic origin. Approximately 80% receives secondary treatment and the resulting effluent enters rivers and canals from over 4500 discharges (Department of the Environment, 1989). At the time of the privatization of United Kingdom public water authorities in September 1989, government figures showed that 22% of sewage treatment works were failing to meet statutory requirements for effluent quality. Sewage effluent is also discharged into United Kingdom coastal waters from approximately 2000 outfalls; 90% of these outfalls discharge less than 100 m beyond low water and 45% discharge untreated sewage, causing gross fouling and microbial contamination of the shoreline (National Rivers Authority, 1991; Water Research Council, 1991). A more or less similar situation exists in many other developed countries. However, there are indications that the situation may improve substantially over the next decade. The United Kingdom Government is committed to improving coastal bathing water quality. This improvement is to be achieved by the construction of new outfalls, the lengthening of existing outfalls and the increased use of primary treatment prior to discharge (National Rivers Authority, 1991). However, since the 'marine treatment' of sewage, defined as 'the use of natural processes in coastal or estuarine water to dilute, disperse and assimilate wastewater following an appropriate level of land based treatment', is to continue, if not increase (Water Research Council, 1991), substantial quantities of nitrogen and phosphate will continue to enter coastal waters. The situation in the United States has improved substantially during the last 20 years. Less than 1% of municipal wastewater is now discharged without secondary treatment and as a result contamination from non-point sources, particularly agricultural and urban run-off, now represent the major threat to surface water quality (Baker, 1992).

Sewage effluent from farm animals also poses a major threat to water quality. Approximately 17% of the total of 24 153 recorded pollution incidents involving discharges into English and Welsh waters in 1988, resulted from farm pollution. The number of such incidents increased 179% between 1979 and 1988. The most frequent causes were releases of animal slurry, discharges of silage effluent and run-off from farm yards, particularly during heavy rainfall (Water Authorities Association, 1988). Marine and freshwater fish farms also contribute to the nutrient loading of water bodies. In the United Kingdom, production from freshwater fish farms has increased steadily from a total of around 2000 tonnes in 1976 to 17 000 tonnes of rainbow trout production alone in 1988 (British Ecological Society, 1990). The effluent from an 'average' fish farm, in terms of its BOD and suspended solid loads is approximately equivalent to the treated waste from 10 590 and 46 000 people, respectively (British Ecological Society, 1990).

Agricultural inputs

The use of fertilizers, particularly those supplying nitrogen, has increased dramatically over the last 50 years. This has allowed substantial increases in both plant and animal production (Tivy, 1990). Without the application of 'artificial' inorganic fertilizers, farming would be uneconomic. In the United Kingdom, application of nitrogen to cereal crops can increase yields by 20 kg per kg of nitrogen applied. However, excessive levels of applications, particularly during winter months, produce little further increase in yield.

Losses from grasslands After the atmosphere and the sea, non-living organic material present in soil represent the largest repository of global nitrogen, containing some 1.5×10^{11} tonnes (Jenkinson, 1990). Even well-cultivated soils, which tend to have low organic contents, contain 2000–3000 kg N ha^{-1}, most of which occurs within the top 25 cm of the soil profile (i.e. within the plough layer). Ploughing and disturbance of the soil stimulates microbial ammonification of organic nitrogen released from decomposing roots and present in the soil, subsequent nitrification produces nitrate, which in the absence of growing plants is readily leached. Because of the nitrogen-fixing activity of clover, old grass–clover pastures contain substantial amounts of nitrogen. When permanent pastures are ploughed up, losses over 20 years may amount to 4 tonnes N ha^{-1}. It has been estimated that the contribution of ploughing permanent grass during the 1940s to the nitrate content of water draining from agricultural areas of England and Wales ranges from less than 5 to over 50 mg l^{-1}, depending on the amount of grassland disturbed. In many areas precipitation which drained from land surfaces in the 1940s is only now being abstracted from the aquifer (Addiscott *et al.*, 1991).

Nutrient losses from non-grazed pastures, cut for hay or silage, are minimal. Grass growth during the autumn and early spring is sufficient to utilize any nitrogen applied. United Kingdom figures suggest that up to 400 kg N ha^{-1} may be applied without substantial loss of nitrogen — twice that which may be applied to cereal crops. In contrast, grazed systems lose large amounts of nitrogen. Between 75 and 95% of the nitrogen consumed by grazing animals is rapidly returned to the soil as dung and urine, from which nitrogen and phosphate are readily leached. These returns are very concentrated and are deposited over very small areas, e.g. cows urinate approximately 2 l at a time over a 0.4 m^2 area, similarly, faecal deposits, which contain approximately half the nitrogen of urine, also affect a very small area. Typical episodes of dung and urine deposition are equivalent to applications of 400–1200 kg N ha^{-1} of nitrogen fertilizer, greatly exceeding the maximum *annual* application of 400 kg N ha^{-1} which can be retained. Urine and dung are not deposited uniformly over a field: 15–20% of a field will receive inputs from urine, 7–10% from dung. The ability of vegetation and soil receiving dung and urine to assimilate nutrients is greatly exceeded. Surplus nitrogen and phosphate are largely lost via run-off and leaching. Studies using N^{15} labelled cattle urine suggest that initial losses can be high. Substantial amounts of the urine drain rapidly down cracks and through macropores. In one study, approximately 26% of the applied nitrogen was leached, 18% lost via ammonia volatilization and 7% by denitrification (Whitehead and Bristow, 1990). Intensively grazed systems, receiving a nitrogen input of 400 kg N ha^{-1} can lose approximately 75%, the majority as leached nitrate. To dilute this loss sufficiently so that drainage water conforms to the European Commission standard of 50 ppm or 11.3 mg^{-1} nitrate–nitrogen, through drainage water must exceed 2600 mm or 100 inches a year (Addiscot *et al.*, 1991). It should be appreciated that the majority of nitrogen inputs into grazing systems are in the form of feed supplements and not as fertilizers applied directly to the grassland.

Livestock are frequently housed for part of the year, and the removal of the resulting farmyard slurry and sewage represents another major source of water contamination. The run-off from farmyards, leakage from storage tanks and application of slurry to fields frequently cause localized contamination of water. The application of slurry to fields is an important farming practice as the slurry supplies both nutrients and helps to improve and maintain soil structure. Much of the nitrogen applied is not in mineral form and therefore not immediately labile to leaching. As the manure is incorporated into the soil and decomposes, increasing amounts will become available to plants, while some will be lost to the atmosphere as a result of denitrification and volatilization. Animal effluents are extremely variable in composition and the rate at which they produce nitrate. This variability makes it difficult to ensure that excess nutrients are not applied (Addiscot *et al.*, 1991).

Losses from arable systems Losses of nitrate from arable fields during the growth season are low. Several studies indicate that approximately 6% of applied nitrogen is leached, approximately twice this amount is lost by denitrification and 1% remains in the soil after harvest. With an average application of fertilizer (190 kg N ha^{-1}), losses due to leaching will amount to approximately 11 kg ha^{-1}; however, in practice losses are considerably greater than this, in the order of 54 kg N ha^{-1} year. The additional loss occurs during the autumn post-harvest period when soils are both warm and moist. Microbial decomposition of soil organic material releases nitrate, which in the absence of plants, is rapidly leached from the soil. The relationship between fertilizer application and the amount which remains in the soil, is thus of major importance. For cereal, little of the applied fertilizer remains (*c.*1.0%); for overcrops such as potato and oilseed rape a high proportion may remain (Addiscot *et al.*, 1991).

Relatively little nitrogen leaches from natural ecosystems; most nutrients are retained and cycled within the system. However, as with agricultural grassland, disturbance results in extensive leaching of mobile nutrients. In forested and wooded systems, clearing allows greater light penetration to the soil surface and the resulting warming will help to stimulate microbial activity. The prevention of regrowth, or in the case of agroforestry the use of herbicides following tree planting, increases nutrient losses (Begon *et al.*, 1990; Cuttle and Gill, 1991).

It is clear that the leaching of fertilizers is only one agricultural source of nutrients in surface and groundwater but it is nevertheless an important one. However the situation is considerably more complex than frequently perceived. An immediate and drastic reduction in the levels of fertilizer application, would have little immediate impact on nitrate concentration in groundwaters but could reduce future deterioration and would help to reduce the problems associated with the eutrophication of surface waters. There is clearly a need to reduce losses from agricultural systems. This will require not only the more efficient use of fertilizers but also careful management of cultivated land to reduce run-off and leaching.

Relative importance of nitrogen and phosphate sources

The relative importance of individual nitrogen and phosphate sources will vary considerably, depending on the nature of the catchment area and the intensity and density of human activity. At Chesapeake Bay in the United States, between 35 and 39% of the Bay's total nitrogen loading is derived from atmospheric deposition, 12% is from direct deposition of NH^+ and NO_3^- to the surface waters, while the remainder arises from deposition of nitrogen over the catchment, 10% of which is transported by rivers into the Bay (Barnes and Mann, 1991; Baker, 1992). In contrast, at Narragansett Bay, Rhode Island, atmospheric inputs of nitrogen account for less than 1% of the yearly input. The predominant sources of both nitrogen and phosphate are sewage and river inputs. Sewage and river inputs account for 42.5 and 50% of nitrogen input, respectively; corresponding values for phosphate are 30 and 69% (Nixon, 1981; Barnes and Mann, 1991).

Of the total 32 000 tonnes of phosphorus entering Dutch surface waters from inland sources each year, 40.9% is of domestic origin from sewage and detergent use, 36.6% is a result of phosphate fertilizer manufacture and 16.9% is from leaching and run-off. In the case of nitrogen the importance of run-off and leaching is considerably greater, accounting for 68% of the estimated 250 000 tonnes annual input of nitrogen into surface waters. Atmospheric deposition of 20×10^6 kg N yr^{-1} accounts for approximately 8% of the total nitrogen input (Uunk, 1991). Atmospheric deposition of nitrogen resulted largely from fossil fuel combustion, combined with agricultural emissions of ammonia, e.g. volatilization of nitrogen from animal slurry spread over fields. The latter was estimated to be 277 000 tonnes NH_3 yr^{-1} in 1985 (Uunk, 1991). Nitrous oxide emissions from the use of fossil fuels have continued to increase since 1900, and are expected to increase by 40–60% over the next 40 years, contributing further to nutrient loading and the eutrophication of water bodies (Barnes and Mann, 1991; Baker, 1992). The nutrient enrichment of coastal waters will increase as coastal regions become increasingly developed and the combustion of fossil fuels continues. In the United States it is estimated that by the year 2000, over half of the population will live within 100 miles of the coast. However, not all causes of eutrophication are directly anthropogenic in origin. The eutrophication of Hickling Broad in the

United Kingdom during the 1960s can be attributed to an increase in the number of black-headed gulls (*Larus rididundus*) roosting on the lake (Moss, 1978).

References

ADDISCOTT, T.M., WHITMORE, A.P., and POWLSON, D.S. (1991) *Farming, Fertilizers and the Nitrate Problem*. CAB International, Wallingford.

BAKER, L.A. (1992) Introduction to nonpoint source pollution in the United States and the prospects for wetland use. *Ecol. Eng.*, **1**, 1–26.

BARNES, R.S.K. and MANN, K.H. (1991) *Fundamentals of Aquatic Ecology*. Blackwell Scientific Publications, Oxford.

BEGON, M., HARPER, J.L. and TOWNSEND, C.R. (1990) *Ecology, Individuals, Populations and Communities*. Blackwell Scientific Publications, Oxford.

BRITISH ECOLOGICAL SOCIETY (1990) *River Water Quality. Ecological Issues No. 1*. British Ecological Society, London.

COLLIER, R. and EDMONDS, J. (1984) The trace element geochemistry of marine biogenic particulate material. *Prog. Oceanogr.*, **13**, 113–199.

CUTTLE, S.P. and GILL, E.K. (1991) Concentration of nitrate in soil water following herbicide treatment of tree planting in an upland agroforestry system. *Agroforest. Sys.*, **13**, 225–234.

DEPARTMENT OF THE ENVIRONMENT (1989) *Digest of Environmental Protection and Water Statistics*. HMSO, London.

DRING, M.J. (1982) *The Biology of Marine Plants*. Edward Arnold, London.

EDZWARD (1977) Phosphorus in aquatic systems: the role of the sediments. In Suffet, I.H. (ed.) *Fate of Pollutants in the Air and Water Environments*. John Wiley and Sons, Chichester.

ETHERINGTON, J.R. (1975) *Environment and Plant Ecology*. John Wiley & Sons, Chichester.

GESAMP (GROUP OF EXPERTS ON THE SCIENTIFIC ASPECTS OF MARINE POLLUTION) (1990) *The State of the Marine Environment. (Joint Group of Experts on Scientific Aspects of Marine Pollution.)* Blackwell Scientific Publications, Oxford.

GLIDEWELL, C. (1990) The nitrate/nitrite controversy. *Chem. Brit.*, **26(2)**, 137–140.

GRAY, N.F. (1989) *Biology of Wastewater Treatment*. Oxford University Press, Oxford.

HARDY, R.W. and HAVELKA, U.D. (1975) Nitrogen fixation research: a key to world food? *Science*, **188**, 633–642.

HYNES, H.B.N. (1960) *The Biology of Polluted Waters*. Liverpool University Press, Liverpool.

JENKINSON, D.S. (1990) An introduction to the global nitrogen cycle. *Soil Use Manag.*, **6**, 56–61.

KNIGHT, K.M., FORMAN, D., ALDABBAGH, S.A. and DOLL, R. (1987) Estimation of dietary intake of nitrate and nitrite in Great Britain. *Food Chem. Toxicol.*, **25(4)**, 277–285.

LESHT, B.M., FONTAINE III, T.D. and DOLAN, D.M. (1991) Great Lakes total phosphorous model: post audit and regionalized sensitivity analysis. *J. Great Lakes Res.*, **17(1)**, 3–17.

LIJKLEMMA, L. (1991) Response of lakes to the reduction of phosphorus load. *Hydrobiol. Bull.*, **24(2)**, 165–170.

MASON, C.F. (1991) *Biology of Freshwater Pollution*, 2nd edn. Longmans Scientific and Technical, Harlow.

MINISTRY OF AGRICULTURE, FISHERIES AND FOOD (1987) *Food Surveillance Paper No. 20*. HMSO, London.

MOSS, B. (1978) The ecological history of a medieval man-made lake, Hickling Broad, Norfolk, United Kingdom. *Hydrobiologia*, **60**, 23–32.

NATIONAL RIVERS AUTHORITY (1991) *Bathing Water Quality in England and Wales—1990*. Water Quality Series No. 3, National Rivers Authority, Bristol.

NIXON, S.W. (1981) Remineralization and nutrient cycling in coastal marine ecosystems. In Nielson, B.J., Cronin, L.E. (eds) *Estuaries and Nutrients*, pp. 111–138. Humana Press, New Jersey.

ORGANISATION FOR ECONOMIC COOPERATION AND DEVELOPMENT (1982) *Eutrophication of Waters: Monitoring, Assessment and Control*. OECD, Paris.

PAERL, H.W. (1988) Nuisance phytoplankton blooms in coastal, estuarine, and inland waters. *Limnol. Oceanogr.*, **33(4)**, 823–847.

REKLEFS, R.E. (1990) *Ecology*, 3rd edn. W.H. Freeman and Company, New York.

SWIFT, M.J., HEAL, O.W. and ANDERSON, J.M. (1979) *Decomposition in Terrestrial Ecosystems*. Blackwell Scientific Publications, Oxford.

TIVY, J. (1990) *Agricultural Ecology*. Longman Scientific and Technical, Harlow.

UUNK, E.J.K. (1991) *Eutrophication of Surface Waters and the Contribution of Agriculture*. Paper read before the Fertiliser Society, London, 11 April 1991. The Fertiliser Society, Greenhill House, Peterborough.

VOLLENWEIDER, R.A. (1975) Input–output models with special reference to the phosphorus loading concept in liminology. *Schweiz. Z. Hydrol.*, **37**, 53–84.

VOLLENWEIDER, R.A. and KEREKES, J.J. (1982) *Eutrophication of Waters: Monitoring Assessment and Control*. Organization for Economic Cooperation and Development, Paris.

WATER AUTHORITIES ASSOCIATION (1988) *Water Pollution from Farms Waste, England and Wales*. Water Authorities Association, London.

WATER RESEARCH COUNCIL (1991) *Design Guide for Marine Treatment Schemes*. Water Research Council, Watford.

WHITEHEAD, D.C. and BRISTOW, A.W. (1990) Transformations of nitrogen following the application of ^{15}N-labelled cattle urine to an established grass sward. *J. Appl. Ecol.*, **27(2)**, 667–678.

WHITFIELD, M. and TURNER, D.R. (1987) The role of particles in regulating the composition of seawater. In Stumm, W. (ed.) *Aquatic Surface Chemistry: Chemical Processes at the Particle–Water Interface*, pp. 457–493. John Wiley, New York.

WOOD, L.B. (1982) *The Restoration of the Tidal Thames*. Hilger, Bristol.

Further reading

ADDISCOTT, T.M., WHITMORE, A.P. and POWLSON, D.S. (1991) *Farming, Fertilizers and the Nitrate Problem*. CAB International, Wallingford.

BARNES, R.S.K. and MANN, K.H. (1991) *Fundamentals of Aquatic Ecology*. Blackwell Scientific Publications, Oxford.

GESAMP (1990) *The State of the Marine Environment (Joint Group of Experts on Scientific Aspects of Marine Pollution)*. Blackwell Scientific Publications, Oxford.

MANNION, A.M. and BOWLBY, S.R. (1992) *Environmental Issues in the 1990s*. John Wiley and Sons, Chichester.

MASON, C.F. (1991) *Biology of Freshwater Pollution*, 2nd edn. Longman Scientific and Technical, Harlow.

Chapter 9

MICROBIAL REMOVAL OF NITRATES

Introduction

The discharge of waste nitrogen compounds into the air and water from industry and the disposal of solid waste containing nitrogen compounds may result in raised concentrations of nitrates in the environment. This occurs either through the direct discharge of nitrate or through the conversion of other waste nitrogen compounds to nitrate by physical or microbiological action in the biosphere. Industrial sources of waste nitrogen compounds are varied, as are the type of compounds discharged (Robertson and Kuenen, 1992). The photographic industry produces significant amounts of ammonium thiosulphate containing wastewater and a variety of other industries produce high ammonia wastes which are converted to nitrate at treatment plants. The petroleum industry and the burning of fossil fuels release substantial amounts of nitrogen compounds to the biosphere. Ironically, processes designed to lower the discharge of one contaminant to the environment occasionally increase the release of another. Wet lime–gypsum flue gas desulphurization plants are a good illustration of this. Designed to lower the environmental discharge of sulphur, they produce wastewater containing high levels of nitrates, e.g. 150–300 mg l^{-1} N (Holm Kristensen and Jepsen, 1991).

The production of foods also contributes to the nitrate pollution load in the environment. Food processing produces considerable quantities of organic and nitrate waste, e.g. the dairy industry. Animal farming, particularly intensive farming, results in the necessity to dispose of large quantities of high nitrogen organic waste in the form of slurries. In addition, large amounts of industrially produced nitrates are routinely applied to agricultural land as fertilizer to improve crop growth and yield. A relatively large proportion of the nitrate fertilizers leach from the areas of application into surface water and groundwater systems (Ch. 8).

The impact of this pollution is described in Chapter 8; however, in brief, the result is eutrophication of waterways and algal blooms. Groundwater sources from aquifers are also at risk from nitrate pollution. In England and Wales 142 groundwater sources used for public supply had nitrate levels above 11.3 mg N l^{-1} (House of Lords, 1989), the maximum admissible concentration set by the European Community (Council of European Communities, 1980). The United States Environmental Protection Agency (EPA) has set the limit at 10.0 mg l^{-1} nitrate–nitrogen, a value exceeded in many areas within the United States (Mateju *et al.*, 1992). These nominal levels have been set in view of the health risks thought to be associated with consumption of high nitrogen water, i.e. infantile methaemoglobinaemia and stomach cancer.

A further nutrient present in a variety of wastes, e.g. animal waste and sewage, is phosphate, which is also applied to soils as a fertilizer. The release of phosphate has substantial environmental impact through its contribution to eutrophication. The conventional technology for the microbial removal of phosphate is the activated sludge process of sewage plants. This is competently reviewed elsewhere (Horan, 1990) and briefly described in Chapter 7. Several workers have studied enhanced biological phosphate removal (EBPR) in activated sludge. EBPR involves control of environmental conditions, in simple terms sequential anaerobic and aerobic fluxes, within the activated sludge. No pure cultures of bacteria from activated sludge capable of EBPR have been isolated although *Acinetobacter* is thought to be the most likely contributor to the process (Jenkins and Tandoi, 1991). Since EBPR is based on an existing technology, it has not been shown to occur with pure isolates and even *in situ* is a somewhat unreliable process. It is not intended to discuss microbial phosphate removal. An excellent review is supplied by Jenkins and Tandoi (1991). The role of macrophyte systems in phosphate removal is discussed in Chapter 10.

The aim of this chapter is to review the biochemistry and physiology of microbial nitrate removal from waste and groundwaters. In addition, consideration will be given to the existing and possible future technologies for this process.

Microbial removal of nitrate from wastes and groundwater: biocatalysts and mechanisms

The contribution of microorganisms to the nitrogen biogeochemical cycle has been extensively studied. Microorganisms catalysing the conversion of nitrogen compounds that go to make up the nitrogen cycle are from a diverse range of physiological types. These include nitrogen-fixing bacteria, ammonifying bacteria, nitrifying bacteria and denitrifying bacteria. In general terms the role of denitrifiers in the nitrogen cycle is the return of nitrogen to the atmosphere. Thus, denitrifiers convert nitrate in a sequence of reactions (see below) to molecular nitrogen, a gas. Nitrogen gas is largely unavailable to living organisms, except nitrogen-fixing prokaryotes and as a result the deleterious effect of high nitrates in the environment may be mitigated. Heterotrophic dissimilatory nitrate reducers can also contribute to the conversion of nitrate to other nitrogen compounds, nitrite or possibly ammonia. These organisms, however, do not catalyse the release of nitrogen to the atmosphere. One potential route for the control of nitrate pollution involves harnessing the catalytic capacity of bacteria capable of nitrate conversions, particularly the denitrifying bacteria.

Denitrifying bacteria and denitrification

Reaction and enzyme systems Denitrification follows the following sequence of reactions:

$$NO_3^- \rightarrow NO_2^- \rightarrow NO(gas) \rightarrow N_2O(gas) \rightarrow N_2(gas)$$

During denitrification the nitrogen compounds act as terminal electron acceptors instead of oxygen during respiration. Commonly, therefore, though not exclusively (see below), denitrification occurs when oxygen becomes limited for aerobic respiration. After oxygen depletion nitrate is the first compound to be used as a terminal electron acceptor. There are a variety of species from several different genera known to be involved in denitrification. Some bacteria are capable of catalysing the full transformation of NO_3^- to molecular nitrogen, e.g. *Paracoccus denitrificans*. Many other bacteria, however, are only capable of catalysing particular stages in the sequence of reactions (Table 9.1). Denitrification of nitrate to molecular nitrogen, therefore, commonly involves a mixed community of microorganisms with complementary metabolisms. Since the nitrogen oxide gases are thought to contribute to the greenhouse effect any designed denitrification process should use a single organism or a community capable of catalysing complete denitrification.

Each stage of the sequence is catalysed by a different enzyme system: NO_3^- to NO_2^- by nitrate reductase (a membrane bound enzyme), NO_2^- to NO and N_2O by nitrite reductase and nitric oxide (NO) reductase which are membrane bound and cytoplasmic enzymes, and N_2O to N_2 by nitrous oxide (N_2O) reductase, a periplasmic enzyme (Payne, 1981; Hochstein and Tomlinson, 1988; Robertson and Kuenen, 1992). Those species lacking particular enzyme systems are unable to catalyse the full range of reactions. For example, *Pseudomonas aureofaciens* does not possess N_2O reductase and is unable to form N_2 (Robertson and Kuenen, 1992). The enzymes are thought to be inducible rather than constitutive. However, it is unlikely that the induction and repression of the enzymes is identical for the different

Table 9.1 Examples of bacteria capable of denitrification or dissimilatory nitrate reduction

Chemolithotrophs	*Heterotrophs*	*Denitrification reaction in optimum conditions*	
Thiobacillus thiparus	*Hysobacter antibioticum*	NO_3^-	NO_2^-
	Pseudomonas *Achromobacter*	NO_3^-	N_2O
Paracoccus denitrificans *Thiobacillus denitrificans*	*Hyphomicrobium* *Pseudomonas* *Halobacterium* *Alcaligenes eutropha*	NO_3^-	N_2
	Klebsiella aerogenes *Escherichia coli*	NO_3^-	NH_3
	Flavobacterium sp.	NO_2^-	NO_2
	Vibrio succinogenes	NO_2^-	N_2

organisms involved in denitrification given their genetic and metabolic diversity (Hochstein and Tomlinson, 1988; Mateju *et al.*, 1992).

Heterotrophic denitrification The majority of denitrifying bacteria are heterotrophs, e.g. *Pseudomonas* (which appear the dominant denitrifiers), *Alcaligenes* and *Flavobacterium*, and require oxidizable organic substrates to act as electron donors, i.e. energy source, during respiration. A wide variety of organic substrates are utilized during denitrification, e.g. methanol, glucose, acetic acid and industrial waste such as sulphite waste liquor (Mateju *et al.*, 1992). Denitrification by heterotrophic bacteria involves the oxidation of organic compounds and the reduction of nitrate to nitrogen through the sequence shown above. The nitrogen compounds act as the terminal electron acceptors during this respiratory process (Eq. 9.1).

$$2\tfrac{1}{2}CH_2O + 2NO_3^- + 2H^+ \rightarrow N_2 + 2\tfrac{1}{2}CO_2 + 3\tfrac{1}{2}H_2O \quad (9.1)$$

Chemolithotrophic denitrifiers Some denitrifiers are obligate or facultative chemolithotrophs using hydrogen or reduced sulphur as sources of energy, i.e. electron donors, e.g. *Thiobacillus denitrificans* (a facultative chemolithotroph) and *Ferrobacillus ferrooxidans* (an obligate chemolithotroph). Chemolithotrophic denitrification involves the oxidation of a reduced inorganic substrate, such as pyrite (Eq. 9.2).

$$2\tfrac{1}{2}FeS_2 + 7NO_3^- + 2H^+ \rightarrow 3\tfrac{1}{2}N_2 + 5SO_4^{2-} + 2\tfrac{1}{2}Fe^{2+} + H_2O \quad (9.2)$$

Dissimilatory nitrate reduction Bacteria capable of catalysing dissimilatory nitrate reduction possess a nitrate reductase system broadly similar to that of denitrifiers. The enzyme systems between dissimilatory nitrate reduction and denitrification diverge at the second phase of the reaction i.e. transformation of nitrite to NO versus the conversion of nitrite to ammonia (Robertson and Kuenen, 1992). During the first step in dissimilatory nitrate reduction, i.e. NO_3^- conversion to NO_2^-, nitrate appears to act as a terminal electron acceptor. The production of ammonia from nitrite, however, seems to be a mechanism for discarding excess reducing power (Cole, 1987).

Factors affecting denitrification

A variety of factors are known to affect denitrification including oxygen availability, nutrient availability, pH, temperature and the presence of inhibitory compounds.

Oxygen The influence of oxygen availability is substantial with some species automatically shutting down denitrification if oxygen is present. Traditionally, this has been viewed as inevitable given that nitrate is acting as an alternative electron acceptor to oxygen during respiration. It has become evident recently that the effect of oxygen is far more complicated. Aerobic denitrification occurs (Robertson and Kuenen, 1992) with oxygen respiration and denitrification happening simultaneously (Robertson and Kuenen, 1990). The ability to continue to denitrify in the presence of oxygen varies with species (Robertson and Kuenen, 1992). For example, *Thiosphoera pantotropha* and *Alcaligenes* sp. denitrify at oxygen concentrations (as % air saturation) of 80 and 50%, respectively (Robertson and Kuenen, 1990). Denitrification in the presence of oxygen, however, is substantially slower than in its absence.

In addition to differences in the physiological response to oxygen, the physical structure of the environment also has considerable influence on denitrification apparently occurring in aerobic conditions. For example, soil particles are frequently formed into aggregates which are penetrated by water-filled pores. These pores are often anaerobic representing a micro-environment where rapid denitrification can occur, even though the pores between the aggregates are aerobic.

Energy source: carbon and inorganic compounds It appears that nitrate is preferentially reduced through the dissimilatory nitrate reductase route when the ratio of electron donor, i.e. organic compound, to electron acceptor, i.e. nitrate, is high. If the ratio is reversed then denitrification is favoured (Tiedje *et al.*, 1982). Organic substrates play several roles during denitrification. They act as an electron donor for denitrification and contribute to cell synthesis. Thus, not all the carbon is available for denitrification. Moreover, further carbon may be lost in the presence of oxygen.

Stoichiometric equations for the heterotrophic denitrification of a variety of carbon sources have been developed by several workers and are reviewed by Mateju *et al.* (1992) (Table 9.2). The amount of denitrified nitrogen shows a linear dependence on the concentration of carbon. The C : N ratio (by weight) of denitrification for different carbon sources has been estimated based on experimental results. For example, the C : N ratio for methanol is 0.93 for nitrate and 0.57 for nitrite while those for acetate are 1.32 and 0.83, respectively (Mateju *et al.*, 1992). A C : N ratio of 1.0 has been suggested as necessary for 80–90% denitrification to occur.

Stoichiometric equations for chemolithotrophic denitrification involving the oxidation of various inorganic energy sources have also been developed (Mateju *et al.*, 1992) (Table 9.2). It appears that chemolithotrophic denitrification shows a similar dependence on the concentration of the electron donor as heterotrophic denitrification.

Table 9.2 Stoichiometric relationship of denitrification with various electron donors

Electron donor	*Stoichiometric relationship*
Hydrogen	$2\frac{1}{2}\ H_2 + NO_3^- \rightarrow \frac{1}{2}\ N_2 + 2\ H_2O + OH^-$
Sulphide	$2\frac{1}{2}\ Fe\ S_2 + 7\ NO_3^- + 2H^+ \rightarrow 3\frac{1}{2}\ N_2 + 5SO_4^{2-} + 2\frac{1}{2}\ Fe^{2+} + H_2O$
Ferrous iron	$10\ Fe^{2+} + 2\ NO_3^- \rightarrow N_2 + 10\ FeOOH + 18H^+$
Methane	$2\frac{1}{2}\ CH_4 + 4\ NO_3^- + 4H^+ \rightarrow 2\ N_2 + 2\frac{1}{2}\ CO_2 + 7\ H_2O$
Methanol	$2\frac{1}{2}\ CH_3OH + 3\ NO_3^- \rightarrow 1\frac{1}{2}\ N_2 + 2\frac{1}{2}\ CO_2 + 3\frac{1}{2}\ H_2O + \frac{1}{2}\ O_2$
Acetic acid	$2\frac{1}{2}\ CH_3COOH + 8NO_3^- \rightarrow 2\ CO_2 + 8\ HCO_3^- + 6H_2O + N_2$
Cellulose	$2\frac{1}{2}\ (C_6H_{10}O_5)_n + 12nNO_3^- \quad 3nCO_2 + 6\frac{1}{2}nH_2O + 6_nN_2 + 12nHCO_3^-$

Certainly, the S : N ratio has been found to have a major impact. At low S : N ratios nitrate conversion to N_2O is limited and nitrite accumulates. If thiosulphate is the electron donor for denitrification then the minimum S : N ratio required appears to be 4.3. Above this value nitrite does not accumulate. The ratio changes with sulphur source, e.g. thiosulphate, elemental sulphur, etc. Undoubtedly similar effects occur with other oxidizable inorganic substrates used in denitrification, e.g. H_2 and Fe^{2+}.

Temperature and pH Both pH and temperature affect denitrification. Although denitrification can be shown to occur at 0–5°C, in general the denitrification rate decreases with temperature. This is due to two effects. First, the majority of heterotrophs which dominate denitrification are mesophilic and have slower metabolic rates at lower temperatures. Second, temperature affects oxygen solubility. At high temperatures oxygen solubility decreases which results in an increase in the rate of denitrification. Gauntlett and Craft (1979) have estimated that denitrification may double for every 10°C rise in temperature, up to a maximum above which denitrification does not occur, e.g. 50°C (Holm Kristensen and Jepsen, 1991). Heterotrophic denitrification is commonly optimum at pH 7–8. Denitrification by chemolithotrophs can occur at lower pHs.

Inhibitory substances Denitrification can be inhibited by substances other than oxygen. Acetylene and sulphide inhibit N_2O reductase, thereby preventing reduction of N_2O to N_2 (Kuenen and Robertson, 1987). Nitrate itself acts as an inhibitor of the final stages in the sequence of denitrification reductions with the enzyme catalysing the conversion of NO to N_2O suppressed by NO_3^- (Goering, 1972). The role of sulphur compounds in the inhibition of denitrification appears to be important. Both sulphide and sulphate (Kowalenko, 1979) have been shown to reduce denitrification. Interestingly, the presence of sulphide depresses denitrification but the reduction of nitrate to ammonia is stimulated by sulphide (Kowalenko, 1979). This suggests that dissimilatory nitrate reduction is enhanced by the presence of sulphide. Heavy metals can inhibit nitrate reduction by as much as 50% in concentrations ranging from 25 to 200 μg l^{-1}. The actual concentration and effect varies with the heavy metal. For example the order of toxicity in lake and synthetic water was Cu>Cd>=Pb>Zn (Waara, 1992). The presence of inorganic compounds are not necessarily inhibitory. For example, stable and continuous denitrification was found to occur in high chloride concentrations (30 g l^{-1}) (Holm Kristensen and Jepsen, 1991).

Treatment technologies

The microbiological treatment of water containing high concentrations of nitrate can be divided into two types: (i) the treatment of liquid waste with a high nitrogen content and (ii) the treatment of extractable water for drinking either *in situ* in the groundwater system or post-extraction.

Wastewater denitrification treatment technologies

Conventional treatment Denitrification in conventional biological wastewater treatment systems tends to occur in closed bioreactors (to maintain anaerobic conditions) including anaerobic activated sludge, packed bed reactors and fluidized beds. These have been fully discussed elsewhere (Horan, 1990) and it is intended to only briefly discuss some of the major factors that affect treatment. The effects of environmental conditions on denitrification in these reactors (Wartchow, 1990; Holm Kristensen and Jepsen, 1991) broadly follows the considerations outlined above. The choice of electron donor is related to the physiological type of denitrifier. Thus, sulphur has been used in processes relying on bacteria such as *Thiobacillus denitrificans* as the denitrifier (Batchelor and Lawrence, 1978). A range of organic substrates has been used for heterotrophic denitrification, e.g. acetate and methanol. The choice of electron donor is dominated not only by the requirements of bacterial physiology and achieving efficient conversion (see above), but also by cost. The cheaper substrates are favoured to ensure biological denitrification is economic. The ideal is for the waste itself to contain oxidizable organics that can provide energy.

Bacterial adsorption and entrapment A technology used in both conventional treatment techniques and in novel systems is bacterial immobilization. This is achieved either through bacterial adsorption on surfaces and the development of a biofilm, a complex process fully described by Characklis and Marshall (1990) or through bacterial entrapment within carrier matrices (see Ch. 15 for a brief discussion of immobilization). The advantage of immobilization is that bacteria can be retained within a reactor at flow rates above their growth rates. Moreover, immobilized biomass is densely packed offering space savings (Robertson and Kuenen, 1992).

Airlift and fluidized bed reactors (Jimenez *et al.*, 1990) containing surface immobilized cells are long-standing processes. In solid surface bioreactors the characteristics of the surface affect denitrification. For example, denitrification rates in a fluidized bed reactor using a pumice stone support and a sand support were 1.882 and 0.898 kg N m^{-1} dy^{-1}, respectively (Jimenez *et al.*, 1990). This may not only relate to the process characteristics of the support, but also to the effect of the solid surface on bacterial activity (Fletcher, 1985). Ideally the packing material should be cheap and readily available as well as having appropriate characteristics for use in a reactor (Jimenez *et al.*, 1990).

More novel immobilized reactor types have been applied to denitrification. These include rotating discs (Winkler, 1981), cells trapped in liquid surfactant membranes (Mohan and Li, 1975) and denitrifiers trapped within a composite two-layer structure consisting of a gel layer bounded by a microporous membrane structure (Lemoine *et al.*, 1991). The latter system was used for continuous denitrification by immobilized *Pseudomonas denitrificans*. Acetate was used as the carbon source at a C : N ratio of 3 (by mol.). The high nitrate water (2.5 mM $NaNO_3$) was applied at a flow rate of 10 cm^{-3} hr^{-1}. Denitrification rates between 15 and 25 μg NO_3^- hr^{-1} cm^{-1} membrane surface were reported giving efficiency of nitrate removal between 20 and 50% (Lemoine *et al.*, 1991).

Soil column and anaerobic contact columns containing a range of different solid supports have also been investigated for denitrification. Again the characteristics of the support influence denitrification rates. This is partly due to the physical characteristics of the supports. For example, permeability is an important consideration. Columns packed with charcoal chips are more permeable to influent than columns packed with volcanic ash which shows lowered denitrification rates (Abe *et al.*, 1991).

Denitrifying bacteria have been immobilized through entrapment in matrices such as calcium alginate, polyelectrolytes, carrageenan, chitosan (Robertson and Kuenen, 1992) and agar (Lemoine *et al.*, 1991). An interesting possibility in nitrogen effluent treatment arises from the characteristics of immobilization matrices. Gel immobilization frequently involves bacterial incorporation within gel beads. The environmental characteristics inside a bead are very different from those on its surface. For example, oxygen is limited inside large beads. Hunik and Tramper (1991) have shown that it may be possible to have a layer of nitrifying bacteria on the bead surface while the bead core contains active denitrifying bacteria under oxygen-limited conditions. Nitrifying bacteria are able to oxidize reduced nitrogen compounds potentially to nitrite and nitrate. Obviously, this raises the possibility of the treatment of complex nitrogen effluents using entrapped nitrifiers and denitrifiers in a single matrix.

Land treatment systems Recently, there has been interest in nitrogen removal from wastewater by denitrification in land treatment systems and rapid infiltration (RI) systems (Iskander, 1981). RI systems are commonly contained but uncovered soil systems. These consist of coarse-textured soils together with a small percentage of finer soils. Wastewater is applied to the soil surface at a rate of between 15 and 350 mm dy^{-1}. The influent generally has a total nitrogen content of 10–30 mg l^{-1}. These systems are often alternately flooded and dried, inducing intermittent anaerobic and aerobic conditions which allows nitrification and denitrification to occur. This is a relatively common procedure at many wastewater treatment facilities. As with other denitrification processes, control of a range of environmental conditions is necessary to encourage optimum denitrification, e.g. influent application, the addition of carbon sources and the type of porous medium—in this case soil particles such as sand, ground granite and field soil.

Bacterial communities and denitrification The combination of nitrification and denitrification in conventional waste treatment systems is common and relies on achieving appropriate conditions to encourage both bacterial communities, i.e. denitrifying bacteria and nitrifying bacteria. For example, in activated sludge systems the development of thick biofilms on solid supports results in the bottom layers of the biofilm becoming oxygen limited because of bacterial activity and diffusional limitations. A combination of an attached biomass and a suspended aerated biomass permits simultaneous nitrification and denitrification (Winkler, 1981). The recent evidence that individual bacteria are capable of simultaneously nitrifying and denitrifying (Robertson and Kuenen, 1990, 1992) presents the possibility that a treatment process for high nitrogen effluent could be designed utilizing a single organism. Even though aerobic denitrification is relatively slow a study is in progress to design a bioreactor and optimize process characteristics for single-species nitrification and denitrification (Pot *et al.*, 1991).

A novel co-operation between bacteria has been suggested in order to use methane, a waste product of sludge treatment, as a cheap carbon source for denitrification. Methanotrophic (methane-oxidizing) bacteria oxidize methane to CO_2 and H_2O. Methanol is produced as an intermediate during this reaction. Careful control of environmental conditions can cause methanol accumulation which can then be used as a carbon source for denitrification (Werner and Kayser, 1991).

Denitrification of drinking water

The processes available for the denitrification of drinking water, i.e. water abstracted for consumption, are broadly similar to those described above. A review of above-ground denitrification methods is given by Hiscock *et al.*, (1991). There are, however, certain limitations on the use of bioreactors for drinking water denitrification. These fall into two categories – bacterial contamination and the characteristics of the electron donor used during denitrification.

It is essential to avoid contamination of the drinking water by heterotrophic bacteria since they may constitute a health hazard. This may be overcome to some extent by the use of immobilized cells in denitrification processes. This does not offer complete protection, however. There is some loss of biofilm material to the liquid phase in any bioreactor through sloughing-off, desorption and loss of daughter cells from the biofilm (Characklis and Marshall, 1990). Similarly, cells immobilized in calcium alginate leak into the water (Nilsson and Olsen, 1982). One suggested solution to this problem is the introduction of filters to remove cells from the treated drinking water.

The choice of electron donor is not only important in determining the efficiency of denitrification and because of cost considerations, but also for its possible associated health risks. This can be illustrated by two examples. Methanol and the initial product of methanol oxidation, formaldehyde, are toxic. As a consequence, if methanol, an efficient substrate for denitrification, is used as the electron donor during denitrification its complete oxidation to CO_2 and H_2O must be assured. Alternatively, facilities must be included at the treatment plant to remove residual methanol and formaldehyde from the drinking water (Robertson and Kuenen, 1992). The use of an inorganic electron donor may also be problematic. Sulphur has been suggested as a potentially useful electron donor. It is solid and can therefore act as a support for the chemolithotrophic denitrifiers, which do not present a health hazard. Unfortunately, the end product of sulphur oxidation is sulphate, a laxative (Robertson and Kuenen, 1992).

In situ denitrification of groundwater

The underground denitrification of groundwater has two advantages. First, seasonal temperature variations do not affect the process, making temperature control systems unnecessary. Second, it can be linked to other secondary treatments within the aquifer, e.g. degradation of organic residuals, reaeration and filtration (Hiscock *et al.*, 1991). A commonly encountered problem in underground treatment is the clogging of the pore space within the aquifer (Hijnen *et al.*, 1988). This occurs for two reasons: (i) microbial growth together with possible production of bacterial exopolymers and (ii) blocking of the pores by NO, N_2O, or N_2, the gases produced by denitrification.

There are three different methods for underground denitrification (Hiscock *et al.*, 1991).

1 The dual-well system is the simplest approach, requiring only one well (a recharge well) for nutrient injection into the aquifer and another well (a pumping well) for abstraction of treated water (Fig. 9.1a).

2 The 'Daisy' system involves a single pumping well surrounded concentrically by several injection wells (Fig 9.1b).

3 The combined system links denitrification in an above-ground bioreactor with underground treatment. Untreated water from the aquifer is passed through a surface bioreactor for denitrification. The water is then recirculated underground where further denitrification occurs and the water is purified (Fig 9.1c).

Factors affecting *in situ* groundwater denitrification are the same as those described for bioreactor based systems, particularly those bioreactors based on solid supports, e.g. the effect of carbon additions (Hiscock *et al.*, 1991; Obenhuber and Lawrence, 1991) and permeability. A comparison between the rates and efficiencies of bioreactor treatments and underground treatment is shown in Table 9.3.

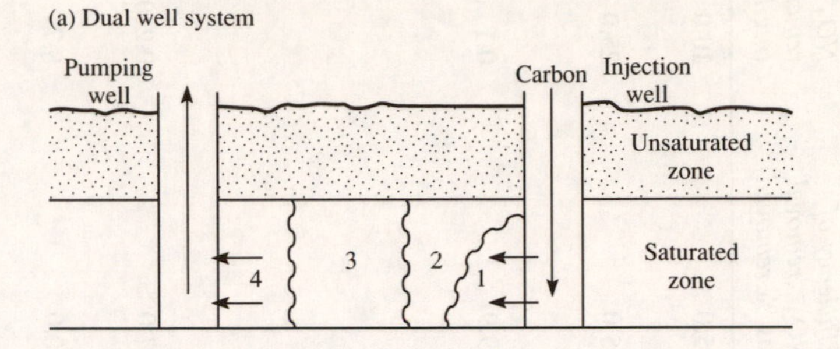

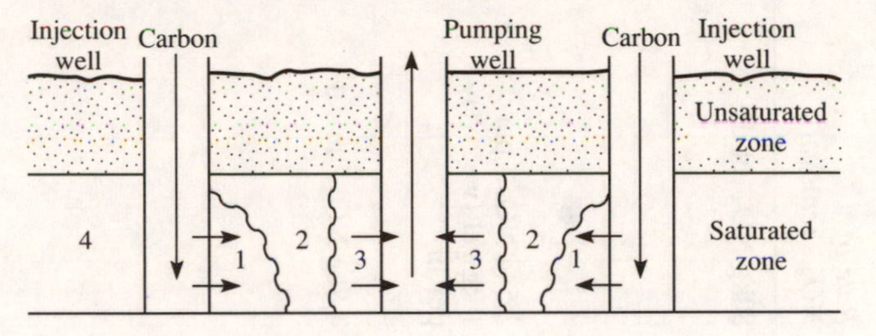

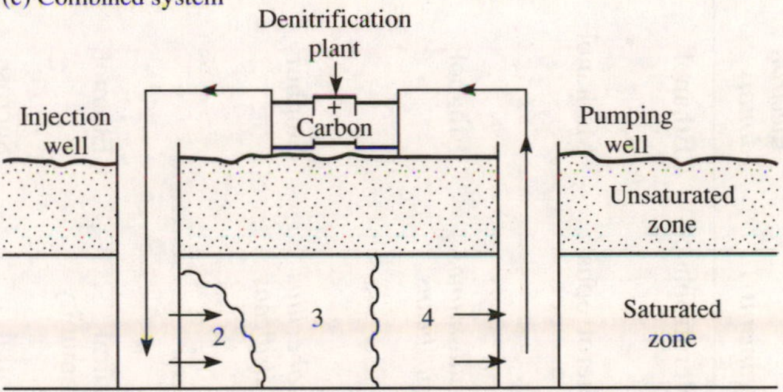

1 represents the zone of natural denitrification, increased by input of carbon and other nutrients via the injection well

2 represents the zone of filtration and reaeration

3 represents the zone of denitrified groundwater

4 represents untreated i.e. high nitrate groundwater

Fig. 9.1. Diagrammatic representation of *in situ* groundwater denitrification systems.

Table 9.3 Comparison of the efficiency and rate of denitrification shown by bioreactors and *in-situ* groundwater treatments

Process	*Organism(s)*	*Electron donor*	*Rate of* NO_3^- *removal*	*Efficiency of* NO_3^- *removal (as % removed)*	NO_3^- *remaining (mg* l^{-1}*)*
Fixed-bed filters	Heterotrophs	Ethanol	80kg NO_3^-N d^{-1}	95.0	10.0
Fluidized sand-bed reactor	Heterotrophs	Methanol	–	75.0	25.0
Encapsulated cells in sequential columns	*Pseudomonas denitrificans*	Ethanol	3.54g NO_3^- per 1 kg gel (wet wt) per hr	99.9	0.1
Upflow reactor sulphur and calcium carbonate supports	*Thiobacillus denitrificans*	Sulphur	4.8kg NO_3^- m^{-3} J^{-1}	–	–
Dual-well system	Natural community	Ethanol	–	97.0	0.2–0.3
Daisy system	Natural community	Sucrose	–	40.0	8.7

See Robertson and Kuenen (1992) for detailed information on the processes, and Hiscock *et al.*, 1991).

References

ABE, K., OZAKI, Y. and MORO-OKA, M. (1991) Advanced treatment of livestock wastewater using an aerobic soil column and an anaerobic contact column. *Soil Sci. Plant Nutrit.*, **37**, 151–157.

BATCHELOR, B. and LAWRENCE, A.W. (1978) Autotrophic denitrification using elemental sulphur. *J. Water Pollut. Control Fed.*, **5**, 1986–2001.

CHARACKLIS, W.G. and MARSHALL, K.C. (eds) (1990) *Biofilms*. Wiley Interscience, Chichester.

COLE, J.A. (1987) Assimilatory and dissimilatory reduction of nitrate to ammonia. In Cole, J.A., Ferguson, S.J. (eds) *The Nitrogen and Sulphur Cycles*, pp. 281–330. Cambridge University Press, Cambridge.

COUNCIL OF EUROPEAN COMMUNITIES (1980) Directive of 15 July 1980 relating to the quality of water intended for human consumption. *EC Official Journal*, L229/11.

FLETCHER, M. (1985) Effect of solid surfaces on the activity of attached bacteria. In *Bacterial Adhesion. Mechanisms and Physiological Significance,* pp. 339–362. Plenum Press, New York.

GAUNTLETT, R.B. and CRAFT, D.G. (1979) Biological removal of nitrate from river water. *Rep. TR 98*. Water Research Centre, Medmenham.

GOERING, J.J. (1972) The role of nitrogen in the eutrophic process. In Mitchell, R. (ed.) *Water Pollution Microbiology*, pp. 43–68. Wiley Interscience, New York.

HIJNEN, W.A.M., KONIG, D., KRUITHOF, J.C. and VAN DER KOOIJ, D. (1988) The effect of biological nitrate removal on the concentration of bacterial biomass and easily assimilable organic carbon compounds in groundwater. *Water Supply*, **6**, 265–273.

HISCOCK, K.M., LLOYD, J.W. and LERNER, D.N. (1991) Review of natural and artificial denitrification in groundwater. *Water Res.*, **25**, 1099–1111.

HOCHSTEIN, L.I. and TOMLINSON, G.A. (1988) The enzymes associated with denitrification. *Ann. Rev. Microbiol.*, **42**, 231–261.

HOLM KRISTENSEN, G. and JEPSEN, S.E. (1991) Biological denitrification of waste water from wet lime–gypsum flue gas desulphurization plants. *Water Sci. Technol.*, **23**, 691–700.

HORAN, N.J. (1990) *Biological Wastewater Treatment Systems. Theory and Operation*. John Wiley and Sons, Chichester.

HOUSE OF LORDS (1989) Nitrate in water. *16th Report Session 1988–1989 of the Select Committee on the European Communities*. HMSO, London.

HUNIK, J.H. and TRAMPER, J. (1991) Abrasion of x-caregeenan gel beads in bioreactors. In Verachtert, H., Verstraete, W. (eds) *Proc. Int. Symp. Environ. Biotechnol.*, Vol. 2, pp. 487–490. The Royal Flemish Society of Engineers.

ISKANDER, I.K. (ed.) (1981) *Modeling Wastewater Renovation: Land Treatment*. John Wiley and Sons, New York.

JENKINS, D. and TANDOI, V. (1991) The applied microbiology of enhanced biological phosphate removal—Accomplishments and needs. *Water Res.*, **25**, 1471–1478.

JIMENEZ, B., BECERRIL, E. and SCOLA, I. (1990) Denitrification in a fluidized bed system using low cost packing material. *Environ. Technol.*, **11**, 409–420.

KOWALENKO, C.G. (1979) The influence of sulphur anions on denitrification. *Can. J. Soil Sci.*, **59**, 221–223.

KUENEN, J.G. and ROBERTSON, L.A. (1987) Ecology of nitrification and denitrification. In Cole, J.A., Ferguson, S.J. (eds) *The Nitrogen and Sulphur Cycles*, pp. 161–218. Cambridge University Press, Cambridge.

LEMOINE, D., JOUENNE, T. and JUNTER, G.-A. (1991) Biological denitrification of water in a two chambered immobilized-cell bioreactor. *Appl. Microbiol. Biotechnol.*, **36**, 257–264.

MATEJU, V., CIZINSKA, S., KREJCI, J. and JANOCH, T. (1992) Biological water denitrification — a review. *Enzyme Microbial Technol.*, **14**, 170–183.

MOHAN, R.R. and LI, N.N. (1975) Nitrate and nitrite reduction by liquid membrane encapsulated whole cells. *Biotechnol. Bioeng.*, **17**, 1137–1156.

NILSSON, I. and OLSEN, S. (1982) Columnar denitrification of water by immobilized *Pseudomonas denitrificans* cells. *Eur. J. Appl. Microbiol. Biotechnol.*, **14**, 86–90.

OBENHUBER, D.C. and LAWRENCE, R. (1991) Reduction of nitrate in aquifer microcosms by carbon additions. *J. Environ. Qual.*, **20**, 255–258.

PAYNE, W.J. (1981) *Denitrification*. John Wiley and Sons, Chichester.

POT, M.A., VAN LOOSDRECHT, M.C., HEIJNEN, J.J. and ROBERTSON, L.A. (1991) Development of a process for removal of ammonia from waste water based on heterotrophic nitrification and aerobic denitrification. In Verachtert, H., Verstraete, W. (eds) *Proc. Int. Symp. Environ. Biotechnol.*, Vol. 2, pp. 583–584. The Royal Flemish Society of Engineers.

ROBERTSON, L.A. and KUENEN, J.G. (1990) Combined heterotrophic nitrification in *Thiosphaera pantotropha* and other bacteria. *Antonie van Leeuwenhoek*, **57**, 139–152.

ROBERTSON, L.A. and KUENEN, J.G. (1992) Nitrogen removal from water and waste. In Fry, F.C., Gadd, G.M., Herbert, R.A., Jones, C.W., Watson-Craik, I.A. (eds) *Microbial Control of Pollution*, pp. 227–267, Cambridge University Press, Cambridge.

TIEDJE, J.M., SEXSTONE, A.J., MYROLD, D.D. and ROBINSON, J.A. (1982) Denitrification: ecological niches, competition and survival. *Antonie van Leeuwenhoek*, **48**, 569–583.

WAARA, K.-O. (1992) Effects of copper, cadmium, lead and zinc on nitrate reduction in a synthetic water medium and lake water from northern Sweden (1992). *Water Res.*, **26**, 355–364.

WARTCHOW, D. (1990) Nitrification and denitrification in combined activated sludge systems. *Water Sci. Technol.*, **22**, 199–206.

WERNER, M., and KAYSER, R. (1991) Denitrification with biogas as external carbon source. *Water Sci. Technol.* **23** (Kyoto), 701–708.

WINKLER, M. (1981) *Biological Treatment of Waste-water*. Ellis Horwood, Chichester.

Further reading

COLE, J.A. and FERGUSON, S.J. (eds) (1987) *The Nitrogen and Sulphur Cycles*. Cambridge University Press, Cambridge.

HISCOCK, K.M., LLOYD, J.W. and LERNER, D.N. (1991) Review of natural and artificial denitrification of groundwater. *Water Res.*, **25**, 1099–1111.

KNOWLES, R. (1982) Denitrification. *Microbiol. Rev.*, **46**, 43–70.

MATEJU, V., CIZINSKA, S., KREJCI, J. and JANOCH, T. (1992) Biological water denitrification—a review. *Enzyme Microbial Technol.*, **14**, 170–183.

PAYNE, W.J. (1981) *Denitrification*. John Wiley and Sons, Chichester.

VERACHTERT, H. and VERSTRAETE, W. (1991) *Proc. Int. Symp. Environ. Biotechnol.*, Vol. 2. The Royal Flemish Society.

Chapter 10

MACROPHYTE SYSTEMS FOR NITRATE AND PHOSPHATE REMOVAL

Introduction

The practice of discharging wastewater into natural wetlands, ponds and onto areas of flat or gently sloping land has been used as a means of disposal since ancient times. Such 'land treatment' can prove very effective where discharges do not exceed the capacity of the receiving system to assimilate, absorb or detoxify contaminants present (Bayes *et al.*, 1989). However, as the human population has grown and become increasingly urbanized, the ability of such systems to cope effectively with domestic sewage has been rapidly exceeded, necessitating the construction of large treatment works and prompting the development of modern treatment methods (Mason, 1991).

In recent years there has been renewed interest in this form of disposal, particularly the use of natural and artificial wetlands. The basic concept is deceptively simple. Pollutants present in wastewater discharged into or onto the wetland, are immobilized and degraded by the normal physical and biological processes which operate in wetland ecosystems (Fig. 10.1). Effective treatment is achieved by the construction or management of wetlands so that environmental

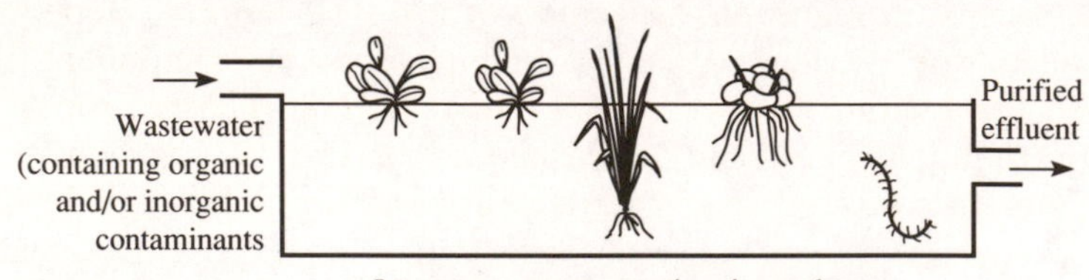

- solids settling
- plant uptake of contaminants
- biochemical and physico-chemical transformations

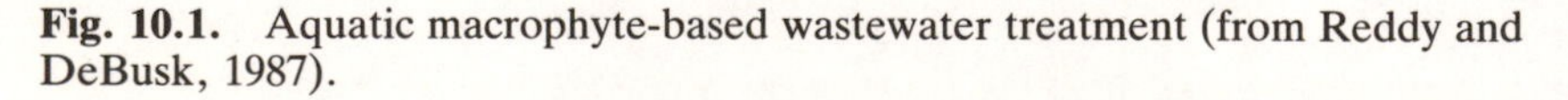

Fig. 10.1. Aquatic macrophyte-based wastewater treatment (from Reddy and DeBusk, 1987).

conditions favour rapid degradation and cleaning of effluent (Reddy and DeBusk, 1987; Reddy and Smith, 1987). In the early 1960s, NASA began actively researching into the use of aquatic plant systems for the treatment of wastewater. Interest was initially centred on the use of aquatic plants as part of a life-support system capable of treating wastewater and maintaining atmospheric CO_2 and O_2 concentrations within an enclosed environment. By 1975, a water hyacinth (*Eichhornia cassipes*)-based treatment system, consisting of a single lagoon with a surface area of 2.02 ha and average depth of 1.22 m, had been constructed at the National Space Technology Laboratories (NSTL), Mississippi. The system has now been in successful operation for well over a decade, dealing with effluent inflows of the order of 600 m^3 dy^{-1} and a biological oxygen demand (BOD_5) loading rate of 33.7 kg ha^{-1}. Following the construction of this sewage system, a similar system was constructed in 1976 for the disposal and treatment of photographic and chemical laboratory wastewater. Like many artificial wetland systems constructed in tropical and subtropical climates, the system originally utilized a lagoon populated with the floating aquatic macrophyte, water hyacinth. The operation of the system, particularly during the colder periods of the year, was improved in 1980 by incorporating an artificial reed bed planted with more cold tolerant, temperate emergent aquatic macrophytes (e.g. *Phragmites communis*, common reed and *Typha latifola*, cattail). The reed bed acts as an effective substrate–plant microbial filter (Wolverton, 1987). In Europe, following the pioneer work conducted in West Germany, particularly at the Max Plank Institute, Krefeld, interest has centred on the use of reed bed systems, utilizing what has become commonly known as the 'root-zone' treatment method (Schiechtl, 1980; Bayes *et al.*, 1989). This approach requires a horizontal flow of wastewater to be established through the rooting media of planted emergent macrophytes, normally *Phragmites communis*. As the wastewater passes through the rhizosphere, microbial activity results in the decomposition of organic material and denitrification of nitrogen sources. Depending on the rooting medium, much of the phosphorus component may become fixed within the soil media (Brix, 1987a). A large, root-zone sewage treatment plant was established as early as 1974 at Othfresen, Germany. This utilized an old settlement pond (area 22.5 ha), planted with *Phragmites communis*, operating at a loading of approximately 5000 pe (i.e. person equivalents) and having an active reed beds area of approximately 5 m^2 pe^{-1} (Brix, 1987b).

Although the root zone method can be very effective, reductions of between 92 and 99% have been claimed for effluent total suspended solids (TSS), BOD, total-N and total-P (Reddy and Smith, 1987), the long-term operation of the plant at Othfresen has highlighted a number of operational problems. Surface run-off of effluent, the development of preferential drainage routes and poor and irregular penetrations of the soil by wastewater have all caused problems. These problems can severely limit the effectiveness of effluent treatment (Brix, 1987b). With careful design, placement of inlet and outlet channels and choice of planting substrate, these problems may be minimized (Alexander and Wood, 1987; Bucksteeg, 1987; Tchobanoglous, 1987).

At present over 19 countries have active research programmes evaluating the use of macrophytes in water pollution control. In 1990, over 300 reed-bed-based treatment systems were operating in Germany (Bucksteeg, 1991), 130 in Denmark (Schierup *et al.*, 1991), and 27 in the United Kingdom (Findlater *et al.*, 1991). Most are based on the root-zone concept of soil beds planted with *Phragmites* and subject to horizontal effluent flow; they may, however, be of various designs and sizes (Fig. 10.2). Artificial wetland systems have been seen as an economically attractive, energy-efficient way of providing high standards of wastewater treatment, particularly for isolated populations, yet capable of enhancing or at least maintaining the conservation or amenity value of an area. In developing countries, they have the additional advantage of representing a 'low technology' solution to the treatment of sewage produced by both large and small dispersed populations. Unfortunately the performance of many of these systems has not lived up to early expectations. To understand why, it is necessary to examine the concept and theory behind their design and operation.

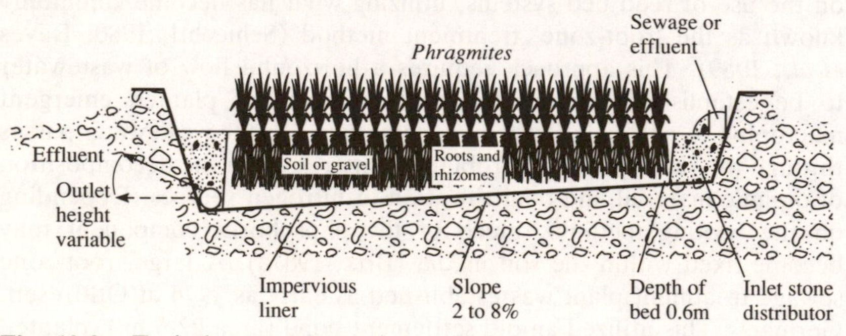

Fig. 10.2. Typical arrangements for a root-zone-type works (from Bayes *et al.*, 1989).

The concept and processes of pollutant removal

Regardless of the type of aquatic macrophyte treatment system (AMATS), whether a natural wetland or an artificial system planted with a monoculture or polyculture of either floating or emergent plants, the processes thought to operate are essentially the same (Fig. 10.1). In addition to the direct uptake and accumulation of contaminates, pollutant removal may be achieved by a complex range of chemical and physical reactions, occurring at the water–sediment, root–sediment and plant–water interfaces (Good and Patrick, 1987; Reddy and DeBusk, 1987; Richardson and Davis, 1987).

Plant accumulation of pollutant

Aquatic macrophytes, particularly floating species, such as the water hyacinth and pennywort are capable of very high rates of growth, e.g. 10 g m^{-2} dy^{-1}. Such growth rates are associated with high levels of nutrient uptake and demand, particularly for nitrogen and phosphorus. Reddy and DeBusk (1987) estimate that water hyacinth plants are capable of removing from the water, by direct uptake, 5850 kg N ha^{-1} yr^{-1}, and of storing within their biomass between 300 and 900 kg N ha^{-1}. Significant quantities of phosphorus can also be removed (350–1125 kg P ha^{-1} yr^{-1}) and accumulated (20–57 kg P ha^{-1}). Despite the ability of rooted emergent macrophytes to store substantial quantities of both nitrogen (200–1560 kg N ha^{-1}) and phosphorus (40–375 kg P ha^{-1}), they are less efficient at lowering wastewater nitrogen and phosphorus contents by direct uptake. This partly reflects their typically lower growth rates and their ability to take up nitrogen and phosphorus from sediment sources in preference to the water body. In addition, frequently more than 50% of accumulated nutrients are stored in underground portions of the plant and are not easily harvested. In order to maximize removal of nitrogen and phosphorus via direct uptake, high growth rates and levels of standing biomass must be achieved. Frequent harvesting may be required to remove accumulated nutrients, encourage new growth and prevent releases from senescent plant material.

Direct uptake and accumulation by plants can also significantly reduce the concentration of other contaminants in wastewater. Natural and artificial wetland systems have been shown to be effective in lowering effluent metal content (e.g. Cu, Zn, Pb, Hg and Ni), partly as a result of both active and passive plant uptake and accumulation (Kleinmann and Girts, 1987; Cooper and Findlater, 1991).

Bacteria mediated decomposition

Organic carbon Bacteria in the sediments associated with plant surfaces and dispersed throughout the water body, will utilize organic carbon present in wastewater as an energy source (Chs 5 and 6). Under oxic conditions, e.g. the upper sediment layers and plant rhizosphere, oxygen is utilized as the electron acceptor during the breakdown of carbon sources. Within biologically active sediments oxygen becomes rapidly depleted, causing the oxygen tension to decline rapidly with depth. As the redox potential declines facultative and obligate anaerobic bacteria increase in abundance, utilizing a sequence of terminal electron acceptors. At redox potentials of 220 mV nitrate is reduced, at 120 mV ferric iron and between -75 and -150 mV sulphate may be utilized (Good and Patrick, 1987). The ability of a water–plant–sediment system to remove organic carbon is enhanced by the presence of natural wetland plant species. Aerenchyma tissue (tissues containing large extracellular air spaces) in the roots and stems of wetland plants allows the transfer of oxygen from the aerial plant portions to submerged parts of the plant (Fig. 10.3). Up to between 50 and 60% of the oxygen reaching submerged plant portions may leak into the surrounding media, creating an aerobic environment for bacteria associated with plant surfaces and sediments close to the root system. Oxygen released in this way will increase the opportunities for aerobic

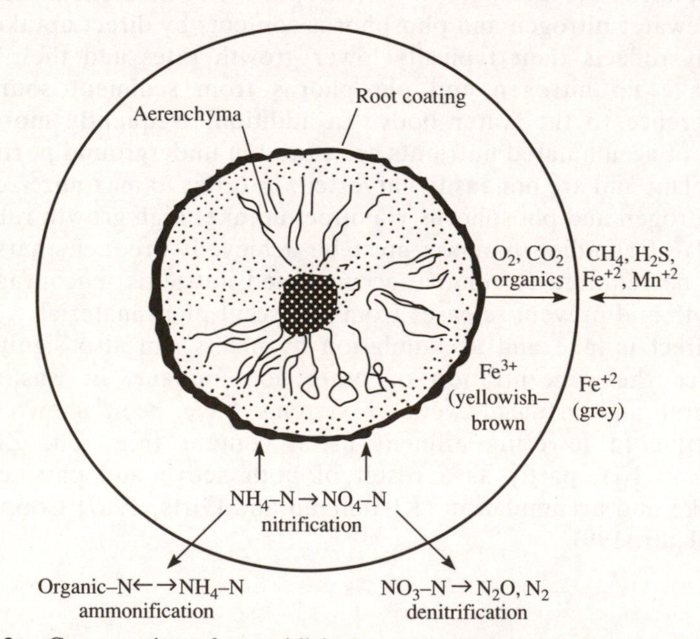

Fig. 10.3. Cross-section of an oxidizing root growing in a reduced sediment. The oxidized rhizosphere is depicted, along with some processes that occur as a result of this aerobic–anaerobic interface (after Mendelssohn and Postek, 1982).

breakdown of carbon sources. Substantial quantities of oxygen may be transported by aquatic macrophytes. Reddy and DeBusk (1987) report values of 3.95 ± 1.86 for *Hydrocotyle umbellata*, 1.29 ± 1.18 for plants of *Eichhornia crassipes* and values of 0.19 ± 0.15 to 1.39 ± 1.49 mg O_2 g^{-1} hr^{-1} for the rooted emergent *Typha* spp. Reddy and DeBusk (1987) have shown that oxygen transported by aquatic plants significantly increased the wastewater oxygen content and rates of organic carbon biodegradation. They report that for a water hyacinth and pennywort system, macrophyte oxygen transfer was responsible for 90% of the reduction in wastewater BOD. Reed *et al.* (1988) estimated that for adequate removal of organic carbon, rooted vegetation needs to transmit an oxygen flux of approximately 1.5 times the organic (BOD) loading.

The ability of wetland plants to transfer oxygen into their submerged portions partly accounts for their tolerance of waterlogging and the subsequent anoxic conditions. It allows the plant to utilize aerobic respiration and provides a mechanism whereby soil toxins, e.g. H_2S and reduced Fe and Mn, may be ameliorated by oxidation in the rhizosphere.

Nitrogen Although some nitrogen removal occurs as a result of direct plant uptake, microbial nitrification and denitrification are normally more important. Nitrification of wastewater ammonia within the rhizosphere and the aerobic sediment layer will yield nitrite (NO_2^-) and nitrate (NO_3^-), some of which may be assimilated by the plants. However, the ultimate removal of nitrogen depends on the subsequent anaerobic denitrification of NO_3^- and NO_2^- to gaseous N_2, which is lost to the atmosphere (Ch. 9). Thus, for efficient removal of wastewater nitrogen, the treatment system must provide a balance of both aerobic and anaerobic microenvironments (Fig 10.3; Bayes *et al.*, 1989).

Although high rates of denitrification in the order of 1 g m^2 dy^{-1} have been reported for AMATS, the extent of nitrogen removal will depend on the design of the system and the form and amount of nitrogen present in the wastewater. For a water hyacinth pond system receiving secondary sewage effluent, denitrification was thought to account for the loss of between 40 and 90% of the input nitrogen. Similar ranges of values have been recorded for freshwater marshes, although here a significant proportion of the nitrogen may be retained by the plant. Bacterial nitrification–denitrification activity associated with the plant rhizosphere will be influenced by the level of nitrogen available. If effluent nitrogen content is low, wetland plants will compete directly with nitrifying and denitrifying bacteria for NH_4^+ and NO_3^-, while high nitrogen, particularly ammonia, will stimulate nitrifying and denitrifying activity (Good and Patrick, 1987).

Physical–chemical processes

Phosphate Phosphate removal is achieved largely as a result of pH-dependent complex formation and co-precipitation reactions involving Ca, Fe and Al. Although plant uptake may be substantial, the sorption of phosphorus by anaerobic reducing sediments appears to be the most important process. The capacity of wetland systems to absorb phosphorus is positively correlated with the sediment concentration of extractable amorphous aluminium and iron. It is suggested that in such sediments gel-like hydrated ferrous hydroxides, along with aluminium complexes, provide sorption sites for orthophosphate-p (Richardson, 1985).

Pathogens Viruses and bacteria present in sewage effluent are removed chiefly by adsorption onto soil particulates and due to the antimicrobial action of the soil–root microflora. For efficient removal of potential pathogens it is important that a slow and even flow of wastewater through the system is achieved (Reddy and Smith, 1987; May *et al.*, 1991).

Design and management of artificial wetlands

Although the basic biological processes responsible for the removal of pollutants by AMATS are well established, there has been little standardization of their design or methods of construction. An empirical 'Black Box' approach has frequently been adopted, small experimental plants being modified and enlarged in the light of experience (Athie and Cerri, 1987; Bayes *et al.*, 1989; Wood, 1991). Unfortunately this approach has not always resulted in the development of efficient systems (Brix, 1987a; Bucksteeg, 1991; Findlater *et al.*, 1991; Schierup *et al.*, 1991). However, based largely on the root-zone concept, a body of general design criteria and principles for the construction of reed bed systems is beginning to emerge (Wood, 1991). Important factors determining the efficiency and pollutant removal capacity of a reed bed system include the following:

- Choice of plant species
- Substrate
- Area of reed bed
- The nature, loading and distribution of effluent

Choice of plant species

Plants within reed bed systems are required to fulfil four major functions: (i) filter solids out of suspension, (ii) provide surfaces for bacterial growth, (iii) translocate oxygen into the root zone, thereby increasing the efficiency of bacterial degradation of pollutants and (iv) maintain the hydraulic permeability of the substrata. Experience suggests that the common reed (*Phragmites communis*) is a particularly suitable species. It combines rapid growth with a deep, extensive rhizome and root system, providing a large surface area for bacterial growth. Root and rhizome penetration help to increase soil porosity. On decay a horizontal network of fine channels remains which serves to stabilize the hydraulic conductivity of the system, allowing good infiltration of the wastewater into the active root–soil matrix.

It is claimed that within 2–5 years of successful reed establishment, rates of hydraulic conductivity within the rhizosphere zone may approach that of coarse sand, regardless of the porosity of the initial soil (Kickuth, 1980).

Although emergent plants, such as *Phragmites*, die back during winter periods, the effectiveness of the system need not be substantially reduced. Dead stems continue to provide an effective route for the movement of oxygen into the rhizosphere and the effects of low atmospheric temperature are partially mitigated by the relatively high temperature of influent wastewater, heat production from microbial activity and the presence of a protective insulating layer of vegetation and litter. Thus a relatively high temperature and rate of microbial activity is maintained. High numbers of active bacteria are associated with the rhizosphere and soil matrix of reed bed systems (Reddy and Smith, 1987; May *et al.*, 1991; Hofmann, 1991).

Substrate

The choice of substrata is critical. In addition to gravel, river sands and pulverized fuel ash, a wide range of soils have been used with varying degrees of success. The medium used is often determined by the availability and range of local substrata. However, these need to be carefully evaluated in terms of their hydraulic permeability and their capacity to absorb nutrients and pollutants (Wood, 1991). The substrate must provide a suitable medium for successful plant growth and allow even infiltration and movement of wastewater. Successful operation requires a hydraulic conductance of approximately 10^{-3} to 10^{-4} m^{-1} s^{-1}. Poor hydraulic conductivity will result in surface flow and channelling of wastewater, severely reducing the effectiveness of the system.

The chemical composition of the substrate will also affect the efficiency of the system. Soils with low nutrient content will encourage direct uptake of nutrients from the wastewater by plants. While, because of the physico-chemical mechanism responsible for the removal of phosphate, substrate with high Al or Fe content will be most effective at lowering wastewater phosphate concentrations. Heavy metal removal will be most effectively achieved with the use of a substrate with a high organic or clay content. In systems constructed for the treatment of agricultural effluents, which are frequently considerably more concentrated than domestic sewage (e.g. BOD up to 3000 mg l^{-1}) and acidic, crushed limestone has been used (Gray *et al.*, 1991). This can provide a highly porous substrate, capable of regulating pH and inducing the precipitation of phosphates.

Area of reed bed

The area required for the traditional land treatment of effluent may be calculated with reference to the 'land-limiting constituent' (LLC), of the effluent (Bayes *et al.*, 1989). This is the pollutant present in the effluent, for which the assimilative capacity of the land is lowest. The area require is then obtained from:

$$\text{Required land area (ha)} = \frac{\text{Supply rate of LLC (kg annum}^{-1})}{\text{Capacity of land to assimilate LLC (kg ha}^{-1}\text{ annum}^{-1})}$$

A similar approach may be used to calculate the approximate area (A_h) of reed bed needed using an empirically derived formula originally presented by Kickuth (1983) for the reduction of BOD_5 in sewage effluent.

$$A_h = KQ_d (\ln C_o - \ln C_t)$$

where A_h = estimated area of reed bed required
K = constant = 5.2
Q_d = the average flow rate of wastewater ($m^3\ d^{-1}$)
C_o = the average BOD_5 of the influent ($mg\ l^{-1}$)
C_t = the average BOD_5 of the effluent ($mg\ l^{-1}$)

The value of $K = 5.2$ was derived for a bed 0.6 m deep and operating at a minimum temperature of 8°C. For less biodegradable wastewaters K values of up to 15 may be appropriate. Using this formula a minimum area of 2.2 $m^2\ pe^{-1}$ is obtained for the treatment of domestic sewage; in practice most designs operate on the basis of 3-5 $m^2\ pe^{-1}$.

The nature, loading and distribution of effluent

Pre-treatment Although AMATS have been used to treat screened primary sewage effluent, their long-term efficiency is improved if the effluent is pre-treated by storage in a settlement tank or pond for 24 hr, prior to discharge onto the active beds or into the treatment lagoons. During storage the BOD of the primary effluent may be reduced by 40% as suspended particles settle. The removal of part of the suspended solids will help to prevent the treatment system from prematurely silting up.

Loading and distribution of effluent The major difference between conventional methods of wastewater treatment and AMATS, is that in conventional systems wastewater is treated rapidly in a highly managed and energy intensive way, whereas AMATS rely on a slow flow of effluent through the system giving long retention times. The discharge and flow of wastewater through AMATS must be regulated so that retention times are sufficiently long for pollutant removal to be efficient. The removal of BOD_5, TOC and COD has been modelled as first order decay functions (Tchobanoglous 1987):

$$C_t = C_o \, e^{-kt}$$

where C_t is the concentration remaining at time t (mg l^{-1}),
C_o the concentration at time t = 0 (mg l^{-1}),
k the specific removal constant for a given constituent at 20°C (d^{-1}), and
t the retention time.

For BOD_5 a *k* has been estimated at approximately 0.3 d^{-1}, giving a 95% reduction in BOD_5 for a retention time (*t*) of 10 days.

The particular value of *k*, will vary with temperature and differ between systems. The effects of temperature may be estimated using the following expression:

$$kt = k_2 \, a^{(T-20)}$$

where *k*t is the removal rate constant at temperature *t* (d^{-1}),
k_2 the removal constant at 20°C (d^{-1}),
a the temperature coefficient, a constant equal to 1.1, and
T the temperature (°C) of the water in the system.

The value of *k*, will be partly determined by the hydraulic conductance of the system. Reed *et al.* (1988) has proposed that for reed bed systems the operational removal rate may be obtained from the above equation and the relation:

$$k_2 = ko \, (37.31 \, b^{4.172})$$

where *ko* is the 'optimum' rate constant for a medium with a fully developed root zone (d^{-1}) and
b is the total porosity of medium expressed as a decimal function.

For typical municipal wastewaters *ko* equals 1.839 d^{-1}, while the BOD_5 removal rates for industrial wastewater with high COD would be considerably slower, *ko* being equal to 0.198 d^{-1}. The effective 95% removal of BOD_5 from such effluent would necessitate a retention time of around 15 days.

The method used to introduce wastewater into the active reed beds depends on whether treatment is achieved by the horizontal or vertical flow of wastewater through the plant root system. In either case effluent must be introduced equally across the whole width of the reed bed, short circuiting and formation of stagnant or 'dead' areas must be avoided. For horizontal systems, trenching and multiple outlets from a number of feeding pipes may be employed. For vertical flow systems a central inlet channel, herring-bone agricultural piping over the surface, and/or spray irrigation systems have been used. In each case the inlet system used must allow the level of water to be maintained at, or just below, the top of the bed. During the establishment of the active vegetation it may be necessary to lower the water table to encourage root and rhizome development. Protective embankments around the system, along with the ability to regulate input flow rate, will be required in order to prevent flooding and uncontrolled discharge to the outlet or surrounding habitats during storm events. During the summer months, the loss of water by evapotranspiration (i.e. the evaporation of water from soil, water and vegetational surfaces) may exceed precipitation, and flow rates will need to be increased in order to prevent drying out (which may damage the vegetation and increase the subsequent hydraulic conductivity of the bed due to the disruption of soil/substrate structure).

Subsequent management

Sludge and straw will accumulate at a rate of between 1.5–2.5 cm yr^{-1}. This accreted material may reduce the efficiency of the system and therefore requires removal. As it accumulates it may cause local variation in the hydrology of the beds, reducing the vigour of reed growth; it may also contain high levels of accumulated heavy metals, which are frequent contaminates of domestic sewage. If the efficiency of the systems declines with age, the bed may require replanting, or flushing with clean water to remove accumulated pollutants. Burning or harvesting of the vegetation may also be required. Both methods are equally effective at removing surplus biomass (i.e. carbon); however, in addition to being environmentally undesirable, burning does not remove nutrients from the system. Nutrients present in the above-ground portion of the vegetation are released back into the system.

The need to harvest has been seen as a potential economic drawback, especially for floating aquatic plant systems, where uptake by the plant represent a major mode of pollutant removal, necessitating frequent removal of biomass. However, this surplus biomass may have significant economic value. Feasibility studies have been undertaken in order to assess the potential economic values of using it as a source of biomass for the production of methane, alcohol, and animal feeds. It is already being used for soil amendment and marketed as an organic compost/fertilizer (Reddy and Smith, 1987).

Evaluation of artificial reed beds

Recent evaluations of some European systems have questioned the effectiveness of the root-zone approach. It is clear that many systems have failed to meet their original design specifications. Although effluent BOD, TSS and TOC are generally reduced effectively, total-nitrogen and phosphorus levels are frequently only lowered by between 25% to 50% (Bucksteeg 1991; Findlater *et al.*, 1991; Schierup *et al.*, 1991). In some cases penetration of wastewater into the soil matrix is extremely poor; as a result, horizontal flow in the rooting zone fails to occur to any significant extent. The majority (90%) of wastewater movement occurs across the surface of the bed within the leaf litter layer (Schierup *et al.*, 1991). In tracer studies, using LiCl, mode retention times of 6.1 and 7.3 days were obtained for two mature Danish reed bed systems. These values agree well with predicted values assuming a soil porosity of 40%. However, lithium was detected in outlet channels within 1-2 hr of its release into the inlet channels. Although lithium continued to appear in outlet channels for 25 days, peak concentrations were recorded during the first two days of the study. Thus, flow through these systems occurs via a combination of surface run-off and subsurface movement, pollutant removal largely resulting from the sedimentation of suspended solids during surface run-off. Where wastewater penetration into the soil matrix is poor, pollutant removal cannot be ascribed to mechanisms operating on the root surfaces or within the soil matrix. Under such conditions the performance of the system will depend on achieving a uniform flow of wastewater through the litter layer (Bucksteeg, 1991). Bucksteeg (1991) suggests that the minimum size of reed beds, irrespective of the substrate used, should be calculated on an assumption of total clogging. Using this 'worst case' approach, Bucksteeg (1991) advocates a minimum size of 5 m^2 pe^{-1}, and where effluents remain visible on the surface, e.g. planted shallow lagoons, an area of at least 10 m^2 pe^{-1}.

The transfer of oxygen into the soil matrix by plants is considered central to the effective operation of reed beds. Total flux values in the order of 5 g m^{-2} dy^{-1} have been recorded for the movement of oxygen into beds planted with *Phragmites* (Gray *et al.*, 1991; Brix and Schierup, 1991). If account is taken of the respiratory consumption of roots and rhizomes, only a small proportion, in the order of 0.02 g m^{-2} dy^{-1}, remains. This quantity would have negligible effects on levels of soil nitrification. However, other studies have obtained considerably higher fluxes of oxygen into the soil matrix. Gray *et al.* (1991) obtained a value of 24.2 g O_2 m^{-2} dy^{-1}, while Hofmann (1991) found that soil redox potentials within a sewage sludge treatment bed planted with *Phragmites* ranged from 94.3 to 170 mV, compared to an unplanted control where values of between -60 and 60 mV were obtained.

It is clear that considerable care must be taken with the choice of substrate and establishment of vegetation if adequate wastewater penetration and oxygen transfer is to be obtained. The uncritical adoption of soil-based horizontal flow systems cannot be justified. Although soil has been favoured because of its potential for greater phosphate and metal retention, the establishment of root and rhizome systems cannot be relied upon to maintain adequate hydraulic conductance or oxygen uptake. The results of some recent studies suggest that improved performance can be obtained with the use of gravel as a substrate and/or the vertical loading of beds with wastewater (Cooper and Findlater, 1991).

Artificial wetlands clearly represent a viable option for wastewater treatment. They have been, and continue to be, successfully used in many countries for the treatment of domestic sewage and removal of heavy metals from contaminated water. However, they cannot, as Bucksteeg (1991) states, be considered to "lie within the generally accepted rules of technology". Their performance cannot, as yet, be reliably predicted. In too many cases pilot plants have been constructed out of expediency, with little regard to previous experience or results of scientific studies. If the undoubted capacity of natural wetland ecosystems to trap and degrade pollutants is to be mirrored in the performance of artificial systems, considerably more fundamental research and testing of small experimental systems is required.

Artificial wetlands may well have a role to play in the control of non-point pollution. It is conceivable that small strips of wetland could, for example, be employed to contain pesticide and fertilizer run-off from agricultural systems and heavy metal run-off from roadways. The potential of wetlands to accumulate and contain synthetic organic pollutants is largely unexplored, as is the possibility of improving the performance of AMATS by selectively introducing desirable bacterial strains capable of specific pollutant decomposition.

References

ALEXANDER, W.V. and WOOD, A. (1987) Experimental investigations into the use of emergent plants to treat sewage in South Africa. *Water Sci. Technol.*, **19(10)**, 51–59.

ATHIE, D. and CERRI, C.C. (1987) The use of macrophytes in water pollution control. *Water Sci. Technol.*, **19(10).**

BAYES, C.D., BACHE, D.H. and DICKSON, R.A. (1989) Land-treatment: design and performance with special reference to reed beds. *Journal of the Institute of Water and Environmental Management*, **3**, 588–598.

BRIX, H. (1987a) Treatment of wastewater in the rhizosphere of wetland plants—the root-zone method. *Water Sci. Technol.*, **19(10)**, 107–118.

BRIX, H. (1987b) The applicability of the wastewater treatment plant in Othfresen as scientific documentation of the root-zone method. *Water Sci. Technol.*, **19(10)**, 19–24.

BRIX, H. and SCHIERUP, H.H. (1991) Soil oxygenation in constructed reed beds: the role of macrophytes and soil–atmosphere interface oxygen transport. In Cooper, P.F., Findlater, B.C. (eds) *Constructed Wetlands in Water Pollution Control*. Pergamon Press, Oxford.

BUCKSTEEG, K. (1987) Sewage treatment in helophyte beds—first experiences with a new treatment procedure. *Water Sci. Technol.*, **19(10)**, 1–10.

BUCKSTEEG, K. (1991) Treatment of domestic sewage in emergent helophyte beds—German experiences and AVT-guidelines H262. In Cooper, P.F., Findlater, B.C. (eds) *Constructed Wetlands in Water Pollution Control.* Pergamon Press, Oxford.

COOPER, P.F. and FINDLATER, B.C. (eds) (1991) *Constructed Wetlands in Water Pollution Control*. Pergamon Press, Oxford.

FINDLATER, B.C., HOBSON, A.J. and COOPER, P.F. (1991) Reed bed treatment systems: performance evaluation. In Cooper, P.F., Findlater, B.C. (eds) *Constructed Wetlands in Water Pollution Control*. Pergamon Press. Oxford.

GOOD, B.J. and PATRICK, JR, W.H. (1987) Root–water–sediment interface processes. In Reddy, K.R., Smith, W.H. (eds) *Aquatic Plants for Water Treatment and Resource Recovery*. Magnolia Publishing Inc., Orlando, Florida.

GRAY, K.R., BIDDLESTONE, A.J., JOB, G. and GALANOS, E. (1991) The use of reed beds for the treatment of agricultural effluent. In Cooper, P.F., Findlater, B.C. (eds) *Constructed Wetlands in Water Pollution Control*. Pergamon Press, Oxford.

HOFMANN, K. (1991) Use of *Phragmites* in sewage sludge treatment. In Cooper, P.F., Findlater, B.C. (eds) *Constructed Wetlands in Water Pollution Control*. Pergamon Press, Oxford.

KICKUTH, R. (1980) Abwassereinigung in Mosaikmatnizen aus aeroben and anaeroben Teilbezirken. *Verhandl.* Verein. *Osterreichischen Cemiken, Abwassertechniches Symp., Graz*, pp. 639–665.

KICKUTH, R. (1983) Einige Dimensionier—ungsgrundsatze fur das Wurzelraumfahren. In Sekovlov, I., Wildere, P. (eds) *Abwasserienigung mit Hilfe van Wasserpflanzen*. Techn. Univ., Hamburg-Harburg.

KLEINMANN, R.L.P. and GIRTS, M.A. (1987) Acid mine water treatment in wetland: an overview of an emergent technology. In Reddy, K.R., Smith, W.H. (eds) *Aquatic Plants for Water Treatment and Resource Recovery*. Magnolia Publishing Inc., Orlando, Florida.

MASON, C.F. (1991) *Biology of Freshwater Pollution*, 2nd edn. Longman Scientific and Technical, Harlow.

MAY, E., BUTLER, J.E., FORD, M.G., ASHWORTH, R., WILLIAMS, J. and BAHGAT, M.M.M. (1991) Chemical and microbiological processes in gravel-bed hydroponic (GBH) systems for sewage treatment. In Cooper, P.F., Findlater, B.C. (eds) *Constructed Wetlands in Water Pollution Control*. Pergamon Press, Oxford.

MENDELSSOHN, I.A. and POSTEK, M.T. (1982) Elemental analysis of deposits on the roofs of *Spartina alterniflora* Loisel. *Amer. J. Botany*, **69**, 904–12.

REDDY, K.R. and DeBUSK, T.A. (1987) State-of-the-art utilization of aquatic plants in water pollution control. *Water Sci. Technol.*, **19(10)**, 61–79.

REDDY, K.R. and SMITH, W.H. (eds) (1987) *Aquatic Plants for Water Treatment and Resource Recovery*. Magnolia Publishing Inc., Orlando, Florida.

REED, S.C., MIDDLEBROOKS,. E.J. and CRITES, R.W. (1988) *Natural Systems for Waste Management and Treatment*. McGraw-Hill, New York.

RICHARDSON, C.J. (1985) Mechanisms controlling phosphorus retention capacity in freshwater wetlands. *Science*, **228**, 1424–1427.

RICHARDSON, C.J. and DAVIS, J.A. (1987) Natural and artificial wetlands ecosystems: ecological opportunities and limitations. In Reddy, K.R., Smith, W.H. (eds) *Aquatic Plants for Water Treatment and Resource Recovery*. Magnolia Publishing Inc., Orlando, Florida.

SCHIERUP, H., BRIX, H. and LORENZEN, B. (1991) Wastewater treatment in constructed reed beds in Denmark—state of the art. In Cooper, P.F., Findlater, B.C. (eds) *Constructed Wetlands in Water Pollution Control*. Pergamon Press, Oxford.

SCHIECHTL, H. (1980) *Bioengineering for Land Reclamation and Conservation*. University of Alberta Press, Alberta, Canada.

TCHOBANOGLOUS, G. (1987) Aquatic plant systems: engineering considerations. In Reddy, K.R., Smith, W.H. (eds) *Aquatic Plants for Water Treatment and Resource Recovery*. Magnolia Publishing Inc., Orlando, Florida.

WOLVERTON, B.C. (1987) Artificial marshes for wastewater treatment. In Reddy, K.R., Smith, W.H. (eds) *Aquatic Plants for Water Treatment with Resource Recovery*. Magnolia Publishing Inc., Orlando, Florida.

WOOD, A. (1991) Constructed wetlands for wastewater treatment—

engineering considerations. In Cooper, P.F., Findlater, B.C. (eds) *Constructed Wetlands in Water Pollution Control*. Pergamon Press, Oxford.

Further reading

ATHIE, D. and CERRI, C.C. (eds) (1987) The use of macrophytes in water pollution control. *Water Sci. Technol.*, **19(10)**

COOPER, P.E. and FINDLATER, B.C. (eds) (1991) *Constructed Wetlands in Water Pollution Control*. Pergamon Press, Oxford.

MASON, C.F. (1991) *Biology of Freshwater Pollution*, 2nd edn. Longman Scientific and Technical, Harlow.

REDDY, K.R. and SMITH, W.H. (eds) (1987) *Aquatic Plants for Water Treatment and Resource Recovery*. Magnolia Publishing Inc., Orlando, Florida.

REED, S.C., MIDDLEBROOKS, E.J., and CRITES, R.W. (1988) *Natural Systems for Waste Management and Treatment*. McGraw-Hill, New York.

SULPHUR AND NITROGEN OXIDES

Chapter 11

ENVIRONMENTAL IMPACTS OF SULPHUR AND NITROGEN OXIDES

Introduction

The potential health risks associated with breathing air polluted by SO_2 and NO_x (nitrous oxides NO, NO_2) are well established and have a long history dating back to the introduction of coal as a common domestic and industrial fuel. Because of the noxious odour, the burning of 'seale coal', which had a high sulphur content, was banned from the City of London in 1306 by Edward I of England. During this time, air quality in London was so poor that the burning of coal while Parliament was in session was completely banned. The death penalty was introduced to reinforce laws governing the burning of coal and it is believed to have been carried out on one unfortunate citizen who burnt coal while Parliament was in session. Sulphur dioxide produced by the combustion of sulphur-containing fossil fuel (particularly coal and oil) is a respiratory irritant. In the atmosphere it is readily transformed to sulphuric acid which frequently becomes concentrated on soot and smoke particles. When inhaled, these particles are highly corrosive and irritating to lung tissue. In London during December 1952, four days of severe smog (a mixture of fog and smoke from coal combustion) resulted in the widespread occurrence and aggravation of respiratory diseases and the death of an estimated 4000 people. This event was largely instrumental in the passing of the 1956 Clean Air Act which regulated the burning of coal by industrial and domestic users by creating smoke-controlled zones, in which only 'smokeless' fuels could be burned. Thankfully today, in most developed countries traditional smog attacks are rare. Health risks are now largely associated with workplace exposure to either high concentrations over short periods or prolonged exposure to low levels. In the United States recommended workplace concentrations of SO_2 are $\leq$ 2 ppm or up to 5 ppm for short-term exposure, although, there is evidence to suggest that long-term exposure to low levels of SO_2 ($<$ 1 ppm) can cause respiratory distress among sensitive individuals with a history of asthma or other respiratory diseases (Goldsmith, 1986).

Currently the major environmental concerns associated with SO_2 and NO_x centre on acidic precipitation and its role in the acidification of soils, surface waters and forest decline. Modern research into 'acid rain' dates from the early 1960s when Scandinavian scientists began to link the occurrence of airborne pollution imported across the North Sea from Britain and mainland Europe, with the acidification of lakes and the decline and disappearance of fish from affected waters. We now know that the impact of SO_2 and NO_x emissions and acid rain on the environment are complex, involving interactions between the physical and biological components of the ecosystem. High H^+ concentrations, in addition to being directly toxic to both animals and plants, inhibit microbial decomposition, nitrogen fixation and increase the solubility and mobility of toxic heavy metals and aluminium within the environment.

Chemistry of acid precipitation

Acid precipitation may be defined as rainwater with a pH less than 5.65. This level corresponds to the pH of distilled water at equilibrium with air containing approximately 340 ppm of CO_2. The pH of the water is acidic because of the formation of carbonic acid:

$$CO_2 + H_2O \rightleftharpoons H_2CO_3$$

Natural rainwater normally has a pH of around 5.6, however, this value may be exceeded locally where trace levels of soil-derived neutralizing cations, e.g. Ca^{2+} and Mg^{2+}, occur in the precipitation. However, over much of eastern America and north-west and central Europe, rainwater pHs range from 4.0 to 5.0. A value of 2.31 has been recorded for rainfall in Pennsylvania (Lynch and Corbett, 1980). Given that pH is measured on a logarithmic scale, these values indicate that the rainwater in these regions is considerably more acidic than expected. The acidity of mist or fog water, due to the small volume of water, is frequently higher than rainwater. In California where rainwater pHs are normally ≥ 4.4, fog water pHs as low as 2.3 have been recorded. Typical fog water pHs range from 3.0 to 3.5 (Freedman, 1989). Increases in rainwater acidity are predominately due to the presence of sulphurous (H_2SO_3), sulphuric (H_2SO_4), nitrous (HNO_2) and nitric (H_2NO_3) acids, which are formed from the oxidation and reactions of SO_2 and NO_x with water in the presence of oxidants, principally hydroxyl radicals OH^- or monatomic oxygen (Fig. 11.1; Mannion, 1992). In north-west Europe sulphuric acid accounts for around 70% of the mean total annual acidity and in eastern North America, 60%. In both cases the remaining 30–40% is largely due to nitric acid. The rate of SO_2 and NO_x conversion can be very rapid. In moist summer conditions, 100% of emissions may be converted to acids within an hour; during the winter, conversion rates are typically lower at around 20% per hour. In dry atmospheres

A. Sulphurous and sulphuric acids

SO_2 is emitted from natural and anthropogenic sources and dissolves in cloud water to produce sulphurous acid:

$$SO_2 + H_2O \longrightarrow H_2SO_3 \rightleftharpoons H^+ + HSO_3^-$$

Sulphurous acid can be oxidized in the gas or aqueous phase by various oxidants

$$SO_2 \xrightarrow{\text{oxidant}} SO_3$$

Aqueous sulphur trioxide forms sulphuric acid:

$$SO_2 + H_2O \longrightarrow H_2SO_4 \rightleftharpoons H^+ + HSO_4^- \rightleftharpoons 2H^+ + SO_4^{2-}$$

B. Nitrous and nitric acids

N_2O is emitted by the process of denitrification and although relatively inert it is a greenhouse gas.

NO and NO_2 (collectively designated as NO_x) are produced by combustion processes and lightning.

They are involved in many chemical processes, some of which damage the ozone layer in the stratosphere:

$$O_3 + NO \longrightarrow NO_2 + O_2$$

Other chemical processes may generate ozone in the troposphere causing photochemical smogs:

$$NO_2 \xrightarrow{\text{light}} NO + O$$
$$O + O_2 \longrightarrow O_3$$

In addition, nitric and nitrous acids may be produced:

$$2NO_2 + H_2O \longrightarrow HNO_3 + HNO_2$$

These acids are components of acid rain along with sulphurous and sulphuric acids

Fig. 11.1. Processes involved in the formation and deposition of acid pollution (from Mannion and Bowlby, 1992).

conversion is slower and relies to a greater extent on photochemical reactions and the presence of oxidizing agents. Summer conversion rates of 16% per day and a winter rate of 3% per day have been recorded (Mason, 1991).

Sources

Anthropogenic emissions of NO_x and SO_2 result primarily from the combustion of coal and oil. Currently, of United Kingdom SO_2

emissions, 60% are derived from power stations and 30% from industrial plants. While sulphur emissions have declined progressively since the early 1970s, emissions of nitrogen are rising, largely as a result of the increased use of motor vehicles. Around 45% of nitrogen emissions result from powers stations, 30% from vehicle exhausts and the remainder originate from a variety of industrial and agricultural sources (Ch. 8; Mason, 1989). Not surprisingly acidic precipitation is most marked in northern industrialized countries. However, acidic depositions within a nation state may well have distant foreign origins. For example, of the sulphur deposited in Norway and Sweden less than 20% derives from domestic sources (McCormick, 1989). One striking aspect of acid precipitation is its regional nature; large areas of terrain are affected. Unlike particulate emissions, depositions do not decline markedly with distance away from major point sources.

Because of the tall smokestacks frequently employed to prevent excessive concentrations of pollutants in the immediate vicinity of major sources, acidic emissions are rapidly transported and dispersed away from the source by wind, atmospheric mixing and dilution. Of the total sulphur emissions ($0.4–0.9 \times 10^6$ Mt yr^{-1}) from a large smelter works at Copper Cliff, Sudbury, Canada, only 1.3% is deposited within 40 km of the 381 m tall smokestack. Within this distance of the works sulphur deposition exceeds normal background levels by only 16% (Freedman, 1989). Distance effects are generally only apparent when total deposition is considered. This includes the contributions made by dry deposition which occurs between episodes of precipitation and results from the direct uptake of gaseous SO_2 and NO_x by vegetation, soil and water surfaces, the gravitational settling, and impaction/filtering of particulate aerosols. Rates of dry and wet deposition show considerable spatial and temporal variations. In general, dry deposition dominates close to the source but declines with distance. The relative magnitude of wet and dry deposition will depend on distance from the source, the nature of the vegetation, prevailing winds and weather conditions. However, even at sites remote from obvious sources of SO_2 and NO_x, dry deposition can account for between 30 and 60% of total nitrogen and sulphur inputs (Eaton *et al.*, 1980; Lindberg *et al.*, 1986). Because of the behaviour of SO_2 and NO_x in aqueous media, the effects of dry deposition onto soil and water will be similar to that of acidic precipitation. This is not true of direct uptake of SO_2 and NO_x by plants, which may account for 60% of total dry depositions. Both gaseous NO_x and SO_2 are toxic to vegetation, capable of reducing plant productivity and causing the loss of sensitive species.

Plant sensitivity to SO_2 and NO_x

Gaseous pollutants enter the plant primarily through the stomata. Because of this, uptake and sensitivity to SO_2 and NO_x vary through the day and with the water status of the plant. During the night or periods of drought, when the stomata are closed, uptake is limited to that which is able to cross the waxy cuticle of the leaf (Mansfield, 1976). The cuticle is normally considered to be impermeable to SO_2 and NO_x, however, acidic precipitation or the presence of a water film may substantially increase the rate and quantities of pollutants entering the plant across the cuticle (Fitter and Hay, 1987). On entry, SO_2 dissolves in the water films bound to the apoplast and mesophyll cell surfaces to form hydrated SO_2 ($SO_2.H_2O$). Hydrated SO_2 acts as a strong acid, dissociating to give HSO_3^- and SO_3^{2-} in proportions determined by the pH of the solution (pK_a for $HSO_3^- \rightarrow SO_3^{2-} + H^+ = 7.2$). The movement of SO_2 into cells appears to be passive. Because of the net negative charge on the cell wall, only the uncharged species $SO_2.H_2O$ actually enters the cell, where it disassociates to produce the phytotoxic SO_3^- ions (pH cytoplasm > 7.0). A portion of the SO_3^- ions may be oxidized, primarily by the chloroplast, to the less toxic sulphate ion. Differences in the ability of plants to regulate their internal pH and to oxidize SO_3^- to SO_4^{2-} are thought to be important factors determining the tolerance of species to SO_2 (Fitter and Hay, 1987; Crawford, 1989). In a similar manner to SO_2, NO_x enters the plant and dissolves in the extracellular water films to give equal proportions of nitrate and the phytotoxic nitrite ions. However, a proportion of both may be converted via the activity of nitrite reductase to ammonia and thus contribute to the normal nitrogen metabolism of the plant.

Primary sites of pollutant action

Three primary sites of action have been identified (Fitter and Hay, 1987; Crawford, 1989).

1 Stomata function: SO_2, NO_x and O_3 can cause rapid disruption of the stoma apparatus, causing increased pollutant uptake and water loss.

2 Chloroplast structure and function: exposure to SO_2, NO_x and O_3 at low levels, insufficient to cause visible damage to the plant, causes gross disruption of the thylakoid membrane system of the chloroplast.

3 CO_2 fixation: exposure to SO_2 and gaseous air pollutants in general cause a rapid decline in rates of carbon fixation (Fig. 11.2). In the case of SO_2, this appears to be due to competition between the sulphite

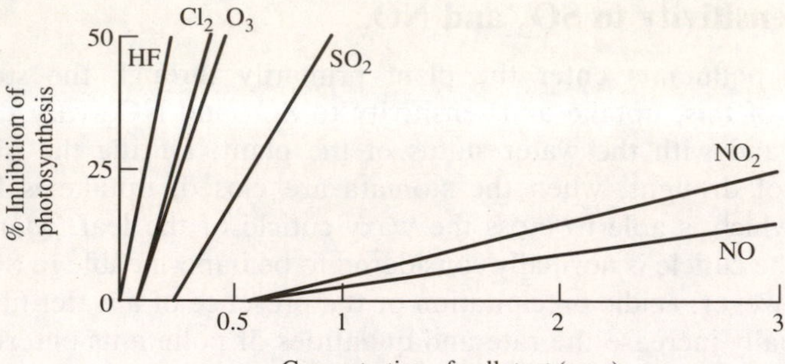

Fig. 11.2. The inhibition of photosynthesis (as measured by CO_2 uptake) in 3–5 week old canopies of barley and oats when exposed to varying levels of gaseous pollutants for 2 hours. During the treatments, the plants were maintained at 24°C, a wind velocity of 1.2–1.6 m s^{-v}, low humidity (45% RH) and high light (40–50 Klux – 280–350 W m^{-1}) (adapted from Bennett and Hill, 1974).

and bicarbonate ions for the CO_2–fixing sites on the enzymes phosphoenolpyruvate carboxylase and ribulose-1,5-biphosphate carboxylase which are responsible for the initial fixation of CO_2 in plants with C4 and C3 photosynthesis, respectively. Studies also suggest that SO_2 may block sulphydryl groups and disrupt cellular pH, while NO_x as a nitrite is capable of interfering with the chloroplast redox systems.

Effects of SO_2 on plant growth

Experimental exposure of plants to SO_2 indicates that low concentrations can significantly reduce yields without visible damage being apparent. However, susceptibility differs markedly between species, among varieties, and with the age and condition of the plant (e.g. nutrient and water status). Short-term exposure experiments (e.g. 10% of growth period) suggest that concentrations > 185 ppb (0.185 ppm) will adversely affect the growth of most crop plants. Tree species are more sensitive being adversely affected by concentrations > 170 ppb (0.170 ppm) (Roberts, 1984). Some species are particularly sensitive. Reduced rates of photosynthesis have been recorded for pea, bean and tomato plants exposed to concentrations as low as 0.03 ppm SO_2 and 0.1 ppm NO_x. Where plants are subject to prolonged exposure, very low concentrations can significantly reduce yields. For example, the yield of *Lolium perenne* (an important agricultural grass species), was depressed by 51% (relative to control plants), when exposed to SO_2 at a concentration of 191 μg m^{-3} (0.067 ppm) for 26 weeks. Control plants were exposed to SO_2 at a concentration of 9 $\mu g\ m^{-3}$ (*c*. 3 ppb) (Bell and Clough, 1973). Roberts (1984), reviewing laboratory and field studies on a wide range of species, concluded that: (i) exposure to 0.076–0.150 ppm for 1–3 months significantly reduced yields in most species, (ii) exposure to 0.038–0.076 ppm for several months reduced the yields of some species and (iii) prolonged exposure to levels < 0.038 ppm variously produced beneficial, no, or only minor reductions in yield. (Beneficial effects occur when soils are deficient in sulphur.)

If the levels at which plant growth is inhibited are compared to typical concentrations of SO_2 and NO_x recorded in United Kingdom rural (SO_2, 0.001–0.05; NO_x, 0.005–0.05 ppm), urban (SO_2, 0.02–0.5; NO_x, 0.02–0.2 ppm) and industrial areas (SO_2 and NO_x ≤ 1.0 ppm), it is clear that the growth of many plants in these areas may be reduced as a result of ambient air quality (Fitter and Hay, 1987). Even in rural areas considered to have clean air (e.g. United States air quality standard for SO_2 is 0.5 ppm) the growth of agricultural crops and natural vegetation may be significantly impaired. In The Netherlands, the distribution and abundance of many wild herbaceous heathland plant species has declined dramatically over the last 20 years. This decline cannot be entirely explained by changes in groundwater levels and habitat disturbance of eutrophication (i.e. nutrient enrichment) caused by agricultural practice. On a 10 km^2 basis, Van Dam *et al.* (1986) have been able to correlate the reductions of eight dry heathland species with the spatial pattern of ambient SO_2 concentrations. More recently Dueck *et al.* (1992) have experimentally determined the threshold level of SO_2 above which heathland vegetation may suffer significant damage. Their results suggest that in order to protect 95% of the species in a heathland community (i.e. prevent damage to the community as a whole), ambient SO_2 concentrations should not exceed 8 $\mu g\ m^{-3}$ (2.7 ppb).

Exposure to concentrations >5 ppm of SO_2 and NO_x result in visible symptoms of damage; SO_2 causes intervienal chlorosis and NO_x the development of irregular brown or black spots on leaf surfaces. Prolonged exposure to these concentrations severely damages vegetation causing the death and displacement of all but the most tolerant species. Tree death, particularly of conifers, and suppression of vegetation cover are frequently evident near major pollution sources. Freedman and Hutchinson (1980), have documented the long-term effects of emissions from the Copper Cliff smelter. The major effects observed are due to SO_2, but acidification and the toxic effects of Cu, Ni and Al play a part. No forest remnants occur within 3 km of the works and vegetation cover is sparse, consisting of isolated clumps of coarse grasses such as *Deschampsia caespitosa* and *Agrostis hyemalis*. At distances of between 3 and 8 km, forest remnants occur in sites sheltered from direct fumigation and capable of providing adequate water, i.e. valleys and lower mesic slopes. These forest remnants are dominated by the more tolerant woody angiosperms such as *Acer rubrum, Quercus rubra,* and *Populus tremuloides.* Tree growth is deformed and stunted with the upper branches frequently dead. Beyond 8 km, forest cover increases, but species diversity is poor and there is little understorey development. It is not until 20 km from the source that a mixed conifer–hardwood forest, typical of the region, occurs.

Acid precipitation, soil acidification and plant–soil interaction

Soil pH is determined by a complex interaction of factors including the following.

1 Carbonic acid concentrations. Because of saprophyte and plant root respiration, CO_2 concentrations are typically high (0.5–1.0% vol./vol.) when compared to that of the atmosphere (0.03%). As a result, soil solution carbonic acid concentrations are high and play an important part in influencing the acidity of pH of soils with pH >6, i.e. with appreciable alkalinity.

2 Plant uptake and the cycling of nitrogen. (i) In acidic soil (pH ≤5.5) the dominant form of inorganic nitrogen is NH_4^+. Plant uptake of NH_4^+ derived from the ammonification of organic nitrogen has no net effect on soil acidity. However, if the NH_4^+ is derived from atmospheric deposition or fertilization, plant uptake of a quantity NH_4^+ contributes an equivalent account of H^+ ions to the soil. The plant excretes H^+ ions in order to maintain electrochemical neutrality. (ii) In soils with a pH >5.5 the major inorganic form of nitrogen is NO_3^-, which is predominately derived by the microbial oxidation of NH_4^+, which yields two units of H^+ ions for every unit of NO_3^- produced. On uptake of NO_3^- plants excrete OH^- ions. If the NH_4 is derived from organic nitrogen present in the soil, there is no net effect on soil acidity. If however, the NH_4^+ results from atmospheric deposition or the use of fertilizers, e.g. $NH_4NO_3^-$, NH_4SO_4 or urea, the assimilation of NH_4^+ will tend to increase soil acidity (Fig. 11.3).

3 Sulphur uptake. The majority of sulphur in soils is present as reduced organic forms, microbial oxidation of which, via various intermediates to SO_4^{2-} and subsequent uptake by plants, has no net effect on soil acidity. However, direct input of SO_4^{2-}, followed by plant uptake can reduce acidity. Under anaerobic conditions SO_4^{2-} is reduced to sulphide, a process involving the uptake of two units of H^+ per unit of SO_4^{2-} reduced. In most circumstances, because of the small amounts

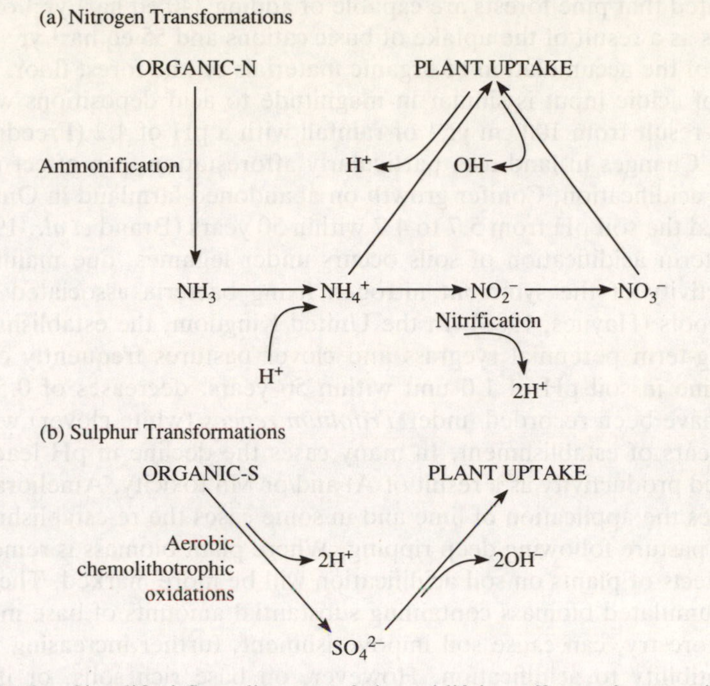

Fig. 11.3. Simplified flow diagram of the acidifying effects of some nitrogen and sulphur transformations in soil.

involved, the sulphur cycle has less of an effect on soil pH than nitrogen (Fig. 11.3).

4 Leaching of ions. In most free-draining soils SO_4^{2-}, Cl^- and NO_3^- are readily leached to the groundwater. The SO_4^{2-} adsorption capacities of northern hemisphere soils are limited and easily saturated by anthropogenic inputs. When anions are leached, soil electrochemical neutrality is maintained by the loss of equivalent quantities of the cation bases Ca^{2+} and Mg^{2+}. In soils with low carbonates contents the loss of Ca^{2+} and Mg^{2+} rapidly reduces the soil's ability to buffer pH changes and aids acidification by reducing the base saturation of the cation exchange capacity. In comparison to water systems, soil is strongly buffered. Carbonate minerals buffer soils over the pH range >8–6.2, silicates from 6.2 to 5.2 and cation exchange capacity from 5.0 to 4.2.

Soil acidification is a well-documented, natural process occurring during the development of soils and vegetation on new or recently exposed strata (primary succession). It results largely from the weathering of soil minerals and the accumulation of organic material on and within the soil matrix (Ricklefs, 1990). However, plant growth and nutrient uptake may affect rates of soil acidification directly. It has been estimated that pine forests are capable of adding 340 eq ha^{-1} yr^{-1} of H^+ to soils as a result of the uptake of basic cations and 55 eq ha^{-1} yr^{-1} as a result of the accumulation of organic materials on the forest floor. This level of acidic input is similar in magnitude to acid depositions which would result from 100 cm yr^{-1} of rainfall with a pH of 4.2 (Freedman, 1989). Changes in land use, particularly afforestation, can effect rates of soil acidification. Conifer growth on abandoned farmland in Ontario lowered the soil pH from 5.7 to 4.7 within 50 years (Brand *et al.*, 1986). Long-term acidification of soils occurs under legumes, due mainly to the activity of the symbiont nitrogen fixing bacteria associated with their roots (Haynes, 1983). In the United Kingdom, the establishment of long-term perennial ryegrass and clover pastures frequently cause a decline in soil pH of 1.0 unit within 50 years; decreases of 0.5 pH units have been recorded under *Trifolium repens* (white clover) within four years of establishment. In many cases the decline in pH leads to reduced productivity as a result of Al and/or Mn toxicity. Amelioration requires the application of lime and in some cases the re-establishment of the pasture following deep ripping. Where plant biomass is removed the effects of plants on soil acidification will be more marked. The loss of accumulated biomass containing substantial amounts of base metals as in forestry, can cause soil impoverishment, further increasing their susceptibility to acidification. However, on base rich soils, or if the interval between harvest is sufficiently long, nutrients lost from the site will be replaced by weathering of the parent soil material.

The influence of anthropogenic and biological sources of acidity are difficult to separate. Although the situation is complex, there is little doubt that the afforestation of catchments on base poor rocks has contributed to the acidification of surface waters and increased their sensitivity to acidic precipitation. It is suggested that biological processes are most important in the litter and upper soil horizons, while anthropogenic inputs play a more pronounced role in the acidification of deeper mineral soil horizons (Tamm and Hallbacken, 1988). Freedman (1989), who provides an excellent review of soil acidification, concludes that there is as yet no "conclusive evidence of soil acidification via acidic deposition - more data is still needed". However, it would seem, at the very least, logical to assume that acidic precipitation will tend to accelerate and increase the likelihood of acidification.

Forest decline

Over the last 30 years there has been increased concern over the global occurrence of what appears to be a new type of forest decline affecting extensive areas of the world's forests (Hinrichsen, 1986; Mueller-Dombois, 1988). Particularly badly affected are European forests and high elevation forests of North America. In 1986, at least 16×10^6 ha of western Europe forests were thought to show symptoms of damage. In western Europe the tree species most affected are the Norway spruce (*Picea abies*), beech (*Fagus sylvaticus*) and to a lesser extent oaks (*Quercus* spp.). In North America the red spruce (*Picea rubens*) and the sugar maple (*Acer saccharum*) have been particularly badly affected. Although symptoms vary between species, frequently observed signs of damage include chlorosis of foliage, abnormal growth, premature leaf loss, the progressive death of branches at the extremities causing a 'stag-head' appearance, root dieback and increased mortality, often as a result of secondary agents such as fungal pathogens and insects (Hinrichsen, 1986; Freedman, 1989). The important point is that although 'natural' forest declines have and do occur for various reasons including disease outbreaks, climate change, etc., they normally affect one or a limited range of species and are not characterized by a consistent set of symptoms. The primary cause of current forest decline

Table 11.1 Proposed mechanism responsible for the link between air pollution and forest decline

1. General stress

   Poor air quality reduces plant growth and photosynthesis and disrupts the movement and allocation of photoassimilate. Assimilate available to roots is reduced, impairing root function

2. Soil acidification

   Acidic precipitation causes leaching of nutrients from the soil and leaf tissue, stressing the plant. Increased Al^{3+} in acidified soil further stresses the plant by inhibiting root growth and function

3. Gaseous pollutant injury

   Decline results from the accumulative effects of prolonged exposure to low concentrations of the phytotoxic air pollutants O_3 and SO_2

4. Magnesium deficiency

   Many of the symptoms shown by a tree subject to decline are similar to those associated with Mg^{2+} deficiency. Exposure to O_3 and acidic precipitation are thought to leach sufficient amounts of Mg^{2+} from both foliage and soil to cause Mg^{2+} deficiency.

5. Excessive nitrogen input

   High levels of NO_x result in excess nitrogen availability which (i) stimulates plant growth increasing demand for other nutrients which are likely to be in short supply, (ii) inhibits mycorrhizas development, (iii) increases sensitivity to frost damage and (iv) causes changes in biomass allocation, e.g. affects root : shoot ratios has not been identified but the regional contamination of airshreds and acidic precipitation are undoubtedly a major factor, although other factors, for example, soil type, heavy metal input (Ch. 13) and ambient O_3 concentrations, may locally play an important role. Five distinct mechanisms have been suggested; these are outlined in Table 11.1. No single mechanism is able to account for all of the symptoms observed, but it is clear that air quality, particularly the concentrations of NO_x are SO_2, are of major importance.

Acidification of surface waters

In contrast to forest decline and soil acidification the importance of acidic precipitation as a prime cause of lake acidification has now been clearly established. Historical records, the long-term monitoring of water chemistry and studies of pH sensitive diatom assemblages in lake sediments all indicate that the onset and rate of acidification is correlated with industrialization and SO_2 emissions (Wellburn, 1988; Mannion, 1989a,b). The pH of Round Loch of Glenhead, Scotland has decreased from 5.5 to 4.4 during the past 130 years. In southern Norway, of 87 lakes surveyed between 1923 and 1949, 21 had pHs below 5.5; by 1980 the number had increased to 41. In addition to long-term shifts in the acidity, acid precipitation can cause marked temporal variations in pH. In the summer, streamwater pH and composition often closely resemble that of the groundwater. However, during winter, with increased streamflow, the pH of the water typically declines as the composition of the water reflects more closely that of precipitation and run off. During snowfall and dry periods of low rainfall, acidic deposits accumulate in the catchment and are released during heavy rain or snow-melt causing pulses of severe acidity (Jacks *et al.*, 1986; Mason, 1991). The sensitivity of water bodies to acidification depends to a large extent on the nature of their catchment. Catchments which contain large amounts of calcium and magnesium carbonate, either in their soils or in the river and lake sediments, will frequently possess sufficient alkalinity to neutralize acidic inputs, thus protecting surface waters from acidification. The most susceptible catchments are those which are dominated by geologically hard slow-weathering oligotrophic minerals such as granite, which give rise to nutrient-poor acidic soil with limited buffering capacity. Land use, particularly forestry is another important factor. The occurrence of large areas of forest, particularly conifer plantations, typically substantially decrease the pH of water draining from the afforested portion of the catchment while substantially increasing the concentration of Al^{3+} and SO_4^{2-} (Mason, 1991).

The weakly buffering capacity of freshwater systems, which is governed by the concentration of bicarbonate ions, makes them particularly prone to acidification. In many respects the process of lake acidification can been seen as analogous to the titration of a dilute alkalinity-buffered system with sulphuric acid (Freedman, 1989). As acidification progresses, three stages can be identified as concentration of bicarbonate ions relative to SO_4^{2-} declines (Fig. 11.4). In the final stages of acidification the pH becomes stabilized, normally below pH 5.0, and because of their increased solubility at low pHs, metal ion (e.g. Al, Cd, Cu, Mn and Zn) concentrations, particularly Al^{3+}, increase (Henriksen, 1989). Not surprisingly, as acidification progresses the ecology of the water body changes dramatically.

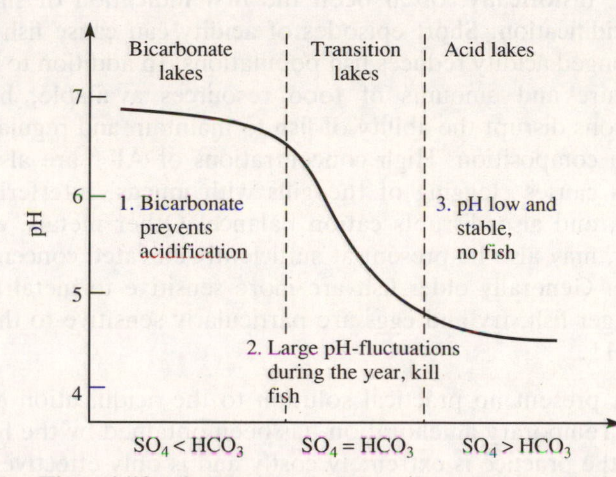

Fig. 11.4. The acidification process in lakes (from Mason, 1991).

1 The species composition changes and diversity of the phytoplankton community decreases. In many acidic lakes the phytoplankton community becomes dominated by dinoflagellates. If sufficient phosphate is available rates of primary production may be maintained or in some cases increased due to reduced grazing pressure from zooplankton and fish.

2 The abundance and diversity of higher plant macrophytes declines. *Sphagnum* mosses become increasingly dominant and because of the high exchange ion capacity of their cell walls may accelerate acidification.

3 As water pH deceases, bacterial activity and nitrogen fixation is inhibited and the importance of fungi increases. The reduced rate of decomposition allows organic material to accumulate. It has been suggested that low rates of decomposition will result in reduced mineralization and nutrient availability.

4 Invertebrate and zooplankton communities become impoverished. The loss of species and changes in community composition are caused by several factors including direct physiological stress imposed by high concentrations of H^+ and Al^{3+} and changes in food chain relationships, e.g. changes in the abundance of edible phytoplankton, predators (fish and invertebrates), bacteria and the amount, form and composition of detritus.

5 Acidified lakes support few or no fish. Records of declining fish stocks and the complete loss of fish from previously productive water bodies has, historically, often been the first indication of significant regional acidification. Short episodes of acidity can cause fish deaths, while prolonged acidity reduces fish populations. In addition to changes in the nature and amounts of food resources available, high H^+ concentrations disrupt the ability of fish to maintain and regulate body fluid cation composition. High concentrations of Al^{3+} are also toxic. Aluminium causes clogging of the gills with mucus, interfering with respiration, and also disrupts cation balance. Other metals, e.g. Cu, Cd and Zn, may also be present at sufficiently elevated concentrations to be toxic. Generally older fish are more sensitive to metal toxicity, while younger fish, fry and eggs are particularly sensitive to the direct effects of H^+.

There is at present no practical solution to the acidification of rivers and lakes. Temporary amelioration has been obtained by the liming of lakes, but the practice is extremely costly and is only effective locally. Acidification is a symptom; the solution lies with attacking the causes, which would require a substantial and sustained reduction in global sulphur emissions.

References

BELL, J.N.B. and CLOUGH, W.S. (1973) Depression of yield in ryegrass exposed to SO_2. *Nature Lond.*, **241**, 47–49.

BENNET, J.H. and HILL, A.C. (1974) Acute inhibition of apparent photosynthesis by phytotoxic air pollutants. In *Air Pollution Effects on Plant Growth*, M. Dugger (ed) American Chemical Society Symposium Series, **3**, 115–127.

BRAND, D.G., KEHOE, P. and CONNORS, M. (1986) Coniferous afforestation leads to soil acidification in central Ontario. *Can. J. Forest. Res.*, **16**, 1289–1391.

CRAWFORD, R.M.M. (1989) Studies in plant survival. Ecological case histories of plant adaptation to adversity. In *Studies In Ecology*, Vol. 11. Blackwell Scientific Publications, Oxford.

DUECK, Th.A., VAN DER EERDEN, L.J. and BERDOWSKI, J.J.M. (1992) Estimation of SO_2 effect threshold for heathland species. *Function. Ecol.*, **6**, 291–296.

EATON, J.S., LIKENS, G.E. and BORMANN, F.H. (1980) Wet and dry deposition of sulfur at Hubbard Brook. In Hutchinson, T.C., Havas, M, (eds) *Effects of Acid Precipitation on Terrestrial Ecosystems*. Plenum, New York.

FITTER, A.H. and HAY, R.K. (1987) *Environmental Physiology of Plants*, 2nd edn. Academic Press, London.

FREEDMAN, B. (1989) *Environmental Ecology. The Impacts of Pollution and Other Stresses on Ecosystem Structure and Function*. Academic Press, London.

FREEDMAN, B. and HUTCHINSON, T.C. (1980) Long-term effects of smelter pollution at Sudbury, Ontario, on forest community composition. *Can. J. Bot.*, **58**, 2123–2140.

GOLDSMITH, J.R. (1986) Effects on human health. In Stern, A.C. (ed.) *Air Pollution*, Vol. 6, 3rd edn, pp. 391–463. Academic Press, New York.

HAYNES, R.J. (1983) Soil acidification induced by leguminous crops. *Grass For. Sci.*, **38**, 1–11.

HENRIKSEN, A. (1989) Air pollution effects on aquatic ecosystems and their restoration. In Ravera, O. (ed.) *Ecological Assessment of Environmental Degradation, Pollution and Recovery*. Elsevier, Amsterdam.

HINRICHSEN, D. (1986) Multiple pollutants and forest decline. *Ambio*, **15(5)**, 258–265.

JACKS, G., OLOFSSON, E. and WERME, G. (1986) An acid surge in a well-buffered stream. *Ambio*, **15(5)**, 282–285.

LINDBERG, S.E., LOVETT, G.M., RICHTER, D.D. and JOHNSON, D. W. (1986) Atmospheric deposition and canopy interactions of major ions in a forest. *Science*, **231**, 141–145.

LYNCH, J.A. and CORBETT, E.S. (1980) Acid precipitation—a threat to aquatic ecosystems. *Fisheries*, **5**, 8–12.

MANNION, A.M. (1989a) Palaeoecological evidence for environmental change during the last 200 yrs. I. Biological data. *Prog. Phys. Geog.*, **13**, 23–46.

MANNION, A.M. (1989b) Palaeoecological evidence for environmental change during the last 200 yrs. II Chemical evidence. *Prog. Phys. Geog.*, **13**, 192–215.

MANNION, A.M. (1992) Acidification and eutrophication. In Mannion, A.M., Bowley, S.R. (eds), *Environmental Issues in the 1990s*, pp. 177–197. John Wiley, Chichester.

MANSFIELD, T.A. (1976) The role of stomata in determining the responses of plants to air pollutants. In Smith, H. (ed.) *Commentaries in Plant Science*, pp. 13–22. Pergamon Press, Oxford.

MASON, C.F. (1991) *Biology of Freshwater Pollution*, 2nd edn. Longman, Harlow.

MASON, J. (1989) The cause and consequences of surface water acidification. In Morris, R., Taylor, E.W., Brown, D.J.A., Brown, J.A. (eds) *Acid Toxicity and Aquatic Animals*, pp. 1–12. Cambridge University Press, Cambridge.

McCORMICK, J. (1989) *Acid Earth: The Global Threat of Acid Pollution*. Earthscan, London.

MUELLER-DOMBOIS, D. (1988) Forest decline and dieback—a global ecological problem. *Trends Ecol. Evol.*, **3(11)**, 310–312.

RICKLEFS, R.E. (1990) *Ecology*. W.H. Freeman, New York.

ROBERTS, T.M. (1984) Effects of air pollution in agriculture and forestry. *Atmos. Environ.*, **18**, 629–652.

TAMM, C.O. and HALLBACKEN, I. (1988) Changes in soil acidity in two forest areas with different acid deposition: 1920 to 1980s. *Ambio*, **17**, 56–61.

VAN DAM, D., VAN DOBBEN, H.F., TER BRUUK, D.F.J. and DeWIT, T. (1986) Air pollution as a possible cause for the decline of some phanerogamic species in The Netherlands. *Vegetatio*, **65**, 47–52.

WELLBURN, A. (1988) *Air Pollution and Acid Rain: Biological Impact*. Longman, Harlow.

Further reading

FREEDMAN, B. (1989) *Environmental Ecology. The Impacts of Pollution and Other Stresses on Ecosystem Structure and Function*. Academic Press, London.

HENRIKSEN, A. (1989) Air pollution effects on aquatic ecosystems and their restoration. In Ravera, O. (ed.) *Ecological Assessment of Environmental Degradation, Pollution and Recovery*. Elsevier, Amsterdam.

HINRICHSEN, D. (1986) Multiple pollutants and forest decline. *Ambio*, **15(5)**, 258–265.

MANNION, A.M. and BOWLEY, S.R. (1992) *Environmental Issues in the 1990s*. John Wiley, Chichester.

McCORMICK, J. (1989) *Acid Earth: The Global Threat of Acid Pollution*. Earthscan, London.

WELLBURN, A. (1988) *Air Pollution and Acid Rain: Biological Impact*. Longman, Harlow.

Chapter 12

DESULPHURIZATION OF COAL AND OILS

Introduction

The formation of acid rain, particularly from sulphur dioxide emissions and the consequent impact on the natural and built environment, e.g. material corrosion, is well known (Ch. 11). A key source of gaseous sulphur emissions is the burning of high-sulphur fossil fuels, often used in the generation of electricity. In addition to the undesirable environmental consequences, there are certain unwelcome technological effects. For example, sulphur compounds reduce the octane number of fuels and lower the effectiveness of antiknock additives in petrol.

At present many industrialized countries are developing strategies and enforcing regulations to limit gaseous sulphur emissions. Inevitably, tighter restrictions on sulphur release requires industry to evaluate and apply methods to reduce sulphur emissions. This may be achieved in several ways, either by reducing the sulphur content in fossil fuels before or during burning, or by treatment of the gaseous waste. There are a number of different conventional treatment technologies aimed at removing sulphur from coal (Maloney and Moses, 1991) and oil (Bhadra *et al.*, 1987) prior to, or during, burning. These are, however, expensive and in the case of heavy oils and bitumen are not cost-effective. A more efficient and less expensive procedure is flue gas desulphurization (FGD). This has been the chosen route of desulphurization for many power plants but still requires high capital investment. The United Kingdom, for example, is investing £2 billion to fit FGD systems to existing power stations (capacity 12 000 megawatts). The cost, however, makes FGD an impractical option for smaller power plants (Maloney and Moses, 1991).

There remains one relatively new technology which may enable efficient and economic removal of sulphur compounds from coal and oils prior to burning, that is microbial desulphurization. The aim of this chapter is to examine the biochemistry and physiology of microbial desulphurization processes and to describe the basic technologies of systems applied to oil or coal microbial desulphurization. It is important to note, however, that the majority of research has been carried out on the microbial desulphurization of coal, by far the largest hydrocarbon reserve material, and inevitably consideration of coal desulphurization will dominate this chapter.

Although gaseous emissions of nitrogen oxides contribute to acidification, sulphur dioxide has a far greater environmental impact (Ch. 11). Some attempt, however, has been made to develop biotechnological procedures for the removal of nitrogen oxides from waste gases. These are in their infancy and appear problematic. For these reasons biological removal of nitrogen oxides from gaseous wastes will not be considered in this chapter.

Composition and structure of coal

Coal is a heterogeneous material consisting of mineral inclusions together with a diverse mixture of carbon compounds, which varies with the type of coal (Klein *et al.*, 1991). The elemental composition and the macromolecular structure of coal is related to its rank (Klein *et al.*, 1991). In general the carbon content increases with rank while water and oxygen content decrease. At the macromolecular level coal consists of aromatic compounds linked by aliphatic or ether bridges. As the rank of coal increases so the content of aromatics within the matrix rises, as does the extent of cross-linkage and the size of matrix structural units. The number of aliphatic and ether linkages, however, decline (Klein *et al.*, 1991).

Sulphur exists in coal in two major forms, organic and inorganic. Organic sulphur compounds are regarded as a structural element in coal and their interactions with other compounds in the matrix are complicated (Boudoui *et al.*, 1987). The dominant heterocyclic sulphur compounds are the thiophenes, e.g. dibenzothiophene (DBT). Other organic sulphur compounds include sulphides, disulphides and thiols (Fig. 12.1). Inorganic sulphur is primarily in the form of iron sulphides, the commonest type being pyrite (FeS_2). The amount of pyrite varies with coal type. For example, bituminous coal, a relatively low rank coal, contains up to 6% pyritic sulphur while lignite contains substantially less. Unlike organic sulphur, iron sulphides are distributed throughout coal as distinct nodules rather than as a component of the matrix. The size of these nodules and their physical distribution depends on the coal type. Inorganic sulphur is also present within coal as sulphate, a chemical oxidation product of pyrite. Sulphate, normally, accounts for only a small and relatively unimportant fraction of the total sulphur content of coal (Boudoui *et al.*, 1987).

The physical structure of coal is important to consider in view of its impact on the accessibility of sulphur compounds to microorganisms or their enzymes. Coal has a porous structure consisting of pores in

Sulphides R – S – R

Disulphides R – S – S – R

Thiols R – SH

Thiophenes

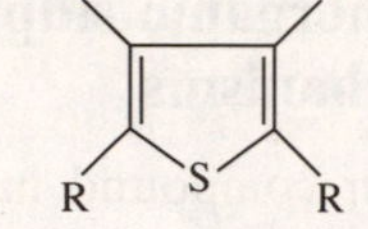

R groups are either hydrogen, alkyl groups or aromatic groups. They may be similar or dissimilar on any one sulphur compound.

Fig. 12.1. Organic sulphur compounds in coal and oils.

four size ranges based on diameters; macropores (50 nm–5 mm), mesopores (2–50 nm), micropores (0.8–2 nm) which form up to 90% of the void volume and submicropores (<0.8 nm). The porous areas are separated by fissures which have diameters in excess of 5 mm (Klein *et al.*, 1991). There is a clear restriction in the accessibility of sulphur compounds to microorganisms through the pore system within coal. In fact, Klein *et al.* (1991) consider that only the surfaces of coal within the macropores are accessible to microorganisms and perhaps mesopores of >20 nm diameter to their enzymes. In practise this means that microbial desulphurization of coal can only be achieved if the coal is finely ground or solubilized.

Composition and structure of oil

Crude oils are complex mixtures of aliphatic, heterocyclic and aromatic hydrocarbons. The precise make-up of oils is exceptionally variable and depends on the extraction site and type of oil. Heavy oils and bitumen are particularly high in sulphur compounds and are highly viscous. Many oil producing countries have large reserves of these crude oils which are likely to become more important as energy sources when more easily extracted oils become depleted. Oils contain both inorganic and organic sulphur. Inorganic sulphur is present as elemental sulphur, metal sulphides and thiosulphates. The major sulphur fraction of oils is organic and composed of a diverse mixture of thiols, thiophenes and substituted benzo- and dibenzothiophenes (Bhadra *et al.*, 1987) (Fig. 12.1). The viscosity of oils means that microbial desulphurization is limited to the oil–water interface. The formation of small oil droplets, e.g. through emulsion technology, therefore, is likely to be advantageous to the microbial removal of sulphur.

Microbial removal of inorganic sulphur from coal and oil: organisms and mechanisms

The main inorganic sulphur compound in many coals is in the form of pyritic sulphur (FeS_2). Although coarse pyrite crystals can be removed economically using physical cleaning procedures, pyrite finely disseminated throughout the coal is less easily removed. Since pyritic sulphur is so common in coals and can present problems for conventional removal technologies, considerable research has been undertaken into microbiological techniques for its removal. These studies, however, have not as yet resulted in any industrial processes being developed.

The removal of inorganic sulphur from coal can be achieved through the microbial oxidation of the sulphur compounds. Certain bacterial metabolic types have potential for removal of inorganic sulphur including obligate chemolithotrophs, e.g. *Thiobacillus ferrooxidans*, and facultative chemolithotrophs, e.g. *Sulfolobus brierleyi* (Kargi, 1986). In addition, heterotrophic microorganisms may contribute to the removal of inorganic sulphur (Rai and Reyniers, 1988).

Mechanisms of inorganic sulphur oxidation by *Thiobacillus*

Thiobacillus species are known to oxidize sulphur during metal (Cu, Zn, U and Ni) leaching from low grade sulphide ores (Ch. 14) under aerobic conditions. Several *Thiobacillus* species, e.g. *T. ferrooxidans* and *T. organoparus*, have been studied for coal desulphurization. These bacteria are aerobic, acidophilic (optimum pH 2–3), mesophilic (optimum temperature 25°C–30°C) chemolithotrophs. Several strains of *T. ferrooxidans*, often isolated from acid mine drainage, in particular have been studied (Kargi, 1988). *Thiobacillus* oxidation of pyritic sulphur occurs through two mechanisms (Kargi, 1982; Hughes and Poole, 1989): (i) direct bacterial oxidation of the sulphur as part of bacterial metabolic processes and (ii) indirect oxidation in which the sulphur is oxidized by acidic solutions of Fe^{3+}, an end product of bacterial metabolism.

The oxidation of pyrite by Fe^{3+} does not involve direct bacterial action and results in the production of elemental sulphur (Eq. 12.1).

$$\underset{}{FeS_2} + \underset{[Fe^{3+}]}{Fe_2(SO_4)_3} \rightarrow \underset{[Fe^{2+}]}{3FeSO_4} + 2S \qquad (12.1)$$

Inorganic sulphur is oxidized bacterially with the production of sulphuric acid (Eq. 12.2).

$$S + 11/2O_2 + H_2O \rightarrow H_2SO_4 \qquad (12.3)$$

This oxidation has the effect of solubilizing the inorganic sulphur and maintains a low pH encouraging the growth of the acidophiles. The Fe^{2+} produced during the reaction described in Equation (12.1) is reoxidized by *Thiobacillus ferrooxidans* forming Fe^{3+} (Eq. 12.3) which contributes to further pyrite oxidation (Eq. 12.1).

$$\underset{[Fe^{2+}]}{2FeSO_4} + 1/2O_2 + H_2SO_4 \rightarrow \underset{[Fe^{3+}]}{Fe_2(SO_4)_3} + H_2O \qquad (12.2)$$

Thiobacillus directly oxidizes pyritic sulphur to ferrous sulphate and sulphuric acid (Eq. 12.4.).

$$FeS_2 + 31/2O_2 + H_2O \rightarrow FeSO_4 + 2H_2SO_4 \qquad (12.4)$$

The ferrous sulphate is reoxidized to ferric sulphate as shown in Equation (12.3).

The overall reaction including both direct and indirect oxidation of pyrite is summarized in Equation (12.5).

$$2FeS_2 + 71/2O_2 + H_2O \rightarrow Fe_2(SO_4)_3 + H_2SO_4 \qquad (12.5)$$

During the sequence of reactions Fe^{2+} and Fe^{3+} are cyclically interconverted. Although these processes occur chemically in the absence of bacteria, there is an increase in the rate of oxidation by a factor of 10^6 in their presence (Hughes and Poole, 1989).

Direct oxidation of inorganic sulphur by *Thiobacillus* species is a membrane-bound reaction requiring direct contact between the substrate and the bacterium. Attachment of *Thiobacillus* to coal particles is therefore exceptionally important (see below). Several different schemes for the cell wall bound interactions have been proposed. The iron is oxidized (Eq. 12.3) at the outer edge of the cell wall and a variety of primary electron acceptors and electron transport systems have been suggested to be involved in the transport of electrons across the cell envelope (Hughes and Poole, 1989; Norris, 1989) for energy production.

The oxidation of inorganic sulphur compounds and production of soluble sulphates in oils is less well understood or studied than in coals. It is thought, however, that the overall mechanism of oxidation is similar to that of pyritic sulphur (Bhadra *et al.*, 1987).

Chemolithotrophic bacteria and inorganic sulphur oxidation

Thiobacillus species efficiently remove sulphur from coal providing particle size is small enough and direct surface contact is ensured (90% pyritic sulphur removed at particle size $<80\ \mu m$) (Kargi, 1986). These organisms are mesophilic, however, and the rate of removal of the sulphur is slow, e.g. 4–5 day residence times in continuous operations (Kargi, 1986). Moderate and extreme thermophiles capable of oxidizing Fe^{2+} and reduced sulphur have also been isolated.

The moderate iron-oxidizing thermophiles show optimum activity between 45 and 50°C although some are active at 30°C. They are similar to strains of *Thiobacillus* (Marsh and Norris, 1983). These organisms, however, are capable of chemolithoheterotrophic growth, e.g. using yeast extract as a substrate, which is linked to increased rates of sulphur oxidation (Marsh and Norris, 1983). The advantages of increased oxidation rates are likely to be counteracted, however, by the increased costs of supplying a complex organic substrate. Since these isolates are active over such a broad temperature range it may not be necessary to control the temperature of bioreactors in which they are catalysing desulphurization, resulting in substantial cost savings.

Extreme thermophiles such as *Sulfolobus* species show optimum growth at 65–80°C and can catalyse desulphurization. These bacteria, commonly isolated from acid hot springs, have a variety of metabolic capabilities. Some are able to metabolize both chemolithotrophically and heterotrophically while others are obligate chemolithotrophs. All are able to oxidize reduced sulphur, but only some can oxidize ferrous iron (Brierley and Brierley, 1986). There is a large variation in their ability to oxidize pyrite with *Sulfolobus brierleyi* apparently the most efficient (Marsh *et al.,* 1983). As with the moderately thermophilic iron-oxidizing bacteria, the addition of yeast extract enhances the rate of oxidation of reduced sulphur and iron by this species. *Sulfolobus acidocaldarius* (optimum growth conditions 60–90°C, pH 1.5–4) has also been used for pyritic sulphur removal from coal (Kargi and Robinson, 1985).

The ability of *Sulfolobus* species to oxidize inorganic substrates at high temperatures may result in several advantages. The rate of reduced sulphur oxidation may be more rapid than for mesophilic bacteria, particularly as non-bacterially mediated oxidation increases substantially at these higher temperatures (Marsh *et al.,* 1983). In addition, the higher temperatures may limit the risk of reactor contamination by other organisms. The sulphur-oxidizing bacteria with chemolithoheterotrophic metabolism may be of particular use in desulphurization since it is possible they could remove not only inorganic sulphur but also organic sulphur from coal and oil. There is some evidence that this is the case, for example, *S. acidocaldarius* can oxidize DBT (Kargi and Robinson, 1984). It is interesting that growth of *Sulfolobus* species can be reduced by compounds leached from coal during desulphurization potentially lowering desulphurization rates (Olsson, *et al.,* 1989).

Crucial to the process of pyritic sulphur oxidation by chemolithotrophic metabolic activity is the attachment of the bacteria to pyrite surfaces. Since oxidation is a cell wall based event and extracellular enzymes are not produced, attachment is essential. A variety of factors affect bacterial attachment to surfaces (Marshall, 1985) including bacterial and solid surface characteristics, nutrient conditions, environmental factors such as pH and temperature, and hydrodynamic factors (Rai and Reyniers, 1988). It is clear that there is a degree of selection in the attachment of chemolithotrophs to coal. For example, *S. acidocaldarius* (Kargi, 1986) and *T. ferrooxidans* (Wainwright, 1988) appear to selectively attach to pyrite surfaces in coal. There has been a mathematical model produced to describe the attachment of *T. ferrooxidans* to coal (Wainwright, 1988). In general, however, little research has been undertaken on the mechanism of chemolithotrophs' attachment to coal surfaces. An understanding of the process would enable optimization of attachment which should result in an improvement in desulphurization.

Heterotrophic microorganisms and inorganic sulphur oxidation

There is some evidence that heterotrophic microorganisms are able to oxidize inorganic sulphur. Species from several different genera of filamentous fungi and yeasts are able to oxidize sulphur (Wainwright and Grayston, 1989). *Aspergillus niger, Fusarium solani* and *Trichoderma* species are among the more active (Faison *et al.*, 1991). Their rates of sulphur oxidation are far lower than those for chemolithotrophic bacteria. Oxidation is thought to be via the polythionate pathway

$$S_0 \rightarrow S_2O_3^{2-} \rightarrow S_4O_6^{2-} \rightarrow SO_4^{2-}$$

It is unclear, however, whether the oxyanions are intermediates or by-products (Marshall, 1985; Bagdigian and Myerson, 1986). There are two potential benefits to fungi arising from sulphur oxidation. Firstly, the fungi may gain energy from chemolithotrophic growth on the reduced sulphur. Secondly, polythionates produced by reduced sulphur oxidation may complex heavy metals reducing their toxic effects.

The oxidation of metal sulphides including cadmium, copper, zinc and lead sulphides, has also been demonstrated for the filamentous fungi *A. niger* and *Trichoderma harzianum* (Bagdigian and Myerson, 1986). The end products of metal sulphide oxidation and the effect of the presence of elemental sulphur varied with metal sulphide type and fungal species (Table 12.1) (Bagdigian and Myerson, 1986). It appears that as with the chemolithotrophic bacteria direct contact between the fungal surface and the metal sulphide is necessary for oxidation to proceed. For example, *T. harzianum* neither adsorbs onto the cell wall nor oxidizes CdS, ZnS or PbS, whereas CuS was both adsorbed and oxidized (Bagdigian and Myerson, 1986). No investigation of FeS_2 oxidation by fungi appears to have been undertaken. The reported rates of fungal sulphur and metal sulphide oxidation are very low and could not compete with chemolithotrophic bacterial oxidation (Marshall, 1985; Bagdigian and Myerson, 1986). It seems unlikely

Table 12.1 Metal sulphide oxidation by *Aspergillus niger*

Metal sulphide	*Elemental sulphur*	$S_2O_3^{2-}$ [(a)]	$S_4O_6^{2-}$ [(a)]	SO_4^{2-} [(a)]	*Thiols* [(b)]
CuS	Present	1.6 (±2.7)	294.4 (±53.0)	601.0(±45.7)	−1.6(±1.0)
CuS	Absent	ND	ND	191.2(±30.3)	8.3(±2.0)
PbS	Present	22.4(±4.1)	ND	211.0(±67.2)	5.9(±0.0)
PbS	Absent	28.1(±10.2)	ND	248.5(±78.1)	4.1(±0.0)

(a) Measured as μg S ml^{-1}.
(b) Measured as μmol ml^{-1}.
ND, not detected.
Figures in parentheses are the standard deviations.

that fungi will be used in pure culture for desulphurization of coals or oils. It is possible, however, that they may contribute in mixed culture processes. Certainly fungi are involved in desulphurization of organic sulphur compounds and have been found to solubilize coal through the oxidation of aromatic compounds to form soluble polar compounds (Cole, 1979; Kargi, 1986).

Heterotrophic bacteria also solubilize metal sulphides (Rai and Reyniers, 1988). Indeed, recently it has been found that heterotrophic bacteria may not only be involved in the removal of organic sulphur (see below) but may also be involved in the removal of inorganic sulphur from oil (Kohler *et al.*, 1984) and coal. *Pseudomonas aeruginosa* and *Pseudomonas putida* appeared to be able to oxidize pyritic sulphur. *P.putida* in particular showed considerable efficiency, removing 69–76% of the inorganic sulphur in Illinois #6 and lignite coal slurries (particle size 147–1397 μm) over a period of 5–7 days. This was the case even though bacterial growth was repressed by the presence of coal and sulphur sources. The end product of the oxidation was sulphate but the mechanism of oxidation has not been determined (Rai and Reyniers, 1988).

General considerations on microbial inorganic sulphur removal

It is thought that assemblages of bacteria, fungi, algae and even protozoa may be involved in metal sulphide leaching in natural environments. The use of mixed cultures of organisms in desulphurization processes may be worthy of consideration. It has, indeed, been found that mixed cultures of 'co-operating' chemolithotrophic microorganisms can be used in inorganic sulphur removal (Table 12.2) (Kargi, 1986).

Removal of inorganic sulphur by bacterial oxidation not only results in the solubilization of sulphur compounds, but also results in the solubilization of metal compounds through the formation of soluble metal sulphates, e.g. $FeSO_4$. Thus, inorganic desulphurization may lead to the removal of metals from oil or coal. Given the growing concerns about metal toxicity (Ch. 13) this suggests an additional advantage in bacterial desulphurization processes.

Table 12.2 Inorganic sulphur removal from coal by mixed cultures of chemolithotrophic bacteria

Mixed culture	*Chemolithotrophic bacteria*	*Role in desulphurization*
A	*Thiobacillus ferrooxidans*	$FeS_2 \rightarrow S^0$
	Thiobacillus thiooxidans	$S^0 \rightarrow SO_4^{2-}$
B	*Heptospirillum ferrooxidans*	$Fe^{2+} \rightarrow Fe^{3+}$
		Fe^{3+} $FeS_2 \rightarrow SO^0$
	Thiobacillus thiooxidans	$S^0 \rightarrow SO_4^{2-}$

Microbial removal of organic sulphur from coal and oil

The removal of organic sulphur from coal and oil has commonly been studied using model organic substrates, particularly DBT. A variety of organisms in mixed and pure cultures have the ability to remove organic sulphur, including heterotrophic bacteria such as *Pseudomonas* species, *Arthrobacter* sp., *Beijerinckia* sp., *Rhizobium* sp. and *Acinetobacter* sp. (Bhadra *et al.*, 1987), heterotrophic fungi, e.g. *Paecilomyces* sp. (Cole, 1979) and faculative chemolithotrophs such as *Sulfolobus* (Bhadra *et al.*,

1987). All the organisms remove sulphur aerobically, however, there is evidence that desulphurization may also occur under anaerobic conditions catalysed by *Desulfovibrio* sp. (Holland *et al.*, 1986).

The mechanisms of organic sulphur removal from oil and coal have been largely described in terms of model substrate breakdown and have not often been verified for the natural material. Organic sulphur is most frequently removed through oxidative processes. The desulphurization process may be through the production of exoenzymes or may be a surface phenomenon with substrate transport to the cell membrane.

Aerobic organic sulphur removal

Several fungi, e.g. *Aspergillus niger,* oxidize thioesters and organic sulphides to their corresponding sulphones and sulphoxides (Laborde and Gibson, 1977). End products for the oxidation of the thiophenes or cyclic thioesters are also sulphoxides and sulphones. Such oxidations can be undertaken by several different fungi (Laborde and Gibson, 1977) and by a variety of bacterial species, e.g. *Pseudomonas* and *Bacillus* (Bhadra *et al.*, 1987). Since dibenzothiophene (DBT) is the dominant heterocyclic sulphur compound in coal and oil and has been used most extensively as a model substrate, the route of desulphurization of this compound will be detailed. The oxidation process suggested for *Pseudomonas* sp. and *Beijerinckia* sp. (Hou and Laskin, 1976; van Afferden *et al.*, 1990) appears to be broadly analogous to the naphthalene oxidative pathway of *Pseudomonas* sp. and is based on the aromatic ring structure being attacked (Bhadra *et al.*, 1987). The organic end product of desulphurization varies with species (Fig. 12.2). More recent research on DBT oxidation by *Brevibacterium* sp. DO (Baldi *et al.*, 1992) has indicated that desulphurization results in the release of sulphite in stoichiometrical quantities. The sulphite is subsequently chemically oxidized to sulphate. Benzoate is the sulphur-free organic end product of the series of sequential oxidations (Fig. 12.3). The mechanism of cleavage of the C–S bond remains unclear. Interestingly, *Brevibacterium* sp. DO was able to co-metabolize benzo[*b*]naphtho[2,1-*d*]thiophene (BNT) with sulphate and sulphur-free 2-naphthoic acid as end products (Klein *et al.*, 1991). It appears that *Sulfolobus* is able to oxidize the sulphur moiety while leaving the aromatic ring structures intact (Bhadra *et al.*, 1987).

Anaerobic organic sulphur removal

Anaerobic desulphurization by *Desulfovibrio* sp. apparently involves several different mechanisms in the reductive splitting of the S–C bond.

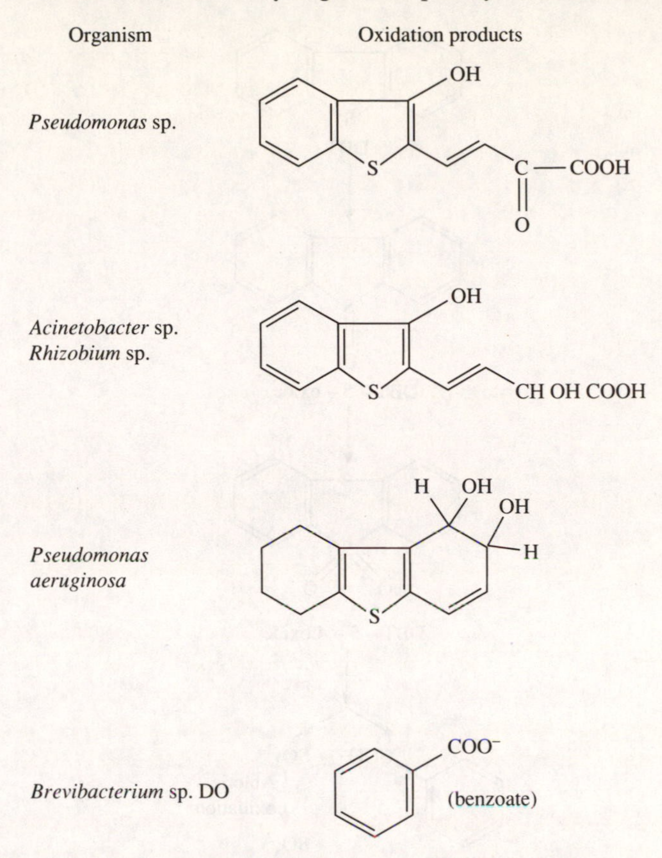

Fig. 12.2. End products of dibenzothiophene (DBT) oxidation.

Desulfovibrio sp. is able to remove sulphur from several sulphur compounds including dibenzyl sulphide and DBT utilizing hydrogen during the process. Hydrogenase activity is, therefore, important in the desulphurization process. The sulphur is removed as H_2S and a variety of organic end products are produced, e.g. reductive desulphurization of dibenzyl sulphide results in the formation of toluene and benzylmercaptan (Holland *et al.*, 1986).

General considerations on microbial organic sulphur removal

Microbial desulphurization of these organic sulphur compounds may be growth associated. This is acceptable, for example, in the case of

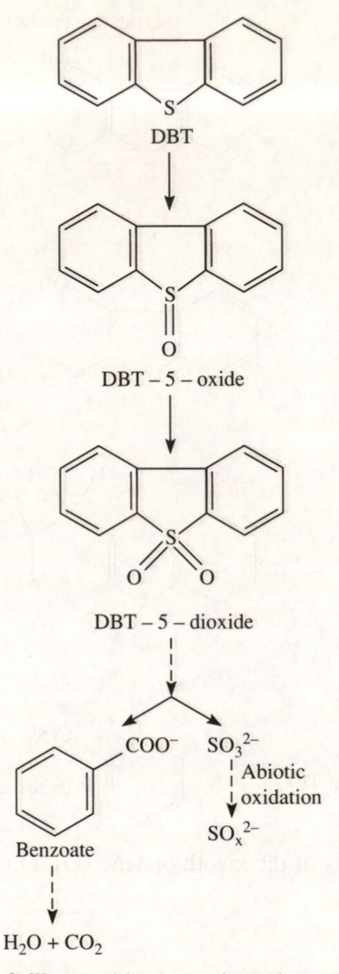

Fig. 12.3. Oxidation of dibenzothiophene (DBT) by *Brevibacterium* sp. I.

Brevibacterium sp. DO which utilizes the sulphur moiety as its sole source of sulphur. In the case of some bacterial species (Bhadra *et al.*, 1987) the organic component itself appears to be utilized as a carbon or energy source or both. This is unlikely to be an acceptable route for desulphurization since the overall calorific value of the coal or oil would be reduced. It is essential that the desulphurization should be targeted on the sulphur moiety.

Given the great variety of organic sulphur compounds in coal and oil it may be advantageous to have a variety of biological desulphurization processes in operation simultaneously perhaps involving mixed cultures or co-metabolism of substrates. Some heterotrophic microorganisms are capable of solubilizing coal (Cole, 1979; Klein *et al.*, 1991) resulting in greater accessibility of both inorganic and organic sulphur compounds. Some organisms may be capable both of solubilization and desulphurization. If so, it is still crucial to ensure that there is no major loss of coal calorific, i.e. energy, value.

Factors affecting the removal of inorganic sulphur from coal

Coal type, pulp density and particle size

The biological removal of pyritic sulphur from coal is affected by several factors. There is clear evidence that the efficiency of removal varies with coal type due to differences in content and distribution of pyrite (Klein *et al.*, 1991) (see above). The nature of the pyrite itself affects the rate of bioleaching. The more chemically reactive pyrites show the fastest bioleaching rates by *T. ferrooxidans*. For example, Queensland pyrite and Pittsburgh-B pyrite show leaching rates of 7.04 and 29.0% loss of Fe dy^{-1} at a pulp density of 0.4% (wt/vol.). The Pittsburgh coal is considered to contain pyrite which is inherently the more active (Bhattacharyya *et al.*, 1990). Although the removal of Fe has been clearly shown in experiments, there does not seem to be any significant release of sulphate, suggesting that sulphur is not metabolized.

Many workers have found that the coal pulp density and the pyrite particle size has a major influence on the rate of sulphur removal. An increase in pulp density from 0.4% to 2.0% (wt/vol.) (particle size 75–150 μm) in Pittsburgh-B pyrite results in a reduction in the rate of bioleaching by *T. ferrooxidans* by approximately a factor of 10. There is also an increase in the lag period before bioleaching occurred from virtually zero to approximately 30 days (batch culture experiments) (Bhattacharyya *et al.*, 1990). These effects may partly be due to increased particle grinding at high pulp densities and oxygen diffusional problems. An increase in particle size decreases the rate of sulphur removal (Andrews *et al.*, 1988; Rai and Reyniers, 1988), e.g. the initial rate of sulphur removal by *S. brierleyi* from 60 and 200 mesh coal was 3.4 and 1.4 mg g^{-1} dy^{-1} respectively (Andrews *et al.*, 1988). This is explicable simply in terms of the requirement for surface attachment (see above); as particle size increases so the surface area available for attachment declines, lowering rates of sulphur removal.

Environmental and nutrient conditions

Other factors affecting the removal of pyritic sulphur include the bacterial type, temperature, pH, availability of CO_2 and O_2, and nutrient conditions. The influence of bacterial type on pyritic sulphur removal has been discussed above and is illustrated in Table 12.3. The effect of temperature and pH is, of course, associated with the physiological type, i.e. thermophile, mesophile and acidophile. Each species will, however, have characteristic optimum temperature and pH for removal of pyritic sulphur. For example, the rates of removal by *S. brierleyi* were proportional to temperature in the range 60–70°C (Andrews *et al.*, 1988).

Table 12.3 Comparison of pyritic sulphur removal by different physiological types of bacteria

Physiological type	*Organism*	*Temperature (°C)*	*pH*	*Residence time (day)*	*Efficiency (% removal)*
Thermophilic facultative chemolithotroph	*Sulfolobus acidocaldarius*	70	2.5	4–6	90
Mesophilic chemolithotroph	*Thiobacillus ferrooxidans*	30	2.4	16	90
Heterotroph	*Pseudomonas putida*	30	7	5–7	67–77

Note: Particle size and pulp density differed in each of the above experiments but were generally considered optimal. The coal type was different in each case.

The presence of organic carbon sources increases the rate of inorganic sulphur removal for those chemolithotrophs capable of heterotrophic growth, i.e. some *Sulfolobus* species and the thermophilic sulphur-oxidizing bacteria (see above). In addition, the availability of CO_2 seems to have a significant impact on desulphurization. The actual effect of CO_2 is somewhat complicated and varies between *Thiobacillus* and *Sulfolobus* species. In thick slurries inorganic sulphur removal by *Thiobacillus* can be lowered by CO_2 limitation. Air sparging of batch culture of 35% (wt/vol.) coal slurries (1% pyrite content) increased sulphur removal by *Thiobacillus* species by as much as 40–50% (Hartdegen *et al.*, 1984). Interestingly, the growth of heterotrophic bacteria in the coal slurries increases desulphurization by *Thiobacillus*. Two reasons are proposed for this. First, organics present in coal and inhibitory to *Thiobacillus* growth are metabolized and removed by the heterotrophs. Second, heterotrophic activity increases the CO_2 content of the slurries (Andrews *et al.*, 1988). Carbon dioxide acts as the carbon source for the chemolithotrophic *Thiobacillus* species and is therefore required for growth. The effect of CO_2 concentration on desulphurization by *Sulfolobus* species is a little more complicated.The rate of removal of sulphur from coal by *S. brierleyi* can be enhanced by slightly increasing the CO_2 content in an air stream. Raising the CO_2 concentration from 7 to 18%, however, lowers the rate of desulphurization from 3.4 to 2.3 mg S g^{-1} coal dy^{-1} (Bhattacharyya *et al.*, 1990). *S. brierleyi* is capable of removing both inorganic sulphur and organic sulphur (up to 15–20%) from the coal. It is possible that increases in CO_2 concentration encouraged chemolithotrophic growth at the expense of heterotrophic removal of organic sulphur.

Factors affecting the removal of organic sulphur from coal and oil

Available reaction surface

The removal or organic sulphur from coal and oil is also influenced by a variety of factors. A key component is the available substrate surface area for exoenzyme attack, or for oils the dissolution and transport of sulphur organic substrates from the oil phase to the cell membrane which may be a rate limiting step (Sagardia *et al.*, 1975). The production of fine coal slurries, therefore, not only aids inorganic sulphur removal, but also organic sulphur removal. In the case of oils the production of emulsions is likely to enhance the rate of organic sulphur removal. Some desulphurizing heterotrophic bacteria are able to produce emulsions, e.g. *P. aeruginosa* PRG1 (Finnerty *et al.*, 1983). The formation of emulsions appears to protect the bacteria against the inhibitory effects of some oil organic compounds (Bhadra *et al.*, 1987). The oil/water ratio has been shown to affect reaction rates probably due to surface effects, e.g. *P. alcaligenes* shows reduced rates of DBT oxidation when the oil weight fraction exceeds 10% (Sagardia *et al.*, 1975).

Environmental and nutrient conditions

The efficiency and rate of bacterial desulphurization of organic sulphur compounds vary with the sulphur substrate and the organism (Table 12.4). Desulphurization of oil usually results in the formation of water-soluble products. Some heterotrophic bacteria utilize the organic sulphur compounds as their sole source of carbon or energy or both (Table 12.4). As

Table 12.4 Oxidation of organic sulphur by selected bacterial species

Organism	*Substrate*	*Intial rate of oxidation*	*Residence time (days)*	*Efficiency (% removed)*	*DBT utilization*
Pseudomonas putida	Coal (74–295μm)	ND	5–7	37.4	ND
Pseudomonas aeruginosa	2% (vol./vol.) DBT in 5% (vol./vol.) oil phase	ND	5	42	ND
Non-growing					
Pseudomonas alcaligenes (DBT 2)		325μ mole h^{-1}			
Pseudomonas stutzeri (DBT 3)	3.5% (vol./vol.) DBT in crude oil	675μ mole h^{-1}	45 hrs	100	Not utilized
Pseudomonas putida (DBT 4)		360μ mole h^{-1}			
Sulfolobus acidocaldarius	Solid DBT in water $\simeq$ 100mg DBT ml^{-1}	4 mg (l^{-1}d^{-1})	ND	80	Carbon source

ND, not determined.
All organisms are heterotrophs except *Sulfolobus acidocaldarius*, a thermophilic facultative chemolithotroph.

discussed above such strains are largely unacceptable for desulphurization processes. A number of microbial species are unable to use organic sulphur compounds in this way and have even been reported to remove sulphur in a non-growing state, acting as biocatalysts, e.g. DBT desulphurization by *P. alcaligenes* DBT2, *P. stutzer*i DBT3 and *P. putida* DBT4 (Nakatani *et al.*, 1968). There is evidence to suggest that DBT removal is linked to possession of a plasmid which encodes enzymes for the degradation of non-sulphur organics, e.g. naphthalene (Finnerty *et al.*, 1983; Bhadra *et al.*, 1987). This may be problematic since compounds other than sulphur organics may be removed from the oil or coal, lowering the energy value of the fuel. The aim of achieving high substrate specificity together with rapid rates of sulphur removal has been addressed not only in the selection of strains from the natural environment but, also, through genetic engineering. *P. alcaligenes* DM220 was genetically modified to increase substrate specificity for DBT desulphurization (Sagardia *et al.*, 1975).

It is implicit that environmental conditions such as temperature, pH and O_2 availability will affect the reaction rate of an aerobic process catalysed by, for example, mesophilic and non-acidophilic, non-alkalophilic heterotrophic bacteria. Control of environmental conditions is likely to be needed for efficient organic sulphur removal, increasing costs. Of particular interest is the necessity for additional growth requirements such as a nitrogen source. Some desulphurizing bacteria require the presence of a complex nitrogen source (rather than a simple salt) such as meat extract before desulphurization will occur (Kodama, 1977). Other workers have found that the presence of a co-substrate, e.g glutamate, is essential for the oxidation of model substrates such as DBT (Beyer *et al.*, 1990). Such additions would substantially raise the costs of any desulphurization process.

Desulphurization process technology

Investigation into the microbial desulphurization of coal and oils has progressed little beyond laboratory scale and is not being exploited on a commercial scale. In particular, the desulphurization of oil has not developed sufficiently for a major consideration of process technology which is likely to be reactor based, e.g. stirred tank reactors, airlift fermentors, etc. It is intended in this section, therefore, to concentrate on and briefly describe the proposed technologies for coal desulphurization.

The influence of process parameters on desulphurization of coal is described by Beyer *et al.* (1990) and includes consideration of factors such as bioreactor type, coal quality, pulp density particle size, etc.

(see above). The economics of microbial desulphurization are also considered in this article (Beyer *et al.*, 1990).

A variety of reactor configurations have been proposed for coal desulphurization ranging from the technologically relatively simple to the more complex. Ideally, microbial desulphurization of coal should be integrated at the mine or conventional coal preparation plants. This may be achieved in several ways. At the mine site it may be possible to desulphurize coal by heap leaching, a technology commonly applied to metal leaching. This technique is simple and cheap. It has limitations, however, since there is little control over environmental conditions, making it impossible to maintain optimum conditions for desulphurization. Conditions may become inhibitory fairly rapidly with rises in temperature and oxygen limitation (Rai and Reyniers, 1988). It is likely to be a slow process for these reasons, with residence times measured in years. Some improvement may be gained by grinding the coal prior to heap desulphurization. The use of shallow lagoons (depth 0.2 m) which can be maintained in controlled environment chambers, aerated and surface agitated has also been suggested as an appropriate technology (Kargi, 1986). This has the advantage of being relatively cheap while allowing the maintenance of optimum desulphurization conditions. Again, ideally ground coal should be used.

Alternative process technologies require a degree of coal preparation and will be more appropriate at coal preparation plants or even at power stations themselves. It is essential that the coal is ground and prepared as slurries before desulphurization. Slurries can then be fed into several different reactor configurations. Airlift or mechanically agitated and aerated reactors may be used for the treatment of coal slurries. These configurations permit control over environmental conditions, in particular the aeration of the system. Airlift reactors are economically most feasible, since mechanical agitation requires considerable energy and may present problems from coal grinding during desulphurization. Pressurized air is used in airlift fermentors which is cheaper and avoids coal grinding (Kargi, 1986). The use of a slurry pipe-line as a plug-flow reactor, i.e. a tubular reactor with a continuous flow of slurry without back mixing, under aerobic conditions has also been suggested as an appropriate technology (Rai and Reyniers, 1988).

A full example of a proposed reactor scheme for the removal of both pyritic sulphur and organic sulphur by *S. acidocaldarius* is outlined by Kargi (1986). The process is two stage, and both reactors are maintained at 70°C and pH 2.5. A ground coal (particle size 50 μm) water slurry is prepared to a pulp density of 20% (wt/vol.) with the addition of mineral salts, e.g. $(NH_4)_2SO_4$ and $MgSO_4$, in a mixing tank. This is transferred to the first reactor and fed with CO_2 and O_2. A residence time within this reactor of 4–6 days allows for the removal of pyritic sulphur with an estimated removal efficiency of 90%. The solids are then separated from the slurry and the effluent is recycled to the mixing tank after the removal of sulphate and pH adjustment. The solids are transferred to another mixing tank and prepared as before. The added mineral salts, however, are sulphate free. The slurry is fed into the second reactor and *S. acidocaldaius* previously grown on DBT is used as the bacterial inoculum. After a residence time of 4 weeks, 40% of the organic sulphur is removed. The coal is then removed by filtration and the liquid, after removal of the reaction products, is recycled to the second mixing tank.

Coals treated by any of the above process technologies will require washing after separation from the slurry liquid, hence ideally the processes should be linked to existing washeries. Any wastewater arising from the desulphurization process will require treatment for safe disposal (Klein *et al.*, 1991). Clearly the recirculation of treatment water in the desulphurization process limits this necessity for wastewater treatment.

References

ANDREWS, G., DARROCH, M. and HANSSON, T. (1988) Bacterial removal of pyrite from concentrated coal slurries. *Biotechnol. Bioeng.*, **32**, 813–820.

BAGDIGIAN, R.M. and MYERSON, E.R. (1986) The adsorption of *Thiobacillus ferrooxidans* on coal surfaces. *Biotechnol. Bioeng.*, **28**, 467–479.

BALDI, F., CLARK, T., POLLACK, S.S. and OLSON, G.G.J. (1992) Leaching of pyrites of various reactivities by *Thiobacillus ferrooxidans*. *Appl. Environ. Microbiol.*, **58**, 1853–1856.

BEYER, M., KLEIN, J., VAUPEL, K. and WIEGAND, E. (1990) Microbial coal desulphurization: calculation of costs. *Bioproc. Eng.*, **5**, 97–101.

BHADRA, A., SCHARER, J.M. and MOO-YOUNG, M. (1987) Microbial desluphurization of heavy oils and bitumen. *Biotech. Adv.*, **5**, 1–27.

BHATTACHARYYA, D., HSIEH, M., FRANCIS, H, KERMODE, R.I., KHAL, A.M. and ALEEM, H.M.I. (1990) Biological desulfurization of coal by mesophilic and thermophilic microorganisms. *Res. Conserv. Recycling*, **3**, 81–96.

BOUDOUI, J.P., BOULEGUE, J., MALECHAUX, L., NIP, M., De LEEUW, J.W. and BOON, J.J. (1987) Identification of some sulphur species in a high organic sulphur coal. *Fuel*, **66**, 1558–1569.

BRIERLEY, C.L. and BRIERLEY, J.A. (1986) Microbial mining using thermophilic microorganisms. In Brock, T.D. (ed.) *Thermophiles: General, Molecular and Applied Microbiology*, pp. 279–305. Wiley, New York.

COLE, M.A. (1979) Solubilization of heavy metal sulphides by heterotrophic bacteria. *Soil Sci.*, **127**, 313–317.

FAISON, B.D., CLARK, T.M., LEWIS, S.N., MA, C.Y., SHARKEY, D.M. and WOODWARD, C.A. (1991) Degradation of organic sulphur compounds by a coal-solubilizing fungus. *Appl. Biochem. Biotechnol.*, **28/29**, 237–251.

FINNERTY, W.R., SHOCKLEY, K. and ATTAWAY, H. (1983) Microbial desulphurization and denitrification of hydrocarbons. In Zajic, J.E., *et al.* (eds) *Microbial Enhanced Oil Recovery*, pp. 83–91. Penwell, Tulsa.

HARTDEGEN, F.J., COBURN, J.M. and ROBERTS, R.L. (1984) Microbial desulphurization of petroleum. *Chem. Eng. Prog.*, **80**, 63–67.

HOLLAND, H.L.L., KHAN, S.H., RICHARDS, D. and RIEMLAND, E. (1986) Biotransformation of polycyclic aromatic compounds by fungi. *Xenobiotica*, **16**, 733–741.

HOU, C.T. and LASKIN, A.I. (1976) Microbial conversion of DBT. *Develop. Ind. Microbiol.*, **17**, 351–362.

HUGHES, M.N. and POOLE, R.K. (1989) *Metals and Micro-organisms*, pp. 303–358. Chapman and Hall, London.

KARGI, F. (1982) Microbial coal desulphurization. *Enzyme Microbial Technol.*, **4**, 13–19.

KARGI, F. (1986) Microbial methods for desulfurization of coal. *TIBTECH*, Nov., 293–297.

KARGI, F. and ROBINSON, J.M. (1984) Microbial oxidation of dibenzothiophene by the thermophilic organism *Sulfolobus acidocaldarius*. *Biotechnol. Bioeng.*, **26**, 687–690.

KARGI, F. and ROBINSON, J.M. (1985) Biological removal of pyritic sulphur from coal by the thermophilic organism *Sulpholobus acidocaldarius*. *Biotechnol. Bioeng.*, **27**, 41–49.

KLEIN, J., VAN AFFERDEN, M., BEYER, M., PFEIFER, F. and SCHACT, S. (1991) Coal in biotechnology. *Biopapers J.*, **11**, 11–17.

KODAMA, K. (1977) Cometabolism of dibenzothiophene by *Pseudomonas jianii* PS. *Agricult. Biol. Chem.*, **41**, 1305–1306.

KOHLER, M., GENZ, I.L, SCHICHT, B. and ECKART, V. (1984) Microbial desulphurization of petroleum and heavy petroleum fractions. 4.Comm. Anaerobic degradation of organic sulphur compounds of petroleum. *Zbl. Mikrobiol.*, **139**, 239–247.

LABORDE, A.L. and GIBSON, D.T. (1977) Metabolism of dibenzothiophene by a *Beijerinekia* species. *Appl. Environ. Microbiol.*, **34**, 783–790.

MALONEY, S. and MOSES, V. (1991) Microbiological desulphurization of coal. In Moses, V., Cape, R.G. (eds) *Biotechnology. The Science and the Business*, pp. 581–590. Harwood Academic Publishers, London.

MARSH, R.M. and NORRIS, P.R. (1983) The isolation of some thermophilic, autotrophic, iron-and sulphur-oxidizing bacteria. *FEMS Microbiol. Lett.*, **17**, 311–315.

MARSH, R.M., NORRIS, P.R. and LE ROUX, N.W. (1983) Growth and mineral oxidation studies with *Sulfolobus*. In Rossi, G., Torma, A.E.

(eds) *Recent Progress in Biohydrometallurgy*, pp. 71–81. Associazione Mineraria Sarda, Iglesias.

MARSHALL, K. (1985) Mechanisms of bacterial adhesion at solid–water interfaces. In Savage, D.C., Fletcher, M. (eds) *Bacterial Adhesion. Mechanisms and Physiological Significance*, pp. 133–161. Plenum Press, New York.

NAKATANI, S., AKASAKI, T., KODAMA, K., MINODA, Y. and YAMADA, K. (1968) Microbial conversion of petrosulphur compounds. II Culture conditions of dibenzothiophene-utilizing bacterial. *Agricult. Biol. Chem.*, **32**, 1205–1211.

NORRIS, P.R. (1989) Mineral oxidising bacteria: metal–organism interactions. In Poole, R.K., Gadd, G.M. (eds) *Microbe Interactions*, pp. 99–117. IRL Press, Oxford.

OLSSON, G., LARSSON, L., HOLST, O. and KARLSSON, H.T. (1989) Microorganisms for desulphurization of coal: the influence of leaching compounds on their growth. *Fuel*, **68**, 1270–1274.

RAI, C. and REYNIERS, J.P. (1988) Microbial desulphurization of coals by organisms of the genus *Pseudomonas*. *Biotechnol. Prog.*, **4**, 225–230.

SAGARDIA, F., RIGAU, J.J., MARTINEZ LAHOZZ, A., FUENTES, F., LOPEZ, C. and FLORES, W. (1975) Degradation of benzothiophene and related compounds by a soil pseudomonads in an oil–aqueous environment. *Appl. Environ. Microbiol.*, **29**, 722–725.

VAN AFFERDEN, M., SCHACHT, S., KLEIN, J. and TRUPER, H.G. (1990) Degradation of dibenzothiophene by *Brevibacterium* sp. DO. *Arch. Microbiol.*, **153**, 324–328.

WAINWRIGHT, M. (1988) Inorganic sulphur oxidation by fungi. In Boddy, L., Marchant, R., Read, D.J. (eds) *Nitrogen, Phosphorous, and Sulphur Utilization by Fungi*, pp. 71–88. Cambridge University Press, Cambridge.

WAINWRIGHT, M. and GRAYSTON, S.J. (1989) Accumulation and oxidation of metal sulphides by fungi. In Poole, R.K., Gadd, G.M. (eds) *Metal–Microbe Interactions*, pp. 119–130. IRL Press, Oxford.

Further reading

BEYER, M., KLEIN, J., VAUPEL, K. and WIEGAND, E. (1990) Microbial coal desulphurization: calculation of costs. *Bioproc. Eng.*, **5**, 97–107.

BHADRA, A., SCHARER, J.M. and MOO-YOUNG, M. (1987) Microbial desulphurization of heavy oils and bitumen. *Biotechnol. Adv.*, **5**, 1–27.

LAWRENCE, R.W., BRANION, R.M.R. and EBNER, H.G. (1986) *Fundamental and Applied Biohydrometallurgy*. Elsevier, Amsterdam.

MONTICELLO, D.J. and FINNERTY, W.R. (1985) Microbial desulphurization of fossil fuels. *Ann. Rev. Microbiol.*, **39**, 371–389.

US DEPARTMENT OF ENERGY/PITTSBURGH ENERGY TECHNOLOGY CENTER (1992) *Proc. 3rd Symp. Biol. Proc. Coal* 4–7 May. US Department of Energy/Pittsburgh Energy Technology Center, Clearwater Beach, Florida.

METAL AND RADIONUCLIDE POLLUTANTS

Chapter 13

ENVIRONMENTAL FATE AND EFFECTS OF METALS AND RADIONUCLIDES

Definitions

True metallic elements, e.g. Cd, Cu, Pb and Zn, are good conductors of electricity, have a lustrous appearance and tend to enter reactions as positively charged cations. A total of 108 elements have these properties and are considered to be true metals. A further seven are referred to as 'semi-metals' or metalloids, e.g. As, Se and Te. They have the physical properties of metals but chemically behave more like non-metal elements.

The chemical, toxicity and environmental fate of the metallic elements is closely related to their position in the periodic table. The term 'heavy metals' has been variously used to refer to the metallic elements low in the periodic table with high atomic weights (>100), or a relative density greater than five. Some 38 elements have a density greater than 5 g cm^{-3}; many are abundant within the earth's crust and were utilized during the evolution of life. Such elements, e.g. Fe, Mn, Cu, Zn and Mo, are today essential nutrients; others which are normally present in low concentrations, e.g. Ag, Cd, Hg and Pb, are highly toxic. The requirement for, and potential toxicity of, heavy metals results from the fact that they are transitional elements, able to form stable co-ordinated compounds with a range of both organic and inorganic ligands. The actinides, elements with atomic numbers 90–103, e.g. plutonium, are all radioactive (Fergusson, 1990; Morgan and Stumm, 1991).

The usefulness of the term heavy metals has been questioned (Hopkins, 1989). Aluminium, which is frequently described as a heavy metal, is not, having an atomic number of only 13; neither is Se, which is often described as a heavy metal and demonstrates similarities in its environmental behaviour to true heavy metals. Alternative terms, e.g. 'heavy elements', have been used to cover the heavy metals and allied elements (Fergusson, 1990). Despite this, the term heavy metal is well established within the literature and will, where appropriate, be used here. However, it is important to appreciate that there are similarities in the environmental fate and behaviour of all the metallic elements reflecting similarities in their basic chemistry, not least being their persistence. Metal elements and their cations, unlike organic compounds, cannot be degraded.

Environmental chemistry

The environmental fate and toxicity of metals and metalloids is largely governed by their chemistry and speciation, i.e. their physico-chemical form. Metals may occur in the environment as hydrated ionic species or they may form a variety of complexes with inorganic and organic ligands. Complex formation may involve electrostatic and/or covalent bonding. Where suitable ligands or binding sites occur on the surfaces of solids, metal elements will become associated with the solid phase. For many metals, particulate surfaces, because of their abundance and the high concentration of suitable binding sites, regulate the bioavailability and concentration of dissolved species.

The chemistry of individual metal elements is related to their position in the periodic table and their ability to behave as a Lewis acid. In simple terms Lewis acids are substances which are able to act as electron acceptors, a Lewis base being a substance capable of acting as an electron donor. Metal co-ordination, i.e. complex formation, results from the interaction of a Lewis acid and base:

$$A + B \rightleftharpoons A{:}B$$

where A represents a Lewis acid, the electron acceptor, and B represents a Lewis base, the electron donor.

Metallic elements have low ionization energies, i.e. the outermost electron(s) is easily lost, resulting in the formation of a positive ion, which may then act as a Lewis acid; reaction with a base effectively neutralizes the charge associated with the metal ion. In the biota, key Lewis bases which frequently co-ordinate with essential trace elements (e.g. Mg, Mn, Fe, Co and Zn) include ligands which contain oxygen (e.g. OH), nitrogen (e.g. NH) and sulphur (e.g. SH).

Ionization energies increase down the periodic table groups, lower members of a group forming cations less readily. The loss of electron(s) results in an excess of positively charged protons within the nucleus of the ion; this imbalance draws the outer electrons inwards towards the nucleus, reducing the atomic radius. Thus, the ionic radius is smaller than the original atomic radius. For example, the atomic radius of Na is 0.186 nm, while the radius of the Na^+ cation is considerably smaller (0.095 nm). The reduced radius affects the charge density or electrostatic energy Z^2/r (where Z is the formal charge, and r the ionic radius) of the ion, because of the relationship between the electrostatic energy of a cation and its radius, for cations carrying the same formal charge, the electrostatic energy will decrease with atomic number as the size of the ion increases. Thus, the ionic species of lighter metals are frequently able to displace heavier metal cations from exchange sites on surfaces where binding is primarily electrostatic, for example, K^+ cations associated with soil cation exchange sites may be readily displaced by the smaller Na^+ ion.

Classification of metallic elements

Periodic table classification Of the 108 known elements, 84 are considered to be metals and 17 are non-metals. The non-metal elements, with the exception of hydrogen, occupy the top right-hand corner of the periodic table. A further seven elements (B, Si, Ge, As, Sb, Se and Te), are classified as semi-metals or metalloids. These elements which separate the main metal and non-metal blocks of the periodic table are intermediate in their chemistry (Fig. 13.1).

Period	Group Ia	Group IIa	Group IIIa	Group IVa	Group Va	Group VIa	Group VIIa	Group VIII			Group Ib	Group IIb	Group IIIb	Group IVb	Group Vb	Group VIb	Group VIIb	Group 0
1	1 H																	2 He
2	3 Li	4 Be											5 B	6 C	7 N	8 O	9 F	10 Ne
3	11 Na	12 Mg											13 Al	14 Si	15 P	16 S	17 Cl	18 Ar
4	19 K	20 Ca	21 Sc	22 Ti	23 V	24 Cr	25 Mn	26 Fe	27 Co	28 Ni	29 Cu	30 Zn	31 Ga	32 Ge	33 As	34 Se	35 Br	36 Kr
5	37 Rb	38 Sr	39 Y	40 Zr	41 Nb	42 Mo	43 Tc	44 Ru	45 Rh	46 Pd	47 Ag	48 Cd	49 In	50 Sn	51 Sb	52 Te	53 I	54 Xe
6	55 Cs	56 Ba	57 La	72 Hf	73 Ta	74 W	75 Re	76 Os	77 Ir	78 Pt	79 Au	80 Hg	81 Tl	82 Pb	83 Bi	84 Po	85 At	86 Rn
7	87 Fr	88 Ra	89 Ac	104 Rf														

Inner transition elements

58 Ce	59 Pr	60 Nd	61 Pm	62 Sm	63 Eu	64 Gd	65 Tb	66 Dy	67 Ho	68 Er	69 Tm	70 Yb	71 Lu
90 Th	91 Pa	92 U	93 Np	94 Pu	95 Am	96 Cm	97 Bk	98 Cf	99 Es	100 Fm	101 Md	102 No	103 Lr

Fig. 13.1. Periodic table of elements (bold line divides metals and metalloids (dashed line) from the non-metals).

The alkali metals, Group Ia, form monovalent cations, while Group IIa, the alkali earth metals, form divalent cations. Groups IIIb through to VIb form the p-block metals, the heavier members of which are commonly referred to as 'heavy metals'. Within the p-block of metals are Al (oxidation state III) which obtains the noble gas electron configuration of Ne when in the Al^{3+} state, Pb with oxidation states of II and IV and Tl with stable oxidation states of I and II. The transition elements occupy rows 4, 5 and 6 of the periodic table. All transition elements, with the exception of Group IIIb metals, which only occur in the III oxidation state, may exist in a wide variety of oxidation states.

The lanthanides, which consist of elements 58 through to 71, along with scandium(21) and yttrium(39) are traditionally referred to as the rare earth metals. Within this group of metals the normal stable oxidation state is III. They tend to show strong ionic bonding and weaker covalent bond characteristics with decreasing radii. As a group they tend to demonstrate type-A behaviour in complex formation (cf. below) (Forstner and Wittman, 1981; Fergusson, 1990; Morgan and Stumm, 1991).

The heaviest members of the periodic table, elements 90–103, are collectively known as the actinides. Amongst these elements the oxidation state II becomes increasingly more stable with increasing atomic number. Actinides with an atomic number of 92 and above are artificial, being the product of the irradiation of uranium or other actinides with neutrons, alpha particles, carbon or nitrogen ions. The presence of these elements in the environment results from the activity of the nuclear and associated industries, or as a consequence of the testing of atomic weapons. All of the actinides are radioactive and thus may pose two distinct threats to public health – direct toxicity and exposure to radiation.

Type-A and type-B metal cations (Lewis hard and soft acid–base behaviour.) A more practical and useful classification of metallic elements may be obtained by considering their behaviour as Lewis acids. Three categories of behaviour may be recognized: type-A, type-B and borderline cations (Table 13.1). The occurrence of type-A or type-B behaviour is governed by the number of electrons present in the outer shell of the cation. Type-A cations possess an inert gas (d^0) electron configuration. They may be visualized as having spherical symmetry. The electron sheath is not easily deformed, and as a result they are referred to as 'hard' Lewis acids. Type-A cations preferentially form complexes with the fluoride ion and ligands which contain oxygen. In solution, type-A metal cations are unable to form stable complexes with sulphur, the HS^- and S^{2-} anions being displaced by OH^-. Chloro and iodo complexes are unstable and tend only to be formed under acidic conditions when OH^- concentrations are low. Type-A cations show a low affinity for chelating agents or ligands which contain only nitrogen or sulphur. In contrast to type-A cations, type-B cations are said to be soft; their electron sheath is readily deformable, i.e. they show high polarizability. Type-B cations co-ordinate preferentially with bases containing iodine, sulphur or nitrogen atoms acting as the electron

Table 13.1 Chemical processes in the environment, relevance of chemical speciation (Pearson, 1963)

Type-A metal cations	*Transition metal cations*	*Type-B metal cations*
Electron configuration of inert gas, low polarizability, 'hard spheres', (H^+), Li^+, Na^+ K^+, Be^{2+}, Mg^{2+}, Ca^{2+}, Sr^{2+}, A^{3+}, Sc^{3+}, La^{3+}, Si^{4+}, Ti^{4+}, Zi^{4+},	One to nine outer shell electrons, not spherically symmetric V^{2+} V^{2+}, Cr^{2+}, Mn^{2+}, Fe^{2+}, Co^{2+}, Ni^{2+}, Cu^{2+}, Ti^{3+}, V^{3+}, Cr^{3+}, Mn^{3+}, Fe^{3+}, Co^{3+}	Electron number corresponds to N_1^O, Pd^O and Pt^0 (10 or 12 outer shell electrons), low electronegativity, high polarizability, 'soft spheres', Cu^+, Ag^+, Au^+, Tl^+, Ga^+, Zu^{2+}, Cd^{2+}, Hg^{2+}, Pb^{2+}, Sn^{2+}, Tl^{2+}, Au^{3+}, In^{3+}, Bi^{3+}

According to PEARSON'S hard and soft acids

Hard acids	Borderline	Soft acids
All type-A metal cations plus Cr^{3+}, Mn^{3+}, Fe^{3+}, Co^{3+}, UO^{2+}, VO^{2+}	All bivalent transition metal cations plus Zn^{2+}, Pb^{2+}, Bi^{3+},	All type-B metal cations minus Zn^{2+}, Pb^{2+}, Bi^{3+},
Also species such as BF_3, BCl_3, SO_3, RSO_2^+, RPO_2^+ CO_2, RCO^+, R_3C^+.	SO_2, NO^+, $B(CH_3)_3$	All metal atoms, bulk metals I_2, Br_2 ICN, I^+, Br^+
Preference for ligand atom:		
N≫P		P≫N
O≫S		S≫O
E≫Cl		l≫F

Table 13.1 continued.

Hard acids	Borderline	Soft acids
Qualitative generalizations on stability sequence:		
Cations:	Cations:	
Stability or (charge/radius)	Irving–Williams order:	
	Mn^{2+}, $<Fe^{2+}$, $<Co^{2+}$, $<Ni^{2+}$,	
	$<Cu^{2+} > Zn^{2+}$	
Ligands:		Ligands:
F>O>N = Cl>Br>I>S		S>I>Br>Cl = N>O>F
$OH > RO > RCO_2$		
$CO_3^{2} \gg NO_3$		
$PO_4^{3} \gg SO_4^{2} \gg ClO_4$		

donors. In comparison with the hard type-A cations these metals show a high affinity for ammonia in water and will form stable chloro and iodo complexes. These metals, along with the transition metal cations, form insoluble sulphide and soluble complexes with S^{2-} and HS^{-} (Forstner and Wittman, 1981; Morgan and Stumm, 1991; Raspor, 1991).

Interactions of metals with particles

Particulate surfaces, whether inorganic (e.g. aluminium silicates) or organic (e.g. algal debris) contain functional groups (–MOH, –ROH, –R–COOH) which readily form complexes with metals. Due to the abundance of hydroxyl groups on surfaces, type-A metals are more particle reactive than type-B cations. Because of this the environmental fate of type-A cations, particularly in aquatic systems, may be regulated by the amounts and behaviour of the metal associated with the particulate phase. It is estimated that 95% of heavy metal transfer from land to sea results from the movement of metals present in the particulate phase. The ocean concentration profiles of some metals, e.g. Am, Pb and Pu, are characterized by a surface maximum which declines rapidly with depth. These profiles result from the scavenging of metals out of the surface waters by sinking particulate material. In marine systems, away from the direct influence of terrestrial run-off particulate material is predominately of biological origin, resulting from the activity of phytoplankton. In freshwater systems inorganic material derived from weathering and soil erosion often accounts for the major portion of particulate present (Whitfield and Turner, 1987; Morgan and Stumm, 1991; Raspor, 1991).

Metal elements react with particulate material of organic and inorganic origin in a similar way, reflecting the similarity in surface physico-chemical properties and the colonization of particle surfaces by bacteria. Inert, non-toxic particulates are rapidly colonized by bacteria, the activities of which influence the ability of particles to accumulate metals, i.e. the physico-chemical properties of bacterial cell walls as well as those of the initial particle will help to determine the extent of metal binding (Whitfield and Turner, 1987).

Formation of carbon complexes

The ability to form carbon complexes is of major importance in determining the environmental fate and toxicity of metals. Metal akyl compounds, e.g. methyl mercury, are of particular concern as they are frequently volatile, are readily accumulated by cells and are toxic, being effective poisons of the central nervous systems. Although stability of metal–carbon complexes in aqueous solutions decreases from Ge (32) through to Pb (82), Hg(80) is able to form stable bonds with carbon, giving rise to true organometallic compounds. The Hg–C bond is not particularly strong (*c*.0.60 kJ $mole^{-1}$), but is significantly stronger that the Hg–O bond. This in part explains the stability and environmental persistence of organomercury compounds (Forstner and Wittman, 1981; Morgan and Stumm, 1991; Raspor, 1991).

Bacterial activity is responsible for the biological alkylation of metals, although for some elements, e.g. Se and Sn, other organisms such as algae may also be important. Biomethylation of Ge, Sn, As, Se, Te, Pb, Pt, Au, Hg and Tl has been reported. The use of alkylated Pb compounds as fuel additives and alkylated Sn compounds in antifouling paints means that for these elements anthropogenic sources greatly exceed those due to biomethylation of these metals (Morgan and Stumm, 1991).

Metals in the atmosphere

The majority of metals present in the atmosphere are associated with particulate material. Because this material is largely inert in nature (i.e. mineral or carbonaceous) and hydroscopic, metals present there occur predominately in the liquid phase, soluble metals being present as ionic species. Only mercury occurs predominately in the gaseous phase, although other metals (Pb, Sn, As, Se and Te) may also occur in a gaseous state (Morgan and Stumm, 1991; Puxbaum, 1991).

Since for most metals anthropogenic sources exceed natural environmental sources, the occurrence of metals in the atmosphere reflects the geographic distribution and concentration of industrial sources. High concentrations are associated with densely populated industrial regions and air masses which originate in such regions. Very low concentrations, not surprisingly, have been found in Antarctica and in the maritime atmosphere over the Pacific Ocean. Higher atmospheric concentrations occur in the Arctic, reflecting the greater extent of industrial development in the northern hemisphere and meterological factors which result in the long distance transport of the polluted Eurasian air masses into the Arctic during the winter months. Even in highly industrialized areas the combined contribution by weight of toxic metals, e.g. Cd, Cr, Cu, Mn, V, Zn and Pb, to total aerosol mass is frequently low (*c*.1.0%). The bulk of the aerosol typically consists of electrolytes (e.g. Na^+, K^+, NH_4^+, Cl^-, NO_3^- and SO_4^{2-}), carbonaceous and mineralic components (e.g. Ca^{2+}, Mg^{2+}, Si and Al compounds) (Morgan and Stumm, 1991; Puxbaum, 1991).

The major source of atmospheric metals is associated with the combustion of fossil fuels and the operation of smelters and metal refineries. Emissions of Be, Co, Mo, Sb and Se are mainly derived from the combustion of coal, and Ni and V are released predominately from oil firing. While As, Cd, Cu and Zn are released from non-ferrous smelters and secondary productions plants, Cr and Mn result from the iron and steel industries. Lead emissions are dominated by releases through the combustion of petrol which contains lead additives (Morgan and Stumm, 1991; Puxbaum, 1991).

Particulate size governs the atmospheric half-life of a particle. Settlement velocity is a complicated function of particle diameter (Fig. 13.2). Because large particles settle most rapidly, dry metal deposition from such particles will be concentrated close to the source. Metals associated with smaller particulates, because of their low settlement velocities, will become dispersed over greater distances. In this instance removal from the atmosphere is largely dependent on wet deposition. The combustion process produces vapour and fine particulates, the condensation and nucleation of which typically results in a bimodal distribution of particulate sizes, which are normally substantially smaller in size than those of natural origin. In ambient air Pb, Cd and Zn occur predominantly within the accumulated mode (0.3–0.8 μm), Ca, Mg and Al are associated with coarse particles (> 3 μm), whereas Mn, V, Cu and Cr are principally associated with particles of an intermediate size (1–5μm).

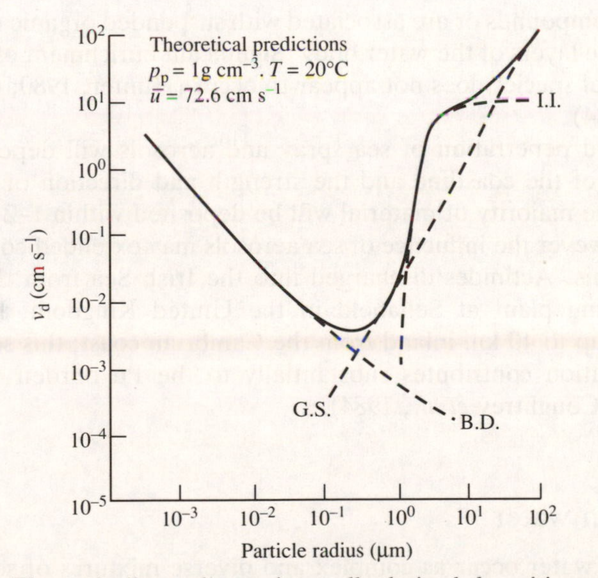

Fig. 13.2. Theoretically- and experimentally-derived deposition velocities as a function of the particle diameter. Contributing processes: BD, Brownian diffusion; GS, gravitational settlement; II, inertial impaction (from NCAR, 1982).

Particle size determines the pattern of deposition in the human respiratory tract. Large coarse particles (10 μm) are preferentially deposited in the nasopharyngeal compartment. Small particles of ≤ 5 μm are able to penetrate the bronchial and alveolar compartments; this range of particle sizes is referred to as the thoracic or respirator fraction. Due to the penetration of this fraction, the metal content of these particles can pose a major health hazard. Although in comparison to anthropogenic sources natural particulates play a relatively minor role in the mass transport of metals, they do represent an important mechanism for the transfer of pollutants. The levels of metals and other pollutants present in seawater are frequently enriched within the sea-surface microlayer as a result of a range of physical and biological processes (e.g. bacterial activity). Enrichment of Pb, Fe, Ni, Cu, fatty acids and chlorinated hydrocarbons within the top 100–105 μm layer of the seawater by factors of 1.5–50 times the bulk water concentration have been reported. Sea spray and aerosol production in the surf from bubble burst and wave and wind action will concentrate pollutants further. For example, studies suggest that during the production of aerosols at the sea surface, Cu may be concentrated by factors within the range of 10^4 to 3×10^5. Thus the formation of aerosols from seawater may represent an important mechanism, especially in coastal regions, for the transfer of pollutants across the land–sea interface. Research suggests that transfer via this route will be most significant for metals which either complex readily with dissolved organic compounds or are associated with suspended organic material in the surface layers of the water body. Significant enrichment of inorganic ionic metal species does not appear to occur (Hunter, 1980; Coughtrey *et al.*, 1984).

The inland penetration of sea spray and aerosols will depend on the exposure of the coastline and the strength and direction of prevailing winds. The majority of material will be deposited within 1–2 km of the coast, however the influence of sea aerosols may extended considerably beyond this. Actinides discharged into the Irish Sea from the nuclear reprocessing plant at Sellafield in the United Kingdom, have been detected up to 40 km inland from the Cambrian coast; this sea-derived contamination contributes substantially to the Pu burden of pasture herbage (Coughtrey *et al.*, 1984).

Metals in water

Metals in water occur as complex and diverse mixtures of soluble and insoluble forms (Fig. 13.3). They may be present as ionic species, inorganic and organic complexes and/or associated with collides and suspended particulate material. Information on total metal concen-

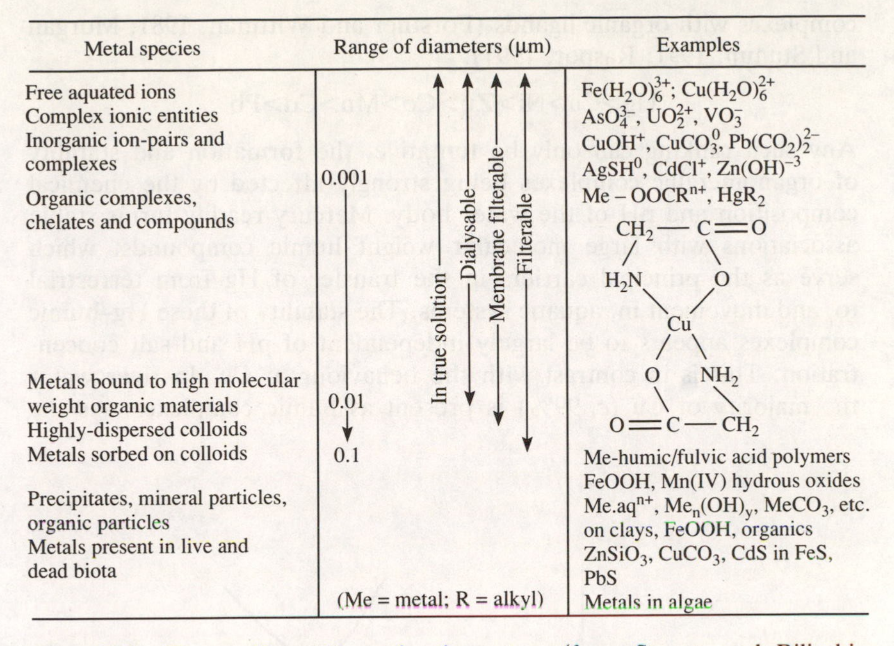

Metal species	Range of diameters (µm)	Examples
Free aquated ions		$Fe(H_2O)_6^{3+}$; $Cu(H_2O)_6^{2+}$
Complex ionic entities		AsO_4^{3-}, UO_2^{2+}, VO_3^-
Inorganic ion-pairs and complexes	0.001	$CuOH^+$, $CuCO_3^0$, $Pb(CO_2)_2^{2-}$, $AgSH^0$, $CdCl^+$, $Zn(OH)^{-3}$
Organic complexes, chelates and compounds		$Me - OOCR^{n+}$, HgR_2
Metals bound to high molecular weight organic materials	0.01	Me-humic/fulvic acid polymers
Highly-dispersed colloids		FeOOH, Mn(IV) hydrous oxides
Metals sorbed on colloids	0.1	$Me.aq^{n+}$, $Me_n(OH)_y$, $MeCO_3$, etc. on clays, FeOOH, organics
Precipitates, mineral particles, organic particles		$ZnSiO_3$, $CuCO_3$, CdS in FeS, PbS
Metals present in live and dead biota		Metals in algae
	(Me = metal; R = alkyl)	

Fig. 13.3. Types of metal species in waters (from Stumm and Bilinski, 1972).

tration is of limited value as the environmental fate and toxicity of the different metal species differs considerably and does not necessarily reflect the overall abundance of the metal in the water body. For example, although particulate forms accounted for less than 1% of the uranium present in seawater off the southern Californian coast, 35% of the uranium adsorbed by seaweeds was found to be particulate in form (Hodge *et al.*, 1979). Unfortunately, often only total metal concentration is known. The analytical problems associated with the determination of metal speciation at very low concentrations in water are complex and have not, as yet, been completely surmounted. Discussion of these problems may be found in the volume edited by Merian (1991), which together with Coughtrey *et al.* (1985) and Fowler (1990) provides reviews of published heavy metal concentrations.

Organometallic compounds in water range in complexity from simple amino acid complexes to those formed with humic substances. Humic compounds frequently constitute the predominant form of dissolved organic material in natural water bodies and may readily form complexes with certain metal cations. Based on empirical studies it is possible to rank metals in terms of their tendency to form stable complexes with organic ligands (Forstner and Wittman, 1981; Morgan and Stumm, 1991; Raspor, 1991):

$$Hg > Cu > Ni > Zn > Co > Mn > Cd > Pb$$

Any such ranking can only be tentative, the formation and stability of organometallic complexes being strongly affected by the chemical composition and pH of the water body. Mercury readily forms stable associations with large molecular weight humic compounds, which serve as the principal carriers in the transfer of Hg from terrestrial to, and movement in, aquatic systems. The stability of these Hg–humic complexes appears to be largely independent of pH and salt concentration. This is in contrast with the behaviour of Cu. In river water the majority of Cu (*c.* 99%) is present as humic complexes. As the

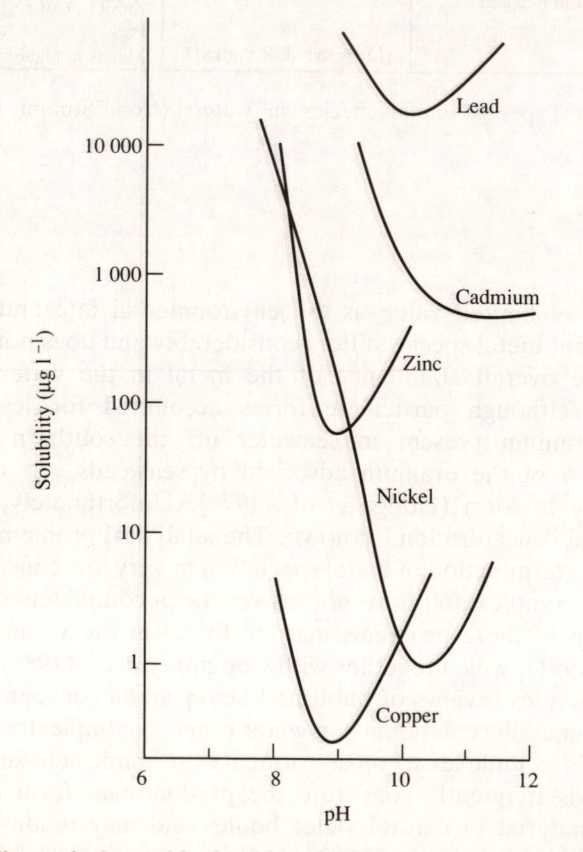

Fig. 13.4. Theoretical metal hydroxide solubility as a function of pH (Laxen, 1983).

river water enters the sea and the salinity increases, the proportion of Cu associated with humic compounds decreases rapidly, the Cu cation being displaced by Ca^{2+} and Mg^{2+}. Inorganic species of Cu, particularly $Cu(OH)_2$, predominate in seawater, humic–Cu complexes accounting for only 10% of the total Cu content present (Burton, 1979). Similarly, monovalent metals, e.g. Cs and Zn ions, may be displaced by Na^+ and K^+, from organic complexes and river particulates entering an estuary. Metal hydroxides are frequently the predominant inorganic metal species present in natural waters and because of the relationship between metal hydroxide solubility and pH, metal cation concentration is often a function of pH. Metal concentrations and mobility increase dramatically at low acidic pHs (Fig. 13.4).

Elements may be classified, on the basis of their ocean depth concentration profiles, into three groups: accumulated, recycled and scavenged elements (Table 13.2) (Whitfield and Turner, 1987). The tendency of metals to form soluble complexes and react with particulate matter differs between these groups. The relative particulate reactivity of an element is reflected in its oceanic residence time (T_y).

$$T_y = \frac{\text{total number of moles of element in ocean}}{\text{rate of addition or removal of element}}$$

The incorporation of elements into sinking particulate material is a major route by which they may be transferred from surface waters. Metals which react weakly with particles have long residence times, while particle reactive metals are characterized by short residence times.

Accumulated elements Accumulated elements are conservative, reacting weakly, usually via ion exchange, with mineral and organic particulate phases and are present predominately as free ions in solution (Bruland, 1983). These elements, which include Cs, Mo and U, have long residence times ($T_y > 10^5$ yr), uniform depth profiles and maintain a constant ratio to one another. Uranium is unique amongst the natural elements in seawater, occurring as an oxocation which reacts strongly with carbonate ions forming a large inert, negatively-charged complex. Some accumulated elements are biologically important (e.g. Na, K, Mg, B, S and Cl), but their concentrations in seawater are too high for their utilization to affect their concentration profiles.

Recycled elements Recycled elements have residence times of between 10^3 and 10^5 yr. They are readily incorporated into biogenic material, particularly phytoplankton, the production of which powers the particulate cycle. These elements are removed from surface waters by plankton and particulates, but are re-released as the biological debris sinks and decomposes. With the exception of C, Si and Ca, transport

Table 13.2 Characteristics of accumulated elements (B to U), recycled elements (Ag to Zn) and scavenged elements (Al to Th) (from Whitfield and Turner, 1987)

Element	*Oxidation state*	*Concentration* *Atlantic* *Surf*	*Atlantic* *Deep*	*Pacific* *Surf*	*Pacific* *Deep*	T_y *(yr)*	*–log R*	*Biological utilization*
				Accumulated Elements				
B	III	.42 mM				1E7	4.7	B
Br	–I	.84 mM				1E8		C
Cl	–I	.53 M				4E8		B
Cs	I	2.3 nM				6E5	<0	D
F	–I	68 μM				4E5		C
K	I	10 mM				5E6	–0.5	A
Li	I	2.6 μM				2E6	0.4	D
Mg	II	53 mM				1E7	2.5	A
Mo	VI	107 nM				6E5		B
Na	I	.47 M				1E8	–0.2	B
Rb	I	1.4 μM				8E5	<0	D
S	VI	28 mM				8E6		A
Tl	I	69 pM				1E4	0.5	D
U	VI	13.5 nM				3E5	1.4	D
				Recycled Elements				
Ag	I			1 pM	23 pM	5E3	–3.3	D
As	V	20 nM	21 nM	20 nM	24 nM	9E4		D
Ba	II	35 nM	70 nM	35 nM	150 nM	1E4	0.4	D
Be	II	10 pM	20 pM	4 pM	25 pM	4E3	5.9	D
C	IV	2.0 mM	2.2 mM	2.0 mM	2.4 mM	8E5		A
Ca	II	10 mM	11 mM	10 mM	11.3 mM	1E6	1.3	B
Cd	II	10 pM	0.35 nM	10 pM	1 nM	3E4	3.3	D
Cr	VI	3.5 nM	4.5 nM	3 nM	5 nM	1E4		C
Cu	II?	1.3 nM	2 nM	1.3 nM	4.5 nM	3E3	5.0	A
Dy	II	5 pM	6.1 pM			300	5.1	D
Er	III	3.6 pM	5.3 pM			400	5.0	D
Eu	III	0.6 pM	1 pM	0.7 pM	1.8 pM	500	5.5	D
Fe	III	2 nM	7 nM	0.2 nM	2 nM	98	5.8	A
Gd	III	3.4 pM	6.1 pM	4 pM	10 pM	300	5.0	D
Ge	IV	1 pM	20 pM	5 pM	100 pM	2E4		D
Ho	III	1.5 pM	1.8 pM	1 pM	3.6 pM		5.0	D
I	V	0.2 μM	0.45 μM	0.35 μM	0.47 μM	3E5		C
La	III	13 pM	28 pM	19 pM	51 pM	200	5.1	D
Lu	III	0.8 pM	1.2 pM	0.35 pM	2.4 pM	4E3	5.1	D
N	V	5 nM	20 μM	5 nM	40 μM	6E3		A
Nd	III	13 pM	23 pM	13 pM	34 pM	500	5.3	D
Ni	II	2 nM	7 nM	2 nM	10 nM	8E4	4.2	C
P	V	50 nM	1.4 μM	50 nM	2.8 μM	1E5		A
Pd	II			0.18 pM	0.66 pM	5E4	5.8	D
Pr	III	3 pM	5 pM	3.2 pM	7.3 pM		5.3	D
Pt	II			0.6 pM	1.4 pM		–1.0	D
Ra	II	8	20	10	35		0	D

Table 13.2 continued.

Element	Oxidation state	*Concentration* Atlantic Surf	Atlantic Deep	Pacific Surf	Pacific Deep	T_y (yr)	–log R	Biological utilization
Sc	III	14 pM	20 pM	8 pM	18 pM	5E3	6.0	D
Se	IV	0.1 nM	0.9 nM	0.07 nM	0.9 nM	3E4		C
Se	VI	0.5 nM	1.5 nM	0.13 nM	1.25 nM			C
Si	IV	1 µM	30 µM	1 µM	150 µM	3E4		C
Sm	III	2.7 pM	4.4 pM	2.7 pM	6.8 pM	200	5.3	D
Sr	II	89 µM	90 µM	89 µM	90 µM	4E6	0.6	D
Tb	III	0.7 pM	1 pM	0.5 pM	1.6 pM		5.3	D
Tm	III	0.8 pM	1 pM	0.4 pM	2 pM		5.2	D
V	V	23 nM		32 nM	36 nM	5E4	2.1	C
Yb	III	3 pM	4.5 pM	2.2 pM	13 pM	400	5.0	D
Zn	II	0.8 nM	1.6 nM	0.8 nM	8.2 nM	5E3	4.7	A
				Scavenged Elements				
Al	III	37 nM	20 nM	5 nM	0.5 nM	150	5.6	D
As	III			0.3 nM	70 pM		4.6	D
Bi	III	0.25 pM		0.2 pM	0.02 pM		6.1	D
Ce	III	66 pM	19 pM	11 pM	4 pM	100	5.0	D
Co	II			0.12 nM	0.02 nM	40	4.1	B
Cr	III			0.2 nM	0.05 nM		6.5	C
Hg?	II	2.5 pM	2.5 pM	1.7 pM	1.7 pM		–3.8	D
I	–I	0.2 µM	10 pM	90 pM	60 pM			C
Mn	II	1.9 nM	1.8 nM	1.9 nM	0.8 nM	50	3.2	A
Pb	II	0.15 nM	20 pM	50 pM	5 pM	50	4.8	D
Sn	IV	20 pM	5 pM			8.4		D
Te	VI	0.9pM	0.4pM	1 pM	0.4 pM		0.2	D
Te	IV	0.4 pM	0.2 pM	0.5 pM	1 pM			D
mTh	IV	87	3		20	50	6.0	D

Notes: (i) Biological utilization. A, essential in all species tested; B, widely but not universally required; C, required in a limited group of species; D, not required. (ii) Concentrations normalized to a salinity of 35. (iii) Ra, Concentration units d.p.m. per 100 kg. (iv) ^{232}Th, Concentration units d.p.m. per 10^6 kg. (v) -log R; Index of scavenging, calculated as where ?$_m$ is the overall 'side-reaction' coefficient and K^{OH} is the first hydrolysis constant of the cation. R represents the balance between solution complexation and adsorption.

out of the surface layers occurs predominantly in the organic phase. The profiles of recycled elements are characterized by pronounced surface depletion, concentrations increasing with depth. Recycled elements include those for which there are specific biological requirements, i.e. both major (e.g. N,P,C and S) and minor (e.g. Se) nutrient elements and the metals Cd, Cr(VI), Cu, Fe and Zn.

Scavenged elements Scavenged elements, which include Al, As, Co, Cr(III), Mn, Pb and Sn, typically have residence times of $<10^3$ yr. Their profiles are characterized by a surface maxima, concentrations decreasing rapidly with depth. These profiles result from additions to the surface layers of the oceans, due partly to anthropogenic activity coupled with rapid removal from the water column. The major removal process is via the scavenging of elements onto the surfaces of sinking particulates.

The behaviour of metals in water may be predicted from a consideration of their apparent formation or stability constants (*b*). (Whitfield and Turner, 1987; Raspor, 1991).

$$b_n = \frac{[ML_n]}{[M]\,[L]^n}$$

where M is the metal, L the ligand and $M + nL \rightleftharpoons ML_n$.

Since particulate surfaces possess a significant number of hydroxyl sites, particle reactivity of metals is related to their stability constants for hydrolysis and electrostatic energies (Z_{2_i/r_i}), which determine the intensity of the ion–water interactions. The involvement of covalent chemical bonding may be included by the use of the index:

$$\Delta\beta = \log\beta_{mF} - \log\beta_{mCl}$$

where the β values represent the stability constants of the monofluoro and monochloro complexes, respectively. Type-A cations ($\Delta\beta > 2$), bond most strongly with hard anions (e.g. F^-, OH^- and SO_4^{2-}) via electrostatic interactions and B-type cations ($\Delta\beta < 2$) bond most strongly with soft anions (e.g. Cl^-, Br^-, HS^-) via covalent interactions. A plot of log (Z_{2_i/r_i}) versus AB yields a complexation field diagram, which enables elements to be grouped according to their chemical speciation patterns in seawater into four categories: A, weakly complexed elements with >1% present as free ion; B, elements dominated by chloride complexes; C, strongly hydrolysed elements and D, fully hydrolysed elements (Fig. 13.5). Adopting a more empirical approach, Fisher (1986) found that phytoplankton metal volume/volume concentration factors (VCF) and sublethal toxicity levels (the metal concentrations required to reduce growth by 50%) were positively correlated with the log of the solubility products of metal hydroxides. He also found that the ocean residence times of metals were negatively correlated with phytoplankton VCF values, while sublethal metal concentrations were positively correlated with VFC values.

The sinking rates of phytoplankton (<2 m dy^{-1}) are too slow to be of major significance in the vertical movement of metals and radionuclides out of the surface layer and across the thermocline into deeper water.

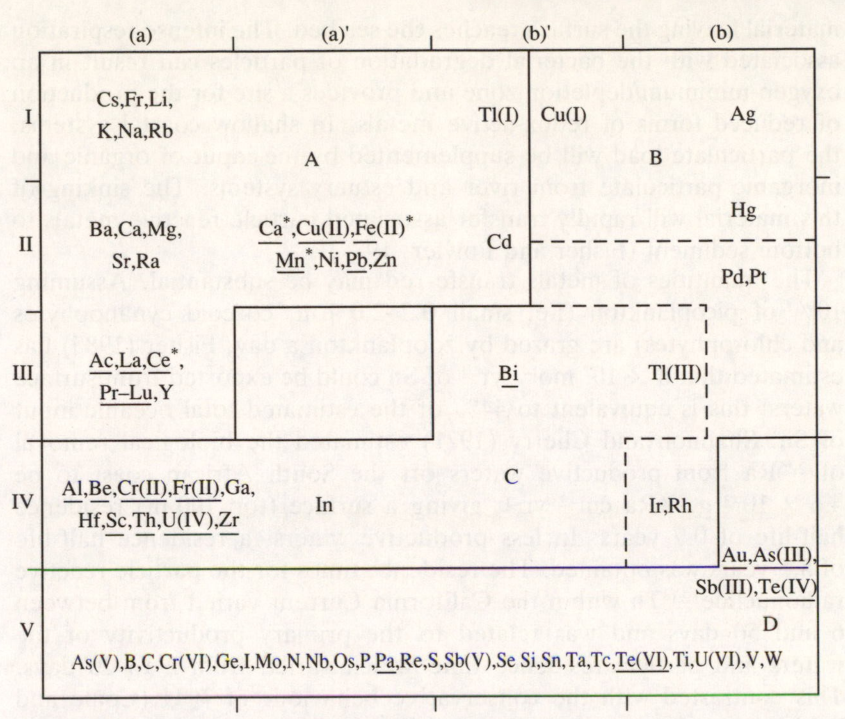

Fig. 13.5. Complexation field (CF) diagram for elements in seawater. The horizontal groupings I, II, IV and V correspond to increasing covalent character (decreasing $\Delta\beta$). The diagram is divided into four sectors: A, weakly complexed elements with >1% present as free cation; B, elements dominated by chloride complexes; C, strongly hydrolysed elements; D, fully hydrolysed elements. The dashed boundaries between sectors B and C reflect the lack of data on the hydrolysis of platinum group metals. Elements that are scavenged in the oceans are underlined and those that are subject to oxidative scavenging are marked with an asterisk (from Whitfield and Turner, 1987).

However, phytoplankton are actively grazed by zooplankton, which effectively repackage the phytoplankton as faecal pellets. These are relatively large particles (>100 μm), with sinking rates of between 100 and 1000 m dy^{-1}. Metals adsorbed onto phytoplankton cell walls are not readily assimilated and become concentrated within zooplankton faecal pellets. Aggregates of phytoplankton cells, particulates including discarded appendicularian houses and zooplankton debris also occur. This 'marine snow' typically has sinking rates of 50–100 m dy^{-1}, and is also involved in the transfer of metals to bottom sediment or deep water. Metals scavenged by this sinking material are released as the material is degraded. The overall effect is to enhance the remineralization of metals in deeper water. In productive open waters, the bulk of organic material is broken down within or just below the thermocline at depths of 500–1000 m, typically <1% of particulate material leaving the surface reaches the sea bed. The intense respiration associated with the bacterial degradation of particles can result in an oxygen-minimum/depletion zone and provides a site for the production of reduced forms of redox-active metals. In shallow coastal systems, the particulate load will be supplemented by the input of organic and inorganic particulate from river and estuary systems. The sinking of this material will rapidly transfer associated particle reactive metals to bottom sediment (Fisher and Fowler, 1987).

The quantities of metals transferred may be substantial. Assuming 10% of picoplankton (i.e. small 0.2–2.0 μm, coccoid cynanophytes and chlorophytes) are grazed by zooplankton a day, Fisher (1985) has estimated that 8×10^{7} mol^{-1} yr^{-1} of Sn could be exported from surface waters; this is equivalent to 44% of the estimated total oceanic input of Sn. Shannon and Cherry (1971) estimated the biological removal of ^{226}Ra from productive waters off the South African coast to be 4.6×10^{-3} g ^{226}Ra cm^{-2} yr^{-1}, giving a surface (top 100 m) residence half-life of 0.7 years. In less productive waters, a residence half-life of 5.5 years was obtained. The residence times for the particle reactive radionuclide ^{234}Th within the California Current varied from between 6 and 50 days and was related to the primary productivity of the waters and particle residence time which varied from 2 to 20 days. This contrasted with the conservative behaviour of ^{238}U (Coale and Bruland, 1985). In the shallow waters of the Irish Sea over 95% of the actinide inventory discharged from the Sellafield waste pipeline rapidly becomes associated with bottom sediments. The majority of Pu and Am released into the seawater rapidly becomes associated with sinking particulates, reducing the load of radionuclide remaining in the water column.

River systems discharge large quantities of metals into coastal waters, the majority of which are associated with suspended particulate material, much of which settles rapidly within the estuary. The potential environmental impact of these discharges depends on the complex set of chemical and physical interactions which occur as river water enters the sea. The mobilization of pollutants associated with this material, harbour dredgings and solid inland wastes dumped at sea represents a significant source of metal contamination (GESAMP, 1990). The mobility and bioavailability of sediment-bound metals is increased by four factors (Forstner and Salomons, 1990).

1 Lowering of pH. Sediment pH may be substantially reduced by bacterial oxidation of sulphur and ferrous iron. Reductions in pH will be most marked among sediments which lack any significant ability to buffer pH. Increased acidity also has the undesirable effect of accelerating the methylation of Hg.

2 Increasing salt concentrations. Competition for sorption sites and formation of soluble chlorometal complexes will favour the release of metals. The formation of soluble $CdCl_4^{2-}$ partly accounts for the mobilization of Cd from particulates, observed as salinity increases.

3 The presence of high concentrations of natural and synthetic complexing agents. The importance of organic chelating agents is likely to be most marked in inland waters. Concern has been expressed over the use of nitroltriacetic acid (NTA), which is used as a substitute for polyphosphates in detergents. Low concentrations of NTA (50–100 $\mu g\ l^{-1}$) reduce the natural adsorption of metals onto particulates.

4 Changing redox conditions. Changes to the redox potential of sediment may result from periodic exposure during the tidal cycle and/or bacterial activity. Anoxic, reducing sediments restrict the mobility of metals as they favour the formation of stable insoluble metal sulphides.

Metals in soils and their mobility The biotransfer of toxic metals from contaminated soils to plants which are subsequently consumed by humans or domestic livestock, represents a route by which human exposure to heavy metals may be substantially increased. Non-tolerant crop plants readily accumulate heavy metals when grown in mildly contaminated soil. The concentrations of Cd, Ni, Cu and Zn in the edible portions of crop plants are substantially increased when grown on soil treated with sewage sludge (Carlton-Smith, 1987). Vegetables grown on soils contaminated by past ore processing contain elevated metal concentrations. In one study, the blood Pb concentrations of women resident in a metal-contaminated Pb mining area were found to be 28% higher among individuals who consumed locally grown vegetables in comparison to women who consumed no locally grown vegetables.

In soils which overlay metal ores very high concentrations of metals may occur. A specialized 'metallophyte' flora is often associated with such sites. The development of metal tolerance among metallophytes occurred early in the evolution of angiosperms. However, the evolution of metal-tolerant ecotypes can occur rapidly in response to localized anthropogenic contamination of sites (Baker, 1987). Metallophytes typically accumulate large quantities of metal. For example, foliage Ni concentrations of 600–10 000 ppm have been observed for plants growing on Serpentine soils containing about 2000 ppm Ni (soil levels of up to 250 000 ppm may be present); this compares with Ni levels of 20–70 ppm in reference samples.

The mobility and bioavailability of metals present in soils depend on physico-chemical properties of both the metal and the soil. In comparison to sandy soils, soils which have high organic and clay mineral contents are able to accumulate large quantities of metals due to an abundance in the soil matrix of cation exchange surfaces and organic ligands capable of complexing with metal ions. The extent and nature of the organic fraction is strongly influenced by the activity of soil microorganisms (bacteria and fungi). In unpolluted soil ecosystems these organisms typically account for more than 90% of the biomass present and for 90% of chemical decomposition of organic material. Where the activity of soil microorganisms is disrupted the ability of soils to accumulate and retain metals will be affected. A reduction in the rate of decomposition will cause organic matter (litter) to accumulate on the soil surface, increasing the number of potential metal binding sites. However, a reduction in the microbial biomass in the soil matrix increases the mobility of metals previously accumulated within and adsorbed onto the surfaces of soil microorganisms. Similarly the mobility of metals bounded to, or complexed with, organic compounds and exopolymers derived from microbial activity will be increased.

Bacterial activity within soils and aquatic sediments is capable of affecting mobility directly via biotransformation and indirectly by generating localized gradients in pH, oxygen and redox potential which may affect the chemical speciation of metals. The ability of certain bacteria to methylate metals is of particular importance given the greater toxicity and bioavailability of such compounds. The uptake and accumulation of mercury by bean plants was increased by a factor of 10 when methyl mercury was supplied in place of inorganic mercury.

Metals in soils are associated with a number of different phases. They may be present as part of the soil minerals, as precipitated compounds, sorbed onto inorganic and organic exchange surfaces or present in the soil solution as soluble inorganic and organic species. Metals present in the different phases are not equally available for plant uptake; the tissue concentration of plants growing in contaminated soils normally correlates most strongly with the metal content of one particular phase, not with total soil concentration (Baker, 1987; Davies *et al.*, 1987). The distribution of metals between the different soil phases is normally assessed by the extraction of metals from soil samples using a range of extractants. Relatively mild extractants, such as ammonium nitrate (1 M solution) are assumed to extract soil solution and exchangeable metal forms. Ethylene diamine tetracetic acid (EDTA, diammonium salts) extracts soil solution and exchangeable and humus-bound forms, while reflux boiling with concentrated nitric acid may be used in the determination of total metal content. Davies *et al.* (1987) using this approach, studied the uptake by radish (*Raphanus sativus*) plants of Pb, Zn, Cu and Cd from a range of contaminated soils. Their results suggest that Pb and Cu were mostly bound to the organic fraction, while the bulk of Cd and Zn were present as easily extracted (5% acetic acid) sorbed forms. Correlations between extract metal and plant leaf concentrations indicate that humus-bound forms of Pb and Cd were the most readily available to plants, while for Zn plant accumulation was correlated with the readily exchangeable fraction. Copper accumulation was poorly correlated with all extract concentrations and total soil content.

Mobility, activity and bioavailability of soil metals are affected by soil moisture, the composition of the soil solution, pH, redox potential and the abundance and distribution of surfaces capable of adsorbing metal ions. These factors may vary over short distances, affecting the concentration and bioavailablity of the metal. The distribution of available plant nutrients in contaminated soils also tends to be spatially heterogeneous. The combined effects of the heterogeneity of toxic metals and nutrient availability are reflected in the growth and distribution of tolerant plant species. Certain tolerant plants are restricted to high nutrient microsites, while others are able to survive in sites which are both nutrient poor and contain high concentrations of toxic metals (Baker, 1987).

The most important factor affecting the proportion of metal available to plants is pH. The mobility/solubility of metals increases with decreasing pH. At low pHs hydrogen ions compete with metals for exchange sites on soil particles. Schmitt and Sticker (1991) have summarized the relationship between metal mobility and pH:

Mobility	pH 4.2–6.6	pH 6.7–8.8
Relatively mobile	Cd,Hg,Ni and Zn	As and Cr
Moderately mobile	As,Be and Cr	Be,Cd,Hg and Zn
Slowly/slightly mobile	Cu,Pb and Se	Cu, Pb, and Ni

In mineral soils, under acidic conditions, the ability of metals to displace other sorbed ions follows the sequence Pb>Cu>Zn≥Cd, which corresponds to increasing p*K* values for first hydrolysis, i.e. the sequence reflects relative particle reactivity. A similar sequence, Pb>Cu>Cd>Zn, is obtained for organic soils. As a consequence of the high particle reactivity of Pb, increased Pb concentrations may increase the availability of other less-strongly sorbed metals. High soil Pb content has been shown to increase the concentration (Miller *et al.*, 1977) and subsequent uptake of Cd by plants. Other ions, if present in high concentrations, e.g. Ca^{2+}, Na^{+} and K^{+}, will also displace ions from adsorption sites within the soil matrix. Sodium and potassium will be particularly important where soils are subject to saline influence. Because of the relationship between metal mobility and pH, residence times for metals in calcareous soils will exceed those associated with acidic soils. Acidic precipitation may further reduce soil residence times, especially in acidic soils unable to adequately buffer their pHs. Experiments mimicking the effects of acid rain on contaminated acidic or spruce forest soils show that the 10% residence time (i.e. time required for the soil concentration of metal to be reduced by 10%

of its original level), decreased with increasing pH. Residence times of 10–20 years were obtained for Zn and Cd, about 100 years for Cu and >200 years for Pb. In metal mine tailings the leaching of heavy metals is frequently exasperated by the oxidation of pyrite (FeS_2), which generates considerable acidity (H_2SO_4), reducing the pH which may fall below 3. If the waste material contains significant quantities of pyrite, low pH and high metal mobility may be maintained for many years despite the application of lime during reclamation.

Metals in the soil solution occur at low concentrations (e.g. 10^{-8} mol l^{-1}). Relatively little is known about their speciation. They may be present as soluble organic complexes formed between the metal, humic and fulvic acids. The terms humic and fulvic acids refer to a complex range of organic acids which are released from the breakdown of organic material. They are found in the soil solution and natural water bodies. The stability of metal complexes formed with these organic substances has been established in a number of experimental studies and ranked in the following order (Schmitt and Sticker, 1991):

$$Cr > Fe > Al > Pb >> Cu > Ni >> Cd > Zn >> Mn = Ca = Mg$$

The formation of stable organic complexes can have effects on the chemical behaviour of the metal, preventing the formation of insoluble precipitates and increasing the potential mobility of the metal, which may in this form be leached out of, or down, the soil profile. The importance of these reactions will be limited by the availability of organic ligands and by the high particle reactivities of many metals. In the absence of organic ligands, Co, Mn and Cd exist primarily as free cations, Cu, Pb and Fe are largely complexed, while Zn and Ni are intermediate in their behaviour.

Since metals present in the soil solution and readily extractable fractions are most available to plants, factors increasing the concentration of metals in these phases will greatly increase their bioavailability. Because of the relationship between soil pH and metal mobility, treating contaminated soils with lime should reduce the bioavailability of heavy metals. Generally this appears to be the case; however, Davies *et al.* (1987) report that Zn leaf contents of radish plants grown on contaminated soil which had been limed were positively correlated with soil pH and exchangeable soil Zn content. In this study, although elevated pH reduced the total amount of Zn available to the plant, multiple regression analysis indicated that leaf Zn levels were positively related to *total* Zn content but *negatively* related to pH; the relative proportion of Zn present in a readily exchangeable and therefore, available form, was increased by liming.

Agricultural soils are aerated, freely drained and well oxidized. Marked seasonal variations in redox potentials do not occur. Under such conditions redox potentials and pE values (pE = − log(aqueous free electron activity)) have little influence on the solubility and mobility of soil metals. However, under mildly reducing ($+2<pE\leq+7$) and strongly reducing ($-6.8<pE\leq-2$) conditions, the speciation and mobility of metals, particularly Fe and Mn, but also Cr, Cu, As, Hg and Pb will be affected. Fe^{3+} is only stable under acidic oxidizing conditions; in reducing soils Fe^{2+} predominates,

$$FeOOH \rightleftharpoons e^- + 3H^+ Fe^{2+} + 2H_2O$$

with the formation of $Fe(OH)_2$, $FeCO_3$, FeS and FS_2 occurring. Similar insoluble sulphides of Cd, Zn, Ni, Co, Pb and Sn are formed under reducing conditions. In practice such conditions will normally only occur in waterlogged soils or in lake and river sediments. On draining, immobilized metals will be released as they are oxidized and more soluble metal species produced.

Metal toxicity

The influence of metal toxicity on human affairs has a long and eventful history dating back at least 5000 years to the initial development of metal working. Lead toxicity has been implicated in the decline of the Roman Empire. To stop superior green wine souring, sapa, a Pb-containing ingredient, was added prior to sealing amphorae in the proportions of 1 part sapa to 50 parts wine. This practice was probably obtained by the Romans from Dorian Greeks, who had used it previously for at least a 1000 years. Sapa was made by simmering, in a pure lead pot, a mixture of herbs, spices and fresh grape juice. Modern attempts to make sapa suggest the wine probably contained *c.*1000 ppm of Pb, making it an effective poison. Wine treated in this way would have contained *c.*20 mg Pb l^{-1}. The consumption of 1 l a day of such wine for a year, not an exceptionally high consumption rate among the Roman elite, would have caused overt Pb poisoning. Patterson *et al.* (1987) state that "if they didn't die, they would, among other things, be infertile and recognized as mad by less-poisoned observers". Less wealthy Romans, unable to purchase the superior grades of wine containing sapa, did not escape the effects of Pb toxicity. Their food and drinking water would have been contaminated. Lead guttering and lead-lined channels were used for the collection and storage of water while earthenware pots and lead cooking utensils would have contaminated many foodstuffs. Lead piping and solders containing Pb, used in the jointing of copper pip/e, still represent a potential source of public water contamination (Forstner and Wittman, 1981; Patterson *et al.*, 1987; Fergusson, 1990).

Lead poisoning probably contributed to the loss of Sir John Franklin's North West Passage expedition, which left London in May 1845. The two expedition ships were last seen in Baffin Bay. They subsequently became trapped in the Arctic ice and had to be abandoned. The surviving ship's companies, attempting to reach safety by travelling south overland on foot, all perished. The fate of the expedition was not discovered until 1859, when a record of the expedition up to 25th April 1848 was found in a cairn at Point Victory. Recent autopsies on the remains of some expedition members, preserved in the permafrost, showed very high levels of Pb, sufficient to cause overt signs of lead poisoning. The resulting disorientation and delirium would have considerably exacerbated the already precarious position of the expedition. The expedition was among the first to rely heavily on canned food; unfortunately, the method of manufacture and solders used then heavily contaminated canned food with Pb (Beattie and Geiger, 1987). Even today Pb from the solder of food tins still accounts for approximately 1/3 of the average per capita absorption of Pb (Forstner and Wittman, 1981; Patterson *et al.*, 1987).

Lead is a ubiquitous pollutant due to its presence in smelted metal ores, its use in a wide range of manufacturing processes and products, and its continued use as a petrol additive. Examination of ancient Peruvian bones suggest that current human body levels are 1000-fold above natural levels. Patterson *et al.* (1987) point out that our current understanding of cellular biology and biochemistry is based on the study of systems significantly contaminated with Pb.

In more recent years, industrial effluents have been responsible for several major episodes of chronic and acute poisonings of humans. The discharge of methyl mercury into Japanese rivers in the 1950s and 1960s poisoned many local people. The toxin became concentrated in the food chain, contaminating locally caught fish, an important source of food. Members of the local population developed 'Minamata Disease', named after the site of contamination and first appearance of the disease, Minamata Bay. The symptoms reflect the neurotoxicity of methyl mercury; severe trembling, mental confusion, followed by numbness, disturbed vision and speech, loss of control of body functions, violent thrashing, unconsciousness and in 40% of the original cases death. In 1975, a total of 3500 victims of Minamata Disease had been documented. Another major incidence of mercury poisoning occurred in Northern Iran during 1972. Peasant farmers, supplied with wheat seeds treated with mercurial fungicides (ethyl mercury *p*-toluene sulfonalilide), ate the seed instead of sowing it, causing localized outbreaks of mercury poisoning. In order to prevent this, local government authorities announced that farmers in possession of treated seed would be liable to prosecution and the possibility of incurring the death penalty. As a result the farmers disposed of the seed in nearby rivers and lakes, contaminating fish and water sources. The combined effects of seed consumption and disposal is estimated to have killed between 5000 and 50 000 and caused permanent disability to more than 100 000 people (Forstner and Wittman, 1981; Freedman, 1989).

Similar accounts of acute and chronic poisoning episodes involving other metals, e.g. Cd, Cr, As, Fe, and Zn may be found in the literature. In most cases these poisoning incidents are localized, being largely confined to industrial workers and populations close to centres of manufacture or waste and effluent disposal; the symptoms and the source of the contaminate can be easily identified and steps taken to alleviate the situation. However, the health risks associated with long-term exposure to low levels of contamination are more difficult to quantify, and thus the requirement for remedial action is more difficult, if not impossible to establish.

The potential human toxicity of metal ions can be related to their position in the periodic table. Toxicity decreases with an increase in the stability of the electron configuration. Among the highly electropositive metal ions of Group IA and Group IIA elements which occur in biological systems primarily as free cations, toxicity increases with atomic number:

IA: Na<K<Rb,Cs
IIA: Mg<Ca<Sr<Ba

Among subgroups IB, IIB, IIIA, toxicity increases with electropositivity:

IB: Cu<Ag<Au
IIB: Zn<Cd<Hg
IIIA: A1<Ga<In<Tl

This increase in toxicity reflects the affinity of the elements for amino, imino and sulphydryl groups, which are often associated with enzyme active centres. Thus, in general terms, type-A metals are less toxic than type-B. However, generalizations such as these, must be treated with some caution. For example, the position of Cd in the periodic table would suggest that alkyl-Cd compounds could pose a significant health risk; however, it is known that, unlike alkyl-Hg compounds, alkyl-Cd compounds are unstable in aqueous solutions.

The toxicity of metals is partially dependent on the form in which they occur and the ease with which they are accumulated. At a cellular level, toxic elements which occur in forms analogous to essential nutrient elements are likely to be particularly toxic. The toxicity of Se is partially attributable to its chemical similarity to sulphur, and the occurrence of selenate (SeO_4^{2-}) and selenite (SeO_3^{2-}). In both algae and higher plants, sulphur is actively taken up as sulphate (SO_4^{2-}). Selenium toxicity is thought to result from the active uptake of selenate in place of sulphate and the subsequent incorporation of Se into metabolites in place of sulphur. The cellular accumulation of non-essential metals occurs via passive diffusion across the cell membrane. For hydrate aqueous metal species, the outer cell membrane represents a significant barrier to uptake. Fisher (1986) has demonstrated that phytoplankton sensitivity to toxic metals is related to the particle reactivity of the metal. Passive uptake of metals is a two-stage process. Metals initially bind onto the outer cell wall and then diffuse into the cell down a concentration gradient. Metals which readily associate with particulates, become concentrated on the cellular surface and subsequently diffuse into the cell. Conservative, non-particle reactive cations will not be accumulated as readily.

The bioavailability and hence toxicity of metals may be affected by the presence of chelating ligands. The formation of metal–organic complexes can detoxify metals such as Cu and Pb. Over the concentration range 1–100 pM inorganic Cu is toxic to *Scendesmus quadricanda*, but the Cu–EDTA complex is not. However, the formation of lipophilic organometallic complexes will increase the uptake and toxicity of the metals. The algae *Nitzschia closterium* tolerates concentrations of 500 nM ionic Cu, however a concentration of 30 nM of Cu present as a lipophilic 8-hydroxyquinoline complex is highly toxic. Fish concentration factors for lipophilic metal species are generally in the range of 10^6–10^8 and for ionic metal species concentration factors range from 10^2 to 10^5. The high toxicity normally associated with alkyl–metal compounds is partly accounted for by the ease with which they are taken up and accumulated in the biota. However, it would be wrong to assume that alkylated metals are always the more toxic species. Alkylated forms of Se and As are substantially less toxic than their stable inorganic forms (Forstner and Wittman, 1981; Fergusson, 1990; Morgan and Stumm, 1991).

The environmental cycling and sources of metals

Metals are continuously released into the environment as a result of natural processes (e.g. weathering, volcanoes) and anthropogenic activity. The relative importance of anthropogenic sources of metals may be estimated by calculating the 'atmospheric interference factor' (AFT) (Morgan and Stumm, 1991), which is equal to the ratio of total anthropogenic emissions to total natural emissions and fluxes. High AFT values imply that element cycling has been substantially disrupted by human activity (e.g. Ag, Cd, Hg and Pb). In contrast, low values suggest that natural processes dominate the global mobility of the metal, although severe local and regional disruption due to industrial or mining activity may occur. Low AFT values may also be obtained despite substantial anthropogenic emissions if the element is abundant and natural fluxes exceed those due to human activity, e.g. Al and Fe (Table 13.3).

Table 13.3 Natural and anthropogenic sources of atmospheric emissions[a] (Lantzy & Mackenzie, 1979)

Element	*Continental dust flux*	*Volcanic dust flux*	*Volcanic gas flux*	*Industrial particulate emissions*	*Fossil fuel flux*	*Total emissions industrial plus fossil fuel*	*Atmospheric interference factor (%)*[b]
Al	356 500	132 750	8.4	40 000	32 000	72 000	15
Ti	23 000	12 000	–	3 600	1 600	5 200	15
Sm	32	9	–	7	5	12	29
Fe	190 000	87 750	3.7	75 000	32 000	107 000	39
Mn	4 250	1 800	2.1	3 000	160	3 160	52
Co	40	30	0.04	24	20	44	63
Cr	500	84	0.005	650	290	940	161
V	500	150	0.05	1 000	1 100	2 100	323
Ni	200	83	0.0009	600	380	980	346
Sn	50	2.4	0.005	400	30	430	821
Cu	100	93	0.012	2 200	430	2 630	1 363
Cd	2.5	0.4	0.001	40	15	55	1 897
Zn	250	108	0.14	7 000	1 400	8 400	2 246
As	25	3	0.1	620	160	780	2 786
Se	3	1	0.13	50	90	140	3 390
Sb	9.5	0.3	0.013	200	180	380	3 878
Mo	10	1.4	0.02	100	410	510	4 474
Ag	0.5	0.1	0.0006	40	10	50	8 333
Hg	0.3	0.1	0.001	50	60	110	27 500
Pb	50	8.7	0.012	16 000	4 300	20 300	34 583

[a]From Lantzy and Mackenzie (1979). See original paper for assumptions and discussion of sources of data. All fluxes are in units of 10^8 g per year.
[b]Atmospheric interference factor = [total emissions ÷ (continental + volcanic fluxes)] x 100.

Elements may be divided into two groups depending on their primary mode of transport. Elements for which mass transfer occurs predominantly in the atmosphere are referred to as atmophile elements. Such elements tend to be volatile and have low boiling point metal oxides. Type-B metals are typically atmophiles. Their environmental dispersal is greatly increased if they are able to form volatile methylated compounds. In contrast to atmophile metals, type-A metals are said to be lithophile, mass transport to ocean sinks occurring predominantly via streams and rivers.

The release of metals into the environment, would not pose problems if they were evenly distributed throughout the biosphere. Unfortunately this is not the case, anthropogenic activity frequently results in elevated local metal concentrations, even where dispersal is rapid and efficient localized contamination is normally evident. Dispersal, while helping to ameliorate metal concentrations in the immediate vicinity of the source, spreads the contaminate over a wider area. Metals associated with fine atmospheric particles (<2.5 μm) may remain aloft for several months, travelling considerable distances. Such emissions from the industrial centres of the developed world have contaminated both polar regions. Metal concentrations present in polar ice profiles correlate with past industrial activity. Dispersal and dilution is not a safeguard against undesirable environmental effects. The pollutant may become concentrated within the food chain (Ch. 1).

References

BAKER, A.J.M. (1987) Metal tolerance. *New Phytol.*, **106**(Suppl.), 93–111.

BEATTIE, O. and GEIGER, J. (1987) *Frozen in Time, Unlocking the Secrets of Franklins Expedition*. Western Producers Prairie Books.

BRULAND, K.W. (1983) Trace elements in sea-water. In Riley, J.P., Chester, R. (eds) *Chemical Oceanography*, pp. 157–220. Academic Press, London.

BURTON, J.D. (1979) Physico-chemical limitations in experimental investigations. *Phil. Trans. Royal Soc. Lond. B.*, **286**, 443–456.

CARLTON-SMITH, C.H. (1987) Effects of metals in sludge-treated soils on crops. Environment Technical Report 251. Water Research Council, Marlow.

COALE, K.H. and BRULAND, W.K. (1985) 234 Th : 238 U disequililbra within the California Current. *Limnol. Oceanogr.*, **30(1)**, 22–25.

COUGHTREY, P.J., JACKSON, D., JONES, C.H., KANE, P. and THORNE, M.C. (1983) *Radionucleotide Distribution and Transport in Terrestrial and Aquatic Ecosystems. A Critical Review of Data,* Vol. 3. A.A. Balkema, Rotterdam.

COUGHTREY, P.J., JACKSON, D., JONES, C.H., KANE, P. and THORNE, M.C. (1984) *Radionuclide Distribution and Transport in Terrestrial and Aquatic Ecosystems. A Critical Review of Data,* Vol. 4, pp. 93–115. A.A. Balkema, Rotterdam.

COUGHTREY, P.J., JACKSON, D. and THORNE, M.C. (1985) Radionuclide Distribution and Transport in Terrestrial and Aquatic Ecosystems. A Critical Review of Data, Vol. 6. A.A. Balkema, Rotterdam.

DAVIES, B.E., LEAR, J.M. and LEWIS, N.J. (1987) Plant availability of heavy metals. Special publication, BES No. 6, pp. 267–275. Blackwell, Oxford.

FERGUSSON, J.E. (1990) *The Heavy Elements. Chemistry, Environmental Impact and Health Effects*. Pergamon Press, Oxford.

FISHER, N.S. (1985) Accumulation of metals by marine picoplankton. *Marine Biol.*, **87**, 137–142.

FISHER, N.S. (1986) On the reactivity of metals for marine phytoplankton. *Limnol. Oceanogr.*, **31(2)**, 443–449.

FISHER, N.S. and FOWLER, S.W. (1987) The role of biogenic debris in the vertical transport of transuranic wastes in the sea. In O'Connor T.P., Burt, W.V., Duedall, I.W. (eds) *Oceanic Processes in Marine Pollution*, pp. 197–207. Krieger, Malabar, Florida.

FORSTNER, U. and SALOMONS, W. (1991) Mobilization of metals from sediments. In Merian, E. (ed.) *Metals and their Compounds in the Environment*, pp. 379–395. VCH, Weinheim.

FORSTNER, U. and WITTMAN, C.T.W. (1981) *Metal Pollution in the Aquatic Environment*. Springer-Verlag, Berlin.

FOWLER, S.W. (1990) Critical review of selected heavy metal and chlorinated hydrocarbon concentrations in the marine environment. *Marine Environ. Res.*, **29**, 1–64.

FREEDMAN, B. (1989) *Environmental Ecology. The Impacts of Pollution and Other Stresses on Ecosystem Structure and Function*. Academic Press, San Diego.

GESAMP (1990) *The State of The Marine Environment*. Blackwell Scientific Publications, Oxford.

HODGE, S., KOIDE, M. and GOLDBERG, E.D. (1979) Particulate uranium plutonium and polonium, in the biogeochemistries of the coastal zone. *Nature* (*Lond.*), **277**, 206–209.

HOPKINS, S.P. (1989) *Ecophysiology of Metals in Terrestrial Invertebrates*. Elsevier Applied Science, London, New York.

HUNTER, K.A. (1980) Processes affecting particulate trace metals in the sea surface microlayer. *Marine Chem.*, **9**, 49–70.

LANTZY, R.J. and MACKENZIE, F.T. (1979) Atmospheric trace metals: global cycle and assessment of man's impact. *Geochim. Cosmochim. Acta*, **43,** 511.

LAXEN, D.P.H. (1983) The chemistry of metal pollutants in water. In *Pollution: Causes, Effects and Control*, Harrison, R.M. (ed) Royal Soc. Chem., Special Publ. No. 44, 104–123.

MERIAN, E. (ed.) (1991) *Metals and Their Compounds in the Environment*. VCH, Weinheim.

MILLER, J.E., HASSETT, J.J. and KOEPPE, D.E. (1977) Interactions between lead and cadium on metal uptake and growth of corn plants. *J. Environ. Qual.*, **6**, 18–20.

MORGAN, J.J. and STUMM, W. (1991) Chemical processes in the environment, relevance of chemical speciation. In Merian, E. (ed.) *Metals and their Compounds in the Environment*, pp. 69–103. VCH, Weinheim.

NCAR (National Center for Atmospheric Research) (1982) *Regional Acid Deposition: Models and Physical Processes*. NCAR, Boulder, CO.

PATTERSON, C.C., SHIRAHATA, H. and ERICSON, J.E. (1987) Lead in ancient human bones and its relevance to historical developments of social problems with lead. *Sci. Tot. Environ.*, **61**, 167–200.

PEARSON, R.G. Hard and soft acids and bases. *J. Am. Chem. Soc.*, **85**, 3533.

PUXBAUM, H. (1991) Metal compounds in the atmosphere. In Merian, E. (ed.) *Metals and Their Compounds in the Environment*, pp. 257–277. VCH, Weinheim.

RASPOR, B. (1991) Metals and metal compounds in waters. In Merian, E. (ed.) *Metals and Their Compounds in the Environment*, pp. 233–253. VCH, Weinheim.

SCHMITT, H.S. and STICKER, H. (1991) Heavy metal compounds in the soil. In Merian, E. (ed.) *Metals and Their Compounds in the Environment*, pp. 311–326. VCH, Weinheim.

SHANNON, L.V. and CHERRY, R.P. (1971) Radium-226 in marine phytoplankton. *Earth Planet. Sci. Lett.*, **11**, 339–343.

STUMM, W. and BILINSKI, H. (1972) Trace metals in natural water. Difficulties of interpretation arising from our ignorance of speciation. In S.H. Jenkins (ed) *Advances in Water Pollution Research. Proc. 6th Int. Conf. Jerusalem,* pp. 39–52. Pergamon Press, New York.

WHITFIELD, M. and TURNER, D.R. (1987) The role of particulates in regulating the composition of seawater. In Stumm, W. (ed.) *Aquatic Surface Chemistry. Chemical Processes at the Particle–Water Interface*, pp. 93–124. John Wiley, New York.

Further reading

FERGUSSON, J.E. (1990) *The Heavy Elements. Chemistry, Environmental Impact and Health Effects*. Pergamon Press, Oxford.

FREEDMAN, B. (1989) *Environmental Ecology. The Impacts of Pollution and Other Stresses on Ecosystem Structure and Function*. Academic Press, San Diego.

GESAMP (1990) *The State of The Marine Environment*. Blackwell Scientific, Oxford.

HOPKIN, S.P. (1989) *Ecophysiology of Metals in Terrestrial Invertebrates*. Elsevier Applied Science, London and New York.

MERIAN, E. (ed). (1991) *Metals and their Compounds in the Environment*. VCH, Weinheim.

PATTERSON, C.C., SHIRAHATA, H. and ERICSON, J.E. (1987) Lead in ancient human bones and its relevance to historical developments of social problems with lead. *Sci. Tot. Environ.* **61**:167–200.

Chapter 14

METAL AND RADIONUCLIDE BIOTREATMENT

Introduction

Living organisms require certain elements to grow and maintain their physical structure, metabolic activity and reproductive capacity. They have evolved appropriate uptake mechanisms for the metals essential for their metabolic activity. On occasions toxic metal elements may be taken up which can result in physiological damage to the organism or even death. Even those metal elements required in minor or trace amounts for normal cell functions may cause inhibitory or toxic effects if present in excessive concentrations (Ch. 13). Many organisms have developed detoxification mechanisms to overcome the detrimental effects of metals. In microorganisms these resistance mechanisms take several forms including transformation of the toxic metal resulting in the release of the metal in another physical and/or chemical form, e.g. metal alkylation. Certain metal uptake and transformation processes catalysed by organisms can be applied to metal waste treatment.

Heavy metal accumulation by plants and microorganisms, and the products of plants, microorganisms or animals, offers a mechanism by which metal-contaminated liquid waste may be treated to lower its metal content prior to discharge to the environment. Microorganisms, however, may not only act as biosorbing agents but enter into a range of other interactions with metals. Indeed, microorganisms show a far wider variety of interactions with metals than do plants and animals. For example, certain metals such as iron play a major role in the energy production of chemolithotrophic bacteria which derive energy from the oxidation of inorganic compounds. This has been applied in the bacterial leaching of low grade sulphide or sulphide-containing ores to extract metals such as copper and uranium (see below). The volatilization of metal by microorganisms, e.g. mercury methylation, the precipitation of metals as metal sulphides and the solubilization of metals by microbially produced products, e.g. acid production, are further examples of the diversity of metal–microbe interactions. Many of these processes have potential applications in the treatment of metal-contaminated material.

The application of whole organisms or their products to metal pollution control extends to radionuclides. Radionuclides fall into the following three categories.

1 Radionuclides which have a stable isotope present in the environment such as the radionuclides of iron. These radionuclides will behave chemically and physically in an identical manner to their stable isotopes, the only difference being their radioactivity.
2 Radionuclides which have no stable analogues but occur naturally in the biosphere, e.g. uranium.
3 Radionuclides which are artificially produced during nuclear fission reactions, i.e. in nuclear power plants or from nuclear weapons tests, and do not normally or naturally occur in the biosphere, e.g. plutonium, americium and technetium. These radionuclides are often highly radioactive and frequently highly toxic.

It is easier to predict the interactions that the first two radionuclide types may have with organisms and, hence potential biotreatments, since they or their analogues exist in the biosphere. Little is known about interactions between organisms and the artificial radionuclides. Although it is possible to predict some effective biotreatment systems from a knowledge of the chemistry of the artificial radionuclides, a number of interactions may well be unpredictable. This suggests that a full study of the degree and nature of organisms' reactions with artificial radionuclides is desirable as an initial step in the development of biotreatment processes.

Recently there has been considerable interest in the application of biotreatment to resolve heavy metal and radionuclide contamination problems. Some of the treatment processes use physiologically active organisms while others use non-living biomass or biological products. The possible and actual applications of biotreatment processes range from removal of heavy metals from liquid effluent, bioremediation of metal-contaminated land, metal recovery from low grade ores and the recovery of valuable, rare or strategically important metals/radionuclides from waste.

The aim of this chapter is to review the mechanisms for treatment of metal- or radionuclide-contaminated waste and metal/radionuclide recovery from waste streams or low grade ores by biological processes. The physiology and biochemistry of interactions between heavy metals and organisms or their products will be considered emphasizing those which may be applied to metal treatment.

Precipitation of heavy metals and radionuclides

The precipitation of metals from aqueous solutions can be biologically mediated primarily through the activity of microorganisms. Bacteria in particular produce extracellular products which interact with free or sorbed metal cations forming insoluble metal precipitates.

The major mechanism for bacterial metal precipitation is through the formation of hydrogen sulphide and the immobilization of the metal cations as metal sulphides. The bacteria involved in this interaction are the sulphate-reducing bacteria including members of the genera *Desulfovibrio* and *Desulfotomaculum*. Sulphate-reducing bacteria are widely distributed in anaerobic environments, e.g. sediment and bogs. They are obligate anaerobes. Sulphide production by sulphate-reducing bacteria is a consequence of their energy generating processes. They couple the reduction of oxidized forms of sulphur, e.g. sulphates and sulphur, with the oxidation of reduced carbon in the form of simple organic molecules. The activity of sulphate-reducing bacteria has been attributed with contributing to the formation of metal sulphide minerals such as covellite (CuS) (Ferris *et al.*, 1989). Metals and metal-like elements including copper, lead, zinc, iron, uranium and selenium, can be precipitated in this way and in an appropriately engineered system can be accumulated and removed as sludge. Some other micro-organisms also precipitate metals as metal sulphides, apparently as a detoxification process, e.g. *Klebsiella aerogenes* precipitates cadmium sulphate at toxic cadmium concentrations (Aiking *et al.*, 1982).

Other microbial products can be involved in metal precipitation. A *Citrobacter* sp. produces a phosphatase enzyme which cleaves hydrogen phosphate (HPO_4^{2-}) from organic phosphates. The hydrogen phosphate subsequently precipitates heavy metals and radionuclides including lead, cadmium and uranium as metal phosphates ($MHPO_4$). The metal phosphate is precipitated at the cell wall and is cell bound (Macaskie, 1990) (see below). Other microorganisms cause metal precipitation through the production of hydrogen peroxide (forming insoluble metal oxides) or fungal production of oxalic acid and subsequent metal immobilization as metal oxalate crystals.

Biotransformations of heavy metals and radionuclides

Microorganisms catalyse a range of metal transformations which may be applied to the treatment of metal- or radionuclide-contaminated waste. These transformations include alkylation, reduction and oxidation reactions (Table 14.1).

Oxidation reactions

Ferrous (Fe^{2+}) and manganese ions (Mn^{2+}) can be deposited through oxidation reactions catalysed by bacteria, fungi, algae and protozoa.

Table 14.1 Heavy metal and radionuclide transformations catalysed by microorganisms

Examples of metal/radionuclide	*Transformation*	*Microorganisms*
Fe^{2+} and Mn^{2+}	Oxidation and precipitation of $Fe(OH)_3$ and MnO_2 (singly or together)	Heterotrophic bacteria, e.g. *Leptothrix*, fungi, algae and protozoa
Cd^{2+}, Au^{3+}, Cu^{2+} and $U_2O_2^{2+}$	Oxidation (direct or indirect) and solubilization	Autotrophic bacteria, e.g. *Thiobacillus ferrooxidans*
Hg^{2+}, Fe^{2+}, Mn^{2+} and Se compounds	Reduction	Heterotrophic and autotrophic bacteria and fungi
Sn^{2+}, Se compounds and Pb^{2+}	Alkylation (especially methylation) results in volatilization of some metals, and dealkylation	Heterotrophic bacteria, fungi and algae

The interaction is best studied for the bacteria and fungi. Ferromanganese oxidizing microorganisms are important in the formation of ferromanganese minerals. The oxidation of the metal cations does not apparently involve energy generation through autotrophic growth, but rather is a sequence of reactions resulting in cation oxidation and precipitation of $Fe(OH)_3$ and MnO_2 within a surface-bound exopolymer which is often in the form of a sheath. The biochemistry of the oxidation reaction is poorly understood but appears to involve Mn^{2+} oxidation at a solid surface which acts as a nucleation point for further Mn^{2+} deposition and oxidation, and initiates Fe^{2+} oxidation at these sites (Nealson *et al.*, 1989). Bacteria such as the genera *Leptothrix* and *Sphaerotilus* (sheathed bacteria) are among the ferromanganese depositing bacteria, while *Gallionella* species oxidize ferrous iron. Many isolates from freshwater and marine sediments, e.g *Hyphomicrobium*, can oxidize manganese (Ghiorse, 1984). Manganese oxidation may have a role in metal waste treatment and bioremediation through the co-precipitation of other metals with MnO_2, although no process has yet been developed.

Mineral oxidizing bacteria, e.g. *Thiobacillus ferrooxidans* and *Leptospirillum ferrooxidans*, can solubilize metals from minerals allowing the extraction and recovery of metals such as copper, cadmium, gold and uranium from low grade or refractory ores. This procedure has been developed for metal extraction rather than for the clean-up of metal-contaminated waste. There are, however, environmental implications since metal leaching from mine waste is a natural process involving the oxidation of mineral sulphides and results in 'acid run-off' which contains metals that have been solubilized, especially iron and copper. Increased management of the process for metal recovery is, therefore, inevitably beneficial to the environment through the planned removal of contaminating heavy metals.

Solubilization of metals occurs either through direct oxidation of the metals by the microorganisms or indirectly through chemical oxidation by products of microbial metabolism. In metal leaching systems these two processes probably occur simultaneously.The bacteria involved in sulphide and ferrous iron oxidation are common in sulphide-rich environments and are chemolithotrophic. They obtain carbon from carbon dioxide and energy from the oxidation of inorganic compounds. These chemolithotrophs are extremophiles. All are acidiphiles (pH 2.5–1.5) and some mineral oxidizing genera, e.g. *Sulfolobus*, are also thermophilic (65–80°C)(Norris, 1989).

Thiobacillus ferrooxidans is probably the most extensively studied of the iron-oxidizing bacteria and catalyses the following reactions important in indirect leaching:

$$\underset{\text{(pyrite)}}{2FeS_2} + 7O_2 + 2H_2O \rightarrow 2FeSO_4 + 2H_2SO_4 \quad (14.1)$$

$$4FeSO_4 + O_2 + 2H_2SO_4 \rightarrow 2Fe_2(SO_4)_3 + 2H_2O \quad (14.2)$$

Many metal sulphide ores are oxidized by ferric sulphate releasing soluble metal salts, e.g. $CuSO_4$, and elemental sulphur (S^0), for example:

$$Fe_2(SO_4)_3 + \underset{\text{(covellite)}}{CuS} \rightarrow 2FeSO_4 + S^0 + CuSO_4 \quad (14.3)$$

Ferric sulphate is regenerated (Eq. 14.2) for further indirect metal oxidation and the elemental sulphur is oxidized to sulphuric acid by *T.ferrooxidans* which also acts as a metal leaching agent producing the soluble metal sulphate salt (Brierley, 1978; Hughes and Poole, 1989a). The direct oxidation of the metal minerals is an aerobic process involving the oxidation of insoluble metal sulphides to their sulphate salts (Eq. 14.1).

Reduction reactions

The reduction of oxidized metal compounds can be catalysed by microorganisms either resulting in a partial decrease in the charge of the metal or in the release of the metallic element. Metals and metalloids which can be reduced by microbial activity include mercury, iron, manganese, selenium, tellurium and arsenic. Selenium and mercury reduction will be considered as examples of microbial reductive processes. Although selenium is not regarded as a heavy metal, ^{123}Se is a relatively important radionuclide in radioactive waste.

Selenium exists in several oxidation states (selenide, Se^{2-}; elemental selenium, Se^0; selenite, SeO_3^{2-}; and selenate, SeO_4^{2-}). Bacteria and fungi can reduce oxidized selenium compounds usually resulting in the production of elemental selenium. Both heterotrophic microorganisms and chemolithotroph microorganisms can reduce selenium compounds, which appears to be an intracellular process and a function of metabolic activity. Selenium salt reduction does not require anaerobic conditions and appears to proceed in a step-wise fashion, i.e. selenate → selenite → selenium → selenide. Individual microorganisms may catalyse only one of these steps rather than complete reduction to selenium or selenide, indeed complete reduction may involve a degree of synergistic activity between microorganisms (Doran, 1982). Until relatively recently it was thought that organisms involved in selenium reduction were all involved in the sulphur cycle (sulphur often acting as a competitive inhibitor of selenium reduction), however a bacterium capable of dissimilatory selenium reduction has recently been isolated (Oremland *et al.*, 1989). This organism uses selenate as the terminal electron acceptor for energy generation, i.e. selenium respiration. Selenate is reduced to red elemental selenium. The reduction of the soluble selenium salts (selenate and selenite) to elemental selenium results in the precipitation of the metal and could potentially be utilized for removal of soluble selenium salts from waste streams, i.e. precipitation and removal in sludge. Moreover, it may be possible to apply these processes to bioremediation of selenium-contaminated land, a growing problem, by ensuring the metal is transformed to its unavailable solid form.

Mercury compounds can be reduced by a range of heterotrophic bacteria and even by the chemolithotroph *Thiobacillus ferrooxidans* (Levi and Linkletter, 1989). Reduction of mercury compounds is a detoxification reaction and involves the uptake of Hg^{2+} and its subsequent intracellular reduction by the enzyme mercuric reductase. The product of the reduction is elemental mercury (Hg^0) vapour, thus mercury is volatilized during reduction, diffusing from the microbial cell and escaping to the atmosphere. Mercury resistance is plasmid encoded by the mer operon genes (Mergeay, 1991). The application of mercury reduction to waste treatment and bioremediation would be problematic. Gaseous products are extremely difficult and expensive to collect and loss to the atmosphere would be environmentally unacceptable.

Alkylation reactions

Other metals, e.g. tin, selenium and lead, can be volatilized by microbial action through the production of alkylated metals. It has been proposed that alkylation reactions are metal detoxification mechanisms (Mergeay, 1991). The alkylation process can occur internally and is mediated by the methylating agents methylcoblamin, *S*-adenosylmethionine and methyl tetrahydrofolic acid, e.g. in the methylation of tin, lead, selenium and mercury. Extracellular methylation of metals can be induced by products of metabolism, for example, marine algae and bacteria produce a range of compounds that are capable of alkylating metals including halomethanes, carbimides and haloaromatic compounds.

Dimethyl selenide ($(CH_3)_2Se$) and dimethyl diselenide ($(CH_3)_2Se_2$) are the common products of selenium methylation which can be catalysed by various heterotrophic bacteria, e.g. *Pseudomonas* sp. and *Corynebacterium* sp. and fungi, e.g. *Alternaria alternata*. The proportion of the volatile products formed and the rate of conversion varies with organism, the type of selenium compound (selenate, selenite or elemental selenium) undergoing methylation and environmental conditions (Doran, 1982). Mercury methylation commonly involves the conversion of mercuric ion (Hg^{2+}) to dimethyl mercury ($Hg(CH_3)_2$) (Levi and Linkletter, 1989). Mercury methylation is probably the most extensively studied of the metal alkylating reactions because of the health significance of its production demonstrated so clearly in the tragedy at Minamata Bay in Japan (see Ch. 13). Organometals are often more toxic than their inorganic counterparts. This presents a major problem in utilizing microbial alkylation reactions for metal waste treatment, in addition to gas capturing difficulties. As a consequence, little work has been undertaken in developing alkylation reactions for waste stream treatment or bioremediation.

Intracellular accumulation

Essential metal nutrients such as iron and magnesium are actively taken up and accumulated by plants and microorganisms from the surrounding environment. In land-based plants the absorption of ions is primarily confined to the roots, while aquatic plants show more general absorption. There are three main mechanisms of metal transport into

Table 14.2 Mechanisms and characteristics of intracellular accumulation

Mechanism	*Driving force*	*Carrier protein*	*Specificity*	*Saturation*	*Inhibition*
Free ion or passive diffusion	Electrochemical or concentration gradient	Absent	Non-specific	Absent	Absent
Facilitated diffusion	Electrochemical or concentration gradient	Present	Specific	Present	Present (competitive and non-competitive)
Active transport	Metabolic energy	Present	Specific	Present	Present (competitive)

both prokaryotic and eukaryotic cells: free ion or passive diffusion, facilitated diffusion and active transport (Levi and Linkletter, 1989; Brierley *et al.*, 1989) (Table 14.2).

Free ion diffusion of ions occurs in response to an electrochemical or concentration gradient across the cell membrane with the metal cations moving down the gradient, i.e. from a higher to a lower metal concentration. It is not an active process and does not involve specific membrane binding sites. The rate of cation diffusion by this process is directly related to the magnitude of the gradient and the permeability of the cell wall and membrane to the ion. Passive diffusion is not subject to saturation.

Facilitated diffusion involves the binding of the cation to a carrier protein or permease (which is often inducible) at the outside of the membrane. The carrier–cation complex then moves across the membrane and the ion is released at the inside. Facilitated diffusion does not require the expenditure of metabolic energy but involves a concentration or electrochemical gradient. This process is relatively specific due to the involvement of a carrier molecule and can be saturated, showing Michaelis–Menten kinetics. It is subject to competitive and non-competitive inhibition. The rate of facilitated diffusion is rapid and greater than passive diffusion. This process is relatively common in eukaryotes such as plants and fungi.

The third cation uptake mechanism is active transport which permits the entry of nutrients against a concentration gradient. This process involves the expenditure of metabolic energy and is mediated by a carrier protein in both eukaryotes and prokaryotes. In plants the binding of ions to the carrier complex and subsequent transfer of the metal–carrier complex across the membrane is driven by membrane-bound ATPases. Uptake by this process is specific, shows Michaelis–Menten kinetics, i.e. uptake increases with cation concentration until saturation, and is subject to competitive inhibition. Metals such as zinc and cobalt undergo active transport in plants. In terrestrial plants the active transport system is complicated by the existence of biphasic uptake which involves two separate systems. System I is induced at low cation concentrations and becomes saturated at concentrations of 0.2 mM. System II is saturated

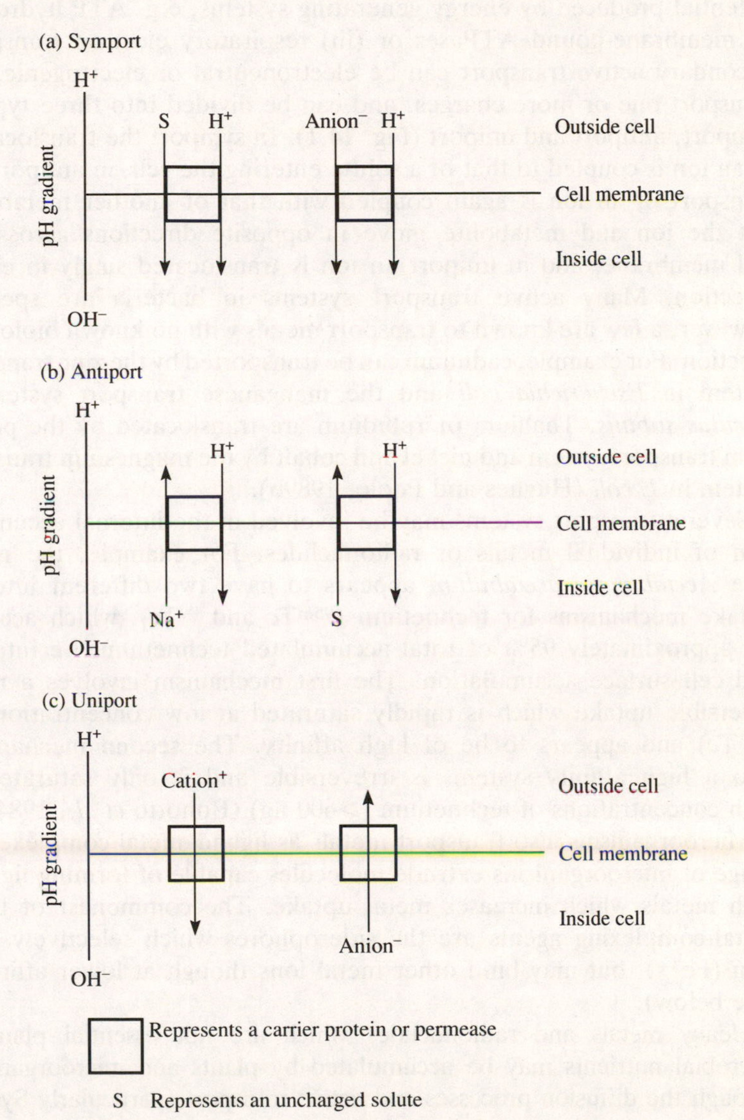

Fig. 14.1. Diagrammatic representation of bacterial secondary active transport systems.

at much higher concentrations (50 mM), lacks the specificity of System I and does not fully follow Michaelis–Menten kinetics.

Prokaryotes also have two types of active transport system which follow Michaelis–Menten kinetics. Primary active transport systems are linked to an enzymatic reaction such as the 'ion-pumps' for Na^+ efflux which are powered through ATP hydrolysis by membrane bound ATPases. Secondary active transport is coupled singly or together to: (i) a previously established concentration gradient, (ii) the electrical potential produced by energy generating systems, e.g. ATP hydrolysis by membrane-bound ATPases or (iii) respiratory electron transport. Secondary active transport can be electroneutral or electrogenic, i.e. transport one or more charges, and can be divided into three types – symport, antiport and uniport (Fig. 14.1). In symport the translocation of an ion is coupled to that of a solute entering the cell, in antiport the transport of an ion is again coupled with that of another metabolite but the ion and metabolite move in opposite directions across the cell membrane, and in uniport an ion is translocated singly in either direction. Many active transport systems in bacteria are specific, however, a few are known to transport metals with no known biological function. For example, cadmium can be transported by the zinc transport system in *Escherichia coli* and the manganese transport system in *Bacillus subtilis*. Thallium or rubidium are translocated by the potassium transport system and nickel and cobalt by the magnesium transport system in *E.coli* (Hughes and Poole, 1989b).

Several transport systems may be involved in the internal accumulation of individual metals or radionuclides. For example, the green alga *Acetabularia acetabulum* appears to have two different internal uptake mechanisms for technetium (^{95m}Tc and ^{99}Tc), which account for approximately 95% of total accumulated technetium, i.e internal and cell surface accumulation. The first mechanism involves a rapid reversible uptake which is rapidly saturated at low concentrations (4 pg Tc) and appears to be of high affinity. The second mechanism, also a high affinity system, is irreversible and is only saturated at high concentrations of technetium (>600 ng) (Bonotto *et al.*, 1984).

Microorganisms also transport metals as ligand–metal complexes. A range of microorganisms extrude molecules capable of forming ligands with metals which increases metal uptake. The commonest of these metal-complexing agents are the siderophores which selectively bind iron (Fe^{3+}), but may bind other metal ions though at lower affinities (see below).

Heavy metals and radionuclides which are not essential plant or microbial nutrients may be accumulated by plants and microorganisms through the diffusion processes and active transport (particularly System II in terrestrial plants and secondary active transport by bacteria). All these uptake processes will occur in those biotechnological systems involving living plants and microorganisms, e.g. meanders and reed beds (Chs 10 and 15). The importance of internal accumulation of heavy metals and radionuclides in the living treatment systems is difficult to assess. It will undoubtedly be affected by environmental conditions, e.g. the presence of other ions and the pH are known to influence uptake rates, an effect which varies between metals. Moreover although many heavy metals are non-essential or even toxic and will not have specific transport systems available, a range of radionuclides will represent essential nutrients and will have specific transport systems for translocation across membranes. This will influence the rate and degree of uptake of certain radionuclides. The physiological state of the organisms will also affect uptake. The requirement for metabolic energy by active transport systems means that any factor which lowers the metabolic activity of the cell will have a subsequent limiting effect on active transport. Therefore, careful management of living systems is essential if active transport is to be considered a major mechanism of metal removal from the waste. Even in biotreatment processes using living systems for metal/radionuclide removal such as reed beds, it is probable that processes other than active uptake may be predominant in metal removal.

Extracellular accumulation

Heavy metals and radionuclides accumulate at cell surfaces through precipitation or binding reactions. Extracellular accumulation is likely to occur in the plant component of the living treatment systems, e.g. reed beds. This is far less significant, however, than surface accumulation by algae, bacteria and fungi. These organisms can be used in treatment processes that permit greater control and the opportunity for metal/radionuclide recovery. It is intended therefore to restrict this section to consideration of extracellular accumulation by microorganisms and unicellular algae. The mechanism of extracellular accumulation, however, will be comparable to those which are likely to occur on plant surfaces in reed beds and meanders.

The sorption of heavy metals and radionuclides by microbial or algal cells is essentially a passive process which does not involve the expenditure of metabolic energy and is primarily controlled by physicochemical factors. The binding of heavy metals and radionuclides by surface sorption can result in considerable accumulation and removal of contaminants from waste streams (Table 14.3). Since the process is passive it occurs to both living and dead cells, as well as to cellular debris. The ability to use dead cells or cellular debris in metal sorption systems for waste treatment offers considerable process advantages (see below).

Table 14.3 Uranium accumulation by microbial sorption process

Organism	*Accumulation capacity (mg g^{-1} dry wt)*	*Initial uranyl ion concentration (mg l^{-1})*	*References*
Fungi			
Apergillus niger	215	100	Mergeay (1991)
Penicillium chrysogeum	70	150	Oremland *et al*. (1989)
Bacteria			
Pseudomonas fluorescens	6	150	Oremland *et al*. (1989)
Waste biomass			
Municipal sludge	12	150	Oremland *et al*. (1989)

Several mechanisms for cell surface sorption have been identified and include cation exchange, complexation or co-ordination, chelation and microprecipitation. These physico-chemical processes may act singly or in combination depending on the metal/radionuclide and the species or even strain of organism. The basis for the differences between types of microorganisms and algae is related to differences in cell wall characteristics (McEldowney, 1990).

Bacterial, fungal and algal cell walls are comprized of complex and highly organized macromolecules. There is a great diversity in eukaryotic cell wall structure. Fungal and yeast cell walls are composed of mannan polysaccharides, galactosamine, chitin, protein and lipid. Common components of algal cell walls include cellulose, gelatinous materials such as alginic acid and fucinic acid, and silicified components. The molecular structure and molecular components of the cell wall varies with algal taxonomic group, although cellulose is the commonest component. The characteristics of algal surfaces vary depending on the cell wall biochemical composition. Most bacteria are Gram-negative, e.g. *Citrobacter*, *Pseudomonas Acinetobacter*, or Gram-positive, e.g. *Bacillus* and *Streptomyces*, and are differentiated on the basis of their cell wall composition and structure. Both bacterial types have the mucopolysaccharide peptidoglycan as a key component, although the amount of this polymer is far higher in Gram-positive bacteria than Gram-negative. Many Gram-positive bacteria have teichoic acids associated with the peptidoglycan. Gram-negative bacteria have lipopolysaccharides, lipids and proteins as major cell wall constituents which lie above the peptidoglycan layer.

In all these cases the cell wall presents a primarily anionic aspect to the surrounding environment. This is due to the presence of functional groups such as carboxyl, hydroxyl, sulphyl and phosphyl groups. These charged groups are particularly important in cation sorption. In addition, cell wall uncharged groups may act as ligands completing the co-ordination number of metal cations, e.g. nitrogen atoms in peptides.

Since the detailed composition and molecular configuration of cell walls, the type of polar group present and the charge distribution within the cell wall is dependent on species, the explanation for metal sorption differences between species and strains is apparent. Moreover, growth conditions affect the characteristics of cell walls, particularly bacterial walls (McEldowney, 1990). Cell surface charge and molecular characteristics can be modified, also, by chemical, e.g. alcohol or KOH, or physical treatments, e.g. heating, which act through denaturation or solvent effects. There is then the potential to manipulate heavy metal or radionuclide sorption capacities and kinetics shown by microbial species so that they are optimal for the particular waste undergoing treatment by several mechanisms including the following:

- the control of growth conditions of a carefully selected biomass type to ensure appropriate cell surface characteristics
- the treatment of biomass by chemical or physical agents
- the genetic manipulation of a chosen microorganism to improve surface metal binding

Many of the sorption systems currently under investigation use waste biomass from other industries, e.g. fermentation waste. This has the advantage of providing a cheap source of biomass, but permits the use of only the second method for manipulation of the cell surface properties to improve sorption characteristics.

The mechanism or mechanisms by which heavy metals or radionuclides are bound to cell surfaces vary with biomass type and the chemistry of the metal/radionuclide undergoing sorption. Examples of different uranium sorption interactions with cell surfaces will be discussed in order to appreciate the range and complexity of the interactions involved. These interactions, however, are not unique to uranium and will be replicated for other metals, e.g. cadmium, lead, copper, zinc and iron.

Uranium is bound to the cell wall by several different mechanisms. Cation exchange reactions, i.e. the exchange of metal cations with counter ions associated with anionic groups on the cell surface appear to be fairly common. Several different anionic active sites, e.g. carboxyl and phosphyl groups, are involved depending on the particular microorganism and the biochemistry of the cell wall. Complexation or co-ordination interactions also occur between cell walls and uranium, although the cell surface ligands involved in the reaction have seldom been elucidated. It is thought that several different active sites and ligands can be involved in the co-ordination of uranyl ions. Some sites such as carboxyl and phosphyl, form primary bonds with the uranium while secondary weaker bonds are formed with surface sites such as hydroxyl and amyl groups. These secondary bonds are thought to augment the primary bonds (Tobin *et al.*, 1984). The secondary sites may only be complexed on saturation of the primary sites. Non-stoichiometric complexation of uranyl ions and levels of accumulation beyond that predicted by availability of cell surface anionic sites may be due to additional metal crystallization at sites of previously complexed uranium. The previously bound metal acts as a nucleation site for the deposition of further metal. This process can act only in conjunction with other surface sorption mechanisms (Beveridge, 1978). A further mechanism of sorption is through precipitation reactions. One such precipitation reaction is shown by a *Citrobacter* sp. which sorbs uranium as insoluble uranium phosphate at the cell wall. Phosphatase exoenzymes produced by the bacterium cleave supplied organic phosphate, producing inorganic phosphate which precipitates with uranium at the cell surface (Macaskie, 1990).

Cation exchange processes, co-ordination reactions or precipitation reactions can occur singly or together, or may be further co-joined with a nucleation reaction. This varies with the organism and the metal under consideration. For example, *Rhizopus arrhizus* has three separate uranyl accumulation mechanisms. These three mechanisms are closely interrelated, although this is not always the case when more than one mechanism is operative. Two mechanisms occur simultaneously and rapidly (to equilibrium within 60 seconds) accounting for 66% of total uptake (0.5 mmol Ug^{-1} cell dry wt). In the first of these two processes, uranyl ions co-ordinate with the amino nitrogen of chitin within the cell wall, and these co-ordination sites immediately on formation act as nucleation sites for crystallization and deposition of further uranium. The third and final process involves the precipitation of uranyl hydroxide within the microcrystalline chitin of the cell wall. This process is slower (to equilibrium >30 minutes) and accounts for 34% of metal uptake. The least significant of these mechanisms with regard to the amount of uranium accumulated is the co-ordination reaction (<3% uptake capacity), however, the remaining two processes are either triggered or assisted by it (Tsezos and Volesky, 1982).

The mechanisms described are probably operative to algal cell walls also. The functional groups in algal cell walls include carboxyl, amyl, hydroxyl, phosphyl, amide, imidazole, thiol and thioether moieties. Several algal genera have been shown to have significant metal sorption capacities including *Chlorella* and *Ulothrix*. The mechanisms are not as well elucidated as with bacteria and fungi, although cation exchange phenomena have been implicated (Gadd, 1992).

There are several criteria for metal/radionuclide biosorption processes to compete on the basis of both efficiency and economy with established conventional techniques for waste stream treatment such as cation-exchange (Volesky, 1987; McEldowney, 1990).

1 The biosorbent biomass should be cheap to grow and/or recover. Clearly it is economically desirable to utilize waste biomass or material from other processes since the production cost will be reduced. Waste microbial biomass is produced from a number of industrial processes (Table 14.4). The use of such waste biomass should not, however, be at the expense of process efficiency. It may be possible to treat waste biomass with chemical or physical agents to improve process characteristics (see above). Algal biomass for metal/radionuclide sorption, in general would have to be produced for the specific purpose of waste treatment since no existing technologies produce waste algal biomass.

Table 14.4 Examples of sources of waste biomass for use in heavy metal or radionuclide effluent treatment

Waste biomass	*Source*
Activated sludge, digested anaerobic sludge	Wastewater treatment
Saccharomyces cerevisiae (yeast)	Brewing
Bacillus subtilis (Gram-positive bacterium)	Enzyme production
Penicillium chrysogenum (fungus)	Penicillin production

2 The biosorbent should have high metal accumulation capacity and accumulation should be sufficiently rapid and efficient to compare with conventional technologies. In order to be competitive a biosorbent should remove >99% of the target metal(s) or radionuclide(s) (Brierley *et al.*, 1986). There is clear evidence that biosorbents can compete in efficiency with existing technologies (Table 14.5), some reaching a sorption capacity of >150 mg g^{-1} dry wt considered essential for an efficient biosorbent system.

The rate of heavy metal accumulation by biosorbents also compares favourably with existing techniques. The time to equilibrium for conventional ion exchange techniques can be several hours, for some

Table 14.5 Percentage increase in efficiency of biosorbent compared to conventional adsorption techniques

Radionuclide	*Biosorbent*	*% Increase in accumulation efficiency of biosorbent over:* (i) ion exchange resin	(ii) activated carbon
Thorium	*Rhizopus arrhizus* (fungus)	200	230
Uranium	*Aspergillus niger* (fungus)	1400	–
Radium	Activated sludge waste biomass	2590	2140

biosorbents it can be as little as minutes or even seconds, e.g. uranium sorption by *Streptomyces longwoodensis*. However, a careful selection of the biomass is necessary since not all show such rapid uptake kinetics with equilibrium reached only after hours of exposure to the metal. Some biosorbents show a biphasic accumulation due to the involvement of more than one accumulation mechanism (Fig. 14.2). This is often demonstrated in an initial rapid uptake which accounts for >50% of uptake capacity, followed by continued sorption over hours, e.g. uranium uptake by *R. arrhizus* (Tsezos and Volesky, 1982).

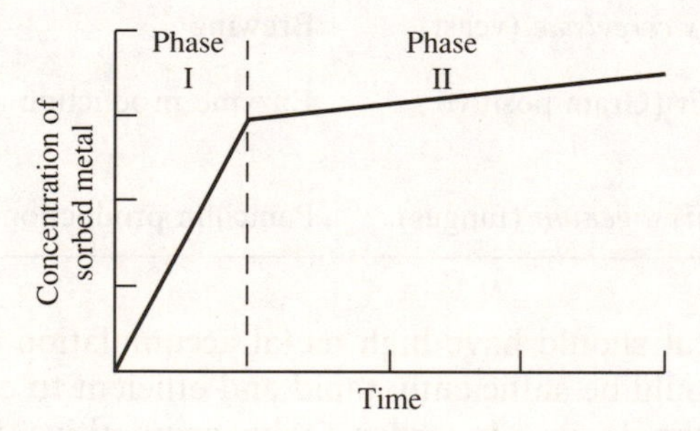

Fig. 14.2. Biphasic heavy metal or radionuclide sorption by microorganisms and algae.

Other parameters in the kinetics of uptake should be considered during the selection of biomass. The effect of the initial radionuclide or heavy metal concentration on the rate accumulation varies with the biomass. Some biosorbents show sorption isotherms which are independent of initial metal concentration with efficient accumulation rates occurring even at very low metal concentrations, a clearly desirable process characteristic. However, this is not always the case and some systems show a strong concentration dependence.

There is a strong linear relationship between initial and equilibrium radionuclide or metal concentrations and biomass sorption until the concentration at which the surface binding sites are saturated is reached. Increases in cell concentration will then increase uptake since, as would be expected, metal uptake at saturation is linearly proportional to cell concentration.

3 Since waste streams are highly variable, the efficiency of biosorption should ideally be unaffected by other waste stream constituents and should be relatively pH stable. This presents one of the biggest challenges in the development of biosorption systems. The effect of pH on metal sorption is often considerable and varies with the biomass and metal. This is true for both microbial and algal biomass. The binding of metals to *Chlorella* is strongly pH dependent. Metal binding affinity has been described as a function of pH in this alga (Darnall *et al.*, 1986).

(a) Metals that are strongly bound at pH 5 or >5 which include Cd^{2+}, Zn^{2+}, Cr^{3+}, Ni^{2+}, Fe^{3+}, Al^{3+}, Cu^{2+}, Pb^{2+} and UO_2^{2+}.
(b) Metals that show strong binding which is pH independent, e.g. Ag^{2+}, Hg^{2+} and $AuCl^{4-}$.
(c) Metal anions that bind strongly at pH 2 or <2 and less strongly at pH 5 such as CrO_4^{2-} and SeO_4^{2-}.

Similar distinctions have been made for microorganisms in their reactions to pH during metal sorption and the effect is known to vary not only with the metal type as indicated for *Chlorella*, but also with the biosorbent type. *Pseudomonas fluorescens* showed uranium sorption independent of pH between pH 2 and 4 while *Streptomyces niveus* showed lower sorption at pH 2 than pH 4. Both these organisms accumulated maximum uranyl ions at pH 4–5. Other microorganisms show greatest accumulation at other pHs. These differences can be explained in terms of the binding interaction, for example, competition for binding sites between the metal and H^+ at low pH or through the effect of pH on metal chemistry influencing subsequent binding, e.g. the formation of hydrolysed uranium or uranyl ions (McEldowney, 1990).

It is evident that pH changes not only affect the capacity but also the rate of accumulation. *Sacchromyces cerevisiae* shows an increasing sorption rate for uranium with increasing pH between pH 2.5 and 5.5. Careful selection of biosorbent may largely solve this problem since there are certainly microorganisms which show sorption efficiency for certain metals unaffected by pH.

The presence of inorganic and organic components other than the target metal or metals is common in waste streams. Such components have often been found to alter the efficiency of sorption. There are several possible mechanisms for this effect (McEldowney, 1990).

(a) Direct competition for cell surface binding sites resulting in lower uptake of the target metal.
(b) Competition between the cell surface ligands and contaminant ligands for complexation with the target metal reducing biosorption.
(c) Binding of the contaminant to the cell surface and subsequently acting as an efficient binding site. This would increase metal sorption.
(d) The contaminant may form a complex with the metal and the resulting complex binding more readily to biomass active sites, increasing metal accumulation.

Examples of all these effects exist for different metals/radionuclides and biosorbent types. The effects of contaminants is further complicated by pH which can significantly alter the influence of inorganic and organic components in sorption efficiency. The effects of three particular components on sorption are worthy of consideration. Iron (Fe^{3+}) is a relatively common component in waste streams and can have a large inhibitory effect on metal sorption by a variety of biomass types for several target metals (Galun *et al.*, 1984). Effluent water hardness will also influence biosorption through the presence of Ca^{2+} and Mg^{2+} which can inhibit metal sorption to many microbial sorbents probably through direct competition for active sites (Lewis and Kiff, 1988). This does not appear to occur with algae such as *Chlorella*. The presence of monovalent cations, e.g. K^+ and Na^+, is also common in waste streams, however generally it does not influence the sorption of divalent or trivalent cations.

It is possible that some of the problems inherent in the presence of a mixture of waste stream components or caused by waste stream pH may be negated by careful biomass selection to ensure efficient metal/radionuclide sorption in the conditions of the particular waste stream. Alternatively, it may be possible to pre-treat biomass to give appropriate uptake characteristics, e.g. through surface modification, or genetically engineer an organism to ensure appropriate characteristics. This is feasible only if the mechanisms of surface binding and the potential mechanisms of interactions between target metals, cell surface and contaminants are known. These modifications and selection procedures will only be effective if the waste stream does not vary. Overcoming the contaminant problems in variable waste streams is more difficult and may require pre-treatment of the waste stream before it enters the biosorption process. Conventional techniques for removing contaminants could be linked with the use of biosorption processes, e.g. Fe^{3+} precipitation. It is worthwhile to realize, however, that waste stream contaminants and pH variability may also have a detrimental impact on the efficiency of conventional techniques.

4 If the aim of the process is the recovery of a strategic or valuable metal (rather than pollution control) then the sorption process should be selective and result in the separation of the target metal from the cocktail of waste stream components. There are many examples where microorganisms and algae show a preferential adsorption series for metals making selective adsorption a real possibility.
Chlorella selective adsorption series (Darnall *et al.*, 1986):

$$UO_2^{2+} > Cu^{2+} > Zn^{2+} > Ba^{2+} = Mn^{2+} > Cd^{2+} = Sr^{2+}$$

Aspergillus niger selective adsorption series (Yakubu and Dudeney, 1986):

$$Fe^{3+} > UO_2^{2+} > Cu^{2+} > Zn^{2+}$$

Metal selectivity may be controlled by selection of the appropriate strain and cell wall characteristics in a similar fashion to minimizing the impact of waste stream contaminants.

5 It should be possible to recover metal ions from biosorbents. This recovery should be rapid, metal selective and cheap.The biomass should be reusable after desorption, indeed for several sorption/desorption cycles. Recovery has several advantages.

(a) Rare or valuable metals/radionuclides can be recovered.
(b) There is no necessity to dispose of metal or radionuclide contaminated biomass which is biodegradable.

It is essential that only low volumes of eluant are applied to the biosorbent for efficient desorption. This ensures a concentrated elution solution for subsequent disposal or further treatment. Certainly, the elution solution should be considerably more concentrated than the original waste stream.

Efficient eluants fall into two categories (McEldowney, 1990): those which form highly soluble salts with metals/radionuclides and those which form soluble complexes with the target metal. Ideally eluants should be common chemicals used by industry, easily obtainable and cheap to further control process costs.

Sulphuric, nitric and hydrochloric acids have been found to be highly efficient eluants for a variety of metals from several biosorbent types forming soluble metal salts. Their efficiency varies with concentration, eluant type and biosorbent (Table 14.6). The main limitation in the use of these eluants, assuming appropriate elution efficiency, is the effect on the biosorbent. Both the structural integrity of the biosorbent and the sorption characteristics (uptake kinetics and capacity) are affected by

Table 14.6 The efficiency of uranium elution from biosorbents

Biosorbent	*Eluant*	*Concentration*	*Percentage efficiency of elution*	*Elution mechanism*
Rhizopus arrhizus	HCL	0.01 N	ineffectual	Formation of soluble salt
		0.1 N	94	
		1.0 N	100	
Rhizopus arrhizus	Na_2CO_3	0.1 N	100	Formation of soluble complexes
	$NaHCO_3$	1 N	100	
Penicillium digitutum	EDTA	0.1 M	61	Formation of soluble complexes
Actinomyces levoris	EDTA	0.01 M	82–88	

these eluants, limiting the subsequent reuse of the biosorbent. Similar considerations apply to complexing agents. For example, carbonate as a complexing agent for uranium is a very effective uranium desorbent (Table 14.6), but damages the biosorbent. This damage appears to occur because of the acidity of carbonate solutions. However, sodium bicarbonate is also highly efficient at uranium desorption from biomass (Table 14.6) and allows extensive reuse of the biosorbent since there are no structural effects or changes in uptake characteristics (McEldowney, 1990).

Biopolymer accumulation

A variety of biopolymers have potential applications as metal-binding agents. These include polysaccharides, proteins and the polyphenolic and related polymers (Hunt, 1986). Metals bind to biopolymers primarily through one or both of the following two processes, depending on the metal and biopolymer: (i) charge interactions – many biopolymers are negatively charged, the density of charge varying with polymer type; (ii) functional groups on the biopolymer may also enter into coordination reactions with the metal, e.g. in thiol groups (–SH) sulphur acts as the donor atom. The solubility of biopolymers varies with the molecular weight of the polymer, the polymer type and environmental conditions such as pH and ionic concentration (Ashley and Roach, 1990).

Whole unpurified polymers tend to be insoluble but are generally highly hydrophilic and permeable. Unpurified insoluble polymers can either be used in metal effluent treatment in fluidized beds or columns, or may be added directly to liquid waste. The use of unpurified polymers ensures that the process is economically competitive with existing technologies. Purification procedures inevitably increase costs and would have to be offset by a considerable increase in sorption efficiency. This, however, does seem to be achievable by treatment of some polymers.

It is possible that solubility may add to the efficiency of the biopolymer in removing metals from liquid waste. Certainly there would be an intimate mixing of waste components if soluble biopolymers were added to liquid effluent. Furthermore the target metals may act to cross-link the polymer chains, particularly in polysaccharides, resulting in their precipitation because of a large increase in molecular weight. Such precipitation may also be achievable through changes in waste stream characteristics such as pH or ionic strength. The precipitated biopolymer–metal complex could then be removed from the waste as sludge.

Polysaccharide biopolymers Polymers consisting of monosaccharide units or monomers, that is polysaccharides, are produced by animals, plants and microorganisms. Certain of these polymers are available in large quantities, for example, the plant cell wall material cellulose, a neutral polysaccharide consisting of β(1–4) glucose monomers. Cellulose is the most abundant natural polymer and, even though neutral, can bind certain metals. For example, copper is bound by cellulose through the replacement of water groups surrounding the copper atom by polymer hydroxyl groups.

The greatest diversity of polysaccharides is produced by microorganisms. A variety of bacteria produces exopolymers which may be loosely associated with the cell wall or may form a distinct gelatinous layer or capsule around the cell. These acidic polysaccharides are normally hydrophilic and polar and bind metals extremely effectively. They consist primarily of pentoses, hexoses, heptoses, amino sugars and hexuronuic acids, depending on bacterial species. In the production of extracellular polymers, both the quantity and molecular structure can be affected by growth conditions. For example, a high C : N ratio stimulates exopolymer production in many bacteria. The degree to which the nature of the polymer can be manipulated by growth conditions varies with bacterial species. Increases in the amount of exopolymer available for metal binding, as would be expected, increases sorption capacity of the polymer. The polymers not only act in metal ion binding, but also act to physically entrap any insoluble particulate metals. Bacterial polymers can be produced on a commercial scale by fermentation, allowing selection of particular polymer characteristics both through strain selection and through choice of appropriate growth conditions.

The presence of exopolymers in activated sludge, primarily produced by members of the bacterial genus *Zoogloea*, is the key factor in determining metal removal from waste streams by activated sludge. *Zoogloea ramigera* produces a heteropolymer consisting of glucose, galactose and pyruvate units, the production of which only occurs at C : N ratios of 10 : 1. The cells and polymer of this organism together have a metal binding capacity of 0.30 and 0.10 g g^{-1} dry wt, respectively, for copper and cadmium. It is possible to desorb the metal by acid treatment and to reuse the biomass/polymer for several treatment cycles. Similar factors as those for metal sorption to cell surfaces influence and control sorption to exopolymers (Hunt, 1986).

There are also animal products which are polysaccharide in nature and offer potential for metal sorption. This includes the second most abundant natural polymer, chitin—a cell wall constituent of fungi and a skeletal material in crustacea. It is cheaply available in bulk (millions of tonnes) as waste from the seafood industry. Chitin consists of β(1–4)-linked *N*-acetyl glucosamine monomers (Fig. 14.3) and has a considerable chain length. It is a highly crystalline polymer and insoluble. Chitin has been shown to be an effective metal biosorbent (Macaskie and Dean, 1990).

Chitosan is a commonly used derivative of chitin, produced through treatment of chitin with boiling alkali (Onsoyen and Skaugrud, 1990). This treatment results in chitin being deproteinized and deacetylated to varying degrees depending on the alkali and heat treatment (Fig. 14.3). This allows a control over the adsorption characteristics of chitosan. Chitosan is a large straight chain polymer of β(1–4)-linked glycans of molecular weight between 10 000 and 1000 000 daltons, depending on

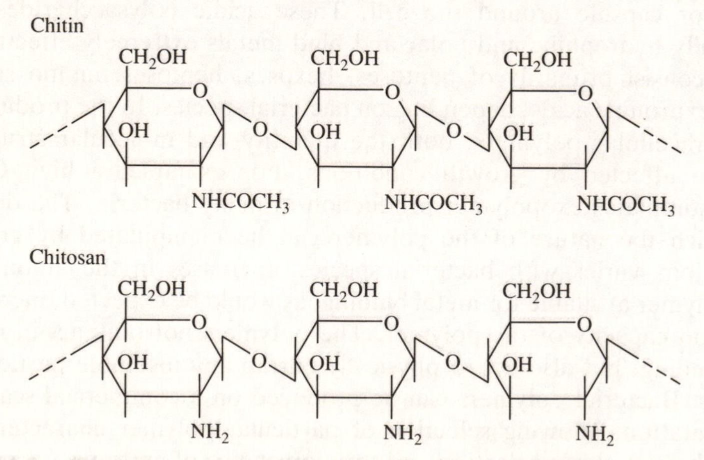

Fig. 14.3. Diagram to show differences in the chemical structure of chitin and chitosan.

the conditions during production. The glycans consist of 2-acetamido-2-deoxy-D-glucose-(glucosamine) (Fig. 14.3). Chitosan has free amino groups and different properties from chitin. Several mechanisms of metal binding to chitosan occur, the mechanism varying with the metal type. Copper is the most fully studied of the metal interactions with chitosan. The binding of copper to chitosan appears to involve the formation of a complex between copper and two hydroxyl groups, an amyl group and a fourth site occupied by water or another hydroxyl group on the chitosan polymer. The efficiency of chitin and chitosan in metal binding varies with the metal or radionuclide. For example, fungi also produce chitin which appears superior in uranium binding characteristics to crustacean chitin probably because of different acetylation characteristics and different associated polymers such as protein. Different invertebrate sources of chitosan show different binding capacities for certain metals, e.g. copper, mercury and nickel. Certainly metal and radionuclide binding to chitin and chitosan can be rapid to saturation levels. Chitin and chitosan can be further treated chemically to change and improve their sorption characteristics. Products such as chitosan phosphates show good metal-binding characteristics.

Another group of naturally produced polymers which have been used industrially for a variety of purposes are those produced by marine algae, e.g. alginates and carrageenans. Marine algae are abundant sources of polysaccharide material which includes both neutral polymers and highly negatively charged polymers which form the mucilage layer surrounding many seaweeds. Algal mucilages have exceptionally high cation binding capacities primarily because of their acidic nature with cation binding to carboxyl and ester sulphate groups.

Polyphenolic biopolymers Lignin is a polymer which consists of cross-linked phenylpropane units (Hunt, 1986). It is a major structural component in plant cell walls and is available in large quantities from the paper industry. During the production of paper, soft wood pulp is treated by the acid–sulphite process which separates out insoluble cellulose from other plant cell wall materials. Lignosulphonic acid, a soluble form of lignin, is produced through this process. It is known that lignosulphonic acid efficiently binds calcium and magnesium. This, together with its availability as an industrial waste product, suggests that lignosulphonic acid may be worthy of study to assess its potential for metal and radionuclide binding.

Metal-binding proteins Metal-binding proteins such as metallothioneins and phytochelatins, appear to be commonly produced by microorganisms (Gadd, 1992). Methallothioneins bind essential and non-essential metals. They are small polypeptides rich in cysteine. It appears that they may be involved in metal resistance. Methallothioneins, because of their metal-binding capabilities, may potentially be used in metal removal.

This is particularly the case, since genetic engineering to increase methallothionein production and to develop metallothioneins specific for certain metals seems feasible (Butt and Ecker, 1987).

Biopolymers, either in their natural or in modified forms clearly offer considerable potential for waste stream treatment. They must, however, comply with the criteria outlined above for extracellular sorption to be able to compete on grounds of efficiency and economy with conventional treatment procedures.

Exoproduct binding The production of metal chelating agents by microorganisms appears to be relatively common (Gadd, 1992). These chelating agents, called siderophores, are catechol or hydroxamate derivatives and are involved in iron uptake by cells. In the environment siderophores solubilize ferric hydroxide making soluble iron available to the cell. Siderophores bind ferric ion at high affinity but can bind other metals though at lower affinities. Siderophores can be produced either microbially or through chemical synthesis, and can be applied to remove specific metals from waste streams. Their efficiency can be increased through chemical modification. For example, the catechol derivative can be modified by substituting Cl^-, Br^-, or NO_2^- on the benzene ring, considerably altering the metal binding of the molecule. Modified siderophores have been used for the removal of metals, e.g. Cd^{2+}, Hg^{2+} and Cu^{2+}, and radionuclides, e.g. Sr^{2+}, Cs^+ and UO_2^{2+}, from mixed cocktails.

The range of reactions entered into by organisms or their products with metals and radionuclides is clearly extensive. Many of these interactions may find application in the development of biotreatment procedures for the removal and recovery of metals and radionuclides from waste streams.

References

AIKING, H., HOK, K., VAN HEERIKHUIZEN, H. and VAN'T RIET, J. (1982) Adaptation to cadmium by *Klebsiella aerogenes* growing in continuous culture proceeds mainly via the formation of cadmium sulfide. *Appl. Environ, Microbiol.*, **44**, 938–944.

ASHLEY, N.V. and ROACH, D.J. (1990) Review of biotechnology applications to nuclear waste treatment. *J. Chem. Technol. Biotechnol.*, **49**, 381–394.

BEVERIDGE, T.J. (1978) The response of cell walls of *Bacillus subtilis* to metal and to electron microscopic stains. *Can. J. Microbiol.*, **24**, 89–104.

BONOTTO, S., GERBER, C.T., GARTER JR., VANDECASTEELE, C.M., MYTTENAERE, C., VAN BAELEN, J., COGNEAU, M. and VAN DER BEN, D. (1984) Uptake and distribution of technetium in several marine algae. In

Cigna, A., Myttenaere, C. (eds) *Int. Symp. Behav. Long-lived Radionuclides Marine Environ.*, pp. 138–396. CEC, Brussels.

BRIERLEY, C.L. (1978) Bacterial leaching. *CRC Crit. Rev. Microbiol.*, **6**, 207–262.

BRIERLEY, C.L., BRIERLEY, J.A. and DAVIDSON, M.S. (1989) Applied microbial processes for metals recovery and removal from waste-water. In Beveridge, F.G., Doyle, R.J. (eds), *Metal Ions and Bacteria*, pp. 359–382. John Wiley and Sons, New York.

BRIERLEY, R.A., GOYAK, G.M. and BRIERLEY, C.L. (1986) Considerations for commercial use of natural products for metal recovery. In Eccles, H., Hunt, S. (eds) *Immobilisation of Ions by Biosorption*, pp. 105–117. Ellis Horwood, Chichester.

BUTT, T.R. and ECKER, D.J. (1987) Yeast metallothionein and applications in biotechnology. *Microbiol. Rev.*, **51**, 351–364.

DARNALL, D.W., GREENE, B., HENZL, M.T., HOSEA, J.M., McPHERSON, R.A., SNEDDON, J. and ALEXANDER, M.D. (1986) Selective recovery of gold and other metal ions from an algal biomass. *Environ. Sci. Technol.*, **20**, 206–208.

DORAN, J.W. (1982) Microorganisms and the biological cycling of selenium. *Adv. Microbial Ecol.*, **6**, 1–32.

FERRIS, F.G., SHOTYK, W. and FYFE, W.S. (1989) Mineral formation and decomposition by microorganisms. In Beveridge, T.J., Doyle, R.J. (eds) *Metal Ions and Bacteria*. John Wiley and Sons, New York.

GADD, G.M. (1992) Microbial control of heavy metal pollution. In Fry, J.C., Gadd, G.M., Herber, R.A., Jones, C.W., Watson-Craik, I.A. (eds) *Microbial Control of Pollution*, pp. 59–88. Cambridge University Press, Cambridge.

GALUN, M., KELLER, P., MALKI, D., FELDSTEIN, H., GALUN, E., SIEGEL, S. and SIEGEL, B. (1984) Removal of uranium (VI) from solution by fungal biomass: inhibition by iron. *Water Air Soil Pollut.*, **21**, 411–414.

GHIORSE, W.C. (1984) Biology of iron-depositing and manganese-depositing bacteria. *Ann. Rev. Microbiol.*, **38**, 515–550.

HUGHES, M.N. and POOLE, R.K. (1989a) *Metals and Microorganisms*, pp. 303–358. Chapman & Hall, London.

HUGHES, M.N. and POOLE, R.K. (1989b) Metal mimicry and metal limitation in studies of metal–microbe interactions. In Poole, R.K., Gadd, G.M. (eds) *Metal–Microbe Interactions*, pp. 1–17. IRL Press, Oxford.

HUNT, S. (1986) Diversity of biopolymer structure and its potential for ion binding applications. In Eccles, H., Hunt, S. (eds) *Immobilisation of Ions by Biosorption*, pp. 15–46. Ellis Horwood Ltd, Chichester.

LEVI, P. and LINKLETTER, A. (1989) Metals, microorganisms and biotechnology. In Hughes, M.N., Poole, R.K. (eds) *Metal and Microorganisms*, pp. 303–358. Chapman and Hall, London.

LEWIS, D. and KIFF, R.J. (1988) The removal of heavy metals from aqueous effluents by immobilized fungal biomass. *Environ. Technol. Lett.*, **9**, 991–998.

MACASKIE, L.E. (1990) An immobilized cell bioprocess for the removal of heavy metals from aqueous flows. *J. Chem. Technol. Biotechnol.*, **49**, 357–381.

MACASKIE, L.E. and DEAN, A.C.R. (1990) Metal-sequestering biochemicals. In Volesky, B. (ed.) *Biosorption of Heavy Metals*, pp. 199–248. CRC Press, Boca Raton.

MCELDOWNEY, S. (1990) Microbial biosorption of radionuclides in liquid effluent treatment. *Appl. Biochem. Biotechnol.*, **26**, 159–180.

MERGEAY, M. (1991) Towards an understanding of bacterial metal resistance. *TIBTECH.*, **9**, 17–24.

NEALSON, K.H., ROSSON, R.A. and MYERS, C.R. (1989) Mechanisms of oxidation and reduction of manganese. In Beveridge, F.G., Doyle, R.J. (eds) *Metal Ions and Bacteria*, pp. 383–412. Wiley.

NORRIS, P.R. (1989) Mineral oxidising bacteria: metal–organism interactions. In Poole, R.K., Gadd, G.M. (eds) *Metal–Microbe Interactions*, pp. 99–117. IRL Press, Oxford.

ONSOYEN, E. and SKAUGRUD, O. (1990) Metal recovery using chitosan. *J. Chem. Technol. Biotechnol.*, **49**, 395–404.

OREMLAND, R.S., HOLLIBAUGH, J.T., MAEST, A.S., PRESSEC, T.S., MILLER, L.G. and CULBERTSON, C.Q. (1989) Selenate reduction to elemental selenium by anaerobic bacteria in sediments and culture: biogeochemical significance of a novel sulphate independent respiration. *Appl. Environ. Microbiol.*, **55**, 2333–2343.

TOBIN, J.M., COOPER, D.G. and NEUFELD, R.J. (1984) Uptake of metal ions by *Rhizopus arrhizus* biomass. *Appl. Environ. Microbiol.*, **47**, 821–824.

TSEZOS, M. and VOLESKY, B. (1982) The mechanisms of uranium biosorption by *Rhizopus arrhizus*. *Biotechnol. Bioeng.*, **24**, 385–401.

VOLESKY, B. (1987) Biosorbents in metal recovery. *Trends Biotechnol.*, **5**, 95–101.

YAKUBU, N.A. and DUDENEY, A.W.L. (1986) Biosorption of uranium with *Aspergillus niger*. In Eccles, H., Hunt, S. (eds) *Immobilization of Ions by Biosorption*, pp. 183–200. Ellis Horwood Ltd, Chichester.

Further reading

BEVERIDGE, T.J. and DOYLE, R.J. (1989) *Metal Ions and Bacteria*. John Wiley and Sons, New York.

GADD, G.M. (1992) Microbial control of heavy metal pollution. In Fry, J.C., Gadd, G.M., Herbert, R.A., Jones, C.W., Watson-

Craik, I.A. (eds) *Microbial Control of Pollution*, pp. 59–88. Cambridge University Press, Cambridge.

HUGHES, M.N. and POOLE, R.K. (1989) *Metals and Microorganisms*. Chapman and Hall, London.

HUTCHINS, S.R., DAVIDSON, M.S., BRIERLEY, J.A. and BRIERLEY, C.L. (1985) Microorganisms in reclamation of metals. *Ann. Rev. Microbiol.*, **40**, 311–336.

POOLE, R.K. and GADD, G.M. (eds) (1989) *Metal–Microbe Interactions*. Special Publication, Society for General Microbiology, Vol. 26. IRL Press, Oxford.

(1990) Papers from the meeting Recovery/removal of metals by biosorption – a chemical reality or a scientist's dream? *J. Chem. Technol. Biotechnol.*, **49**, 329–404.

Chapter 15

BIOTECHNOLOGY FOR METAL AND RADIONUCLIDE REMOVAL AND RECOVERY

Introduction

Biological processes for heavy metal and radionuclide removal from liquid effluent are based on both active, e.g. biotransformation, and passive processes such as biosorption (Ch. 14). These are not mutually exclusive and can occur simultaneously depending on the effluent treatment system in operation. Indeed, several mechanisms functioning at the same time is relatively common in living treatment systems.

Living systems used for the removal of metals or radionuclides from waste are either natural or man-made. Natural systems or natural-setting systems are based on pre-existing whole ecosystems including wetland, i.e. bogs and marshy areas, or aquatic, i.e. lakes and ponds, ecosystems. The biocatalysts are the resident plants, algal and microbial communities. Man-made living systems include two distinct types of facilities. First, those built systems which provide the basis for artificially produced complex ecosystems similar to the natural-setting systems. These include man-made meanders and impoundments. Second, purpose-built systems which contain living microbial communities or individual microbial populations. These include sewage treatment facilities consisting of complex microbial communities and various biocontactor designs often based on individual microbial species.

Non-living biological treatment processes for metal removal include microbial and algal biosorbents, and complexing agents arising from biological sources. Contact between the metal/radionuclide effluent and the biocatalyst is achieved within bioreactors or contactors of several different designs, e.g. packed fixed bed or dispersed bed contactors (see below).

Biological treatment technologies for heavy metal and radionuclide removal from liquid waste should not be viewed as necessarily operating in isolation. It may be desirable to combine a variety of biological systems to achieve maximum efficiency of metal removal. Moreover it may also be appropriate to combine these systems with existing conventional treatments to produce the best quality effluent.

The aim of this chapter is to review the technology of these emerging biological treatment processes and assess their application to metal and radionuclide effluent treatment. The key mechanisms of metal removal for each process will be indicated and the efficiency of the various processes assessed. It is not intended to describe the technology involved in the bioleaching of metal ores.

Non-living biological processes for metal effluent treatment

The key mechanism in heavy metal and radionuclide effluent treatment by non-living biocatalysts is metal sorption (Ch. 14). The use of non-living biomass or materials of biological origin has several advantages over the use of living systems in the treatment of metal- or radionuclide-containing liquid waste. In particular the problems that might occur through metal toxicity to living organisms are obviated. Heavy metal or radionuclide toxicity to living organisms may have several potential impacts on accumulation.

1 The metabolic activity of the organisms may be reduced resulting in a reduction in active accumulation or transformation reactions. In addition, changes in cell surface characteristics which will alter metal sorption may result from changes in metabolic activity.

2 Cell-bound enzyme efficiency may be reduced, affecting such processes as, for example, *Citrobacter* sp. phosphatase metal-binding reaction (Ch. 14) (Macaskie, 1990).

Non-living processes are unaffected by high metal concentrations or variations in waste stream characteristics other than through their impact on the passive accumulation of the metal/radionuclide (Ch. 14). Non-living systems have a further advantage in that there is no necessity to control growth conditions, e.g. O_2 tension, nutrient or growth factor availability, necessary to maintain either actively metabolizing or resting organisms. Nor is there any need to remove and dispose of surplus nutrients or metabolic waste products from the treatment solution. The lack of metabolic waste products avoids the potential interactions between these and metals or radionuclides which may alter metal sorption characteristics. Such interactions, for example, may result in the presence of possibly significant amounts of metabolite-complexed metal in the treated waste stream (Ch. 14) which did not interact with the biocatalyst.

Process technologies: microbial and algal biosorbents

It is essential in any biosorption process to ensure efficient contact between the solid-biosorbent phase and the liquid-metal containing phase. This can be achieved in batch, semi-continuous or continuous flow procedures in a variety of process arrangements, e.g. stirred tank and column contactors. There are constraints, however, on process design imposed by native microbial and algal biomass. Native biomass can only be used in stirred tank reactors. This is because of three characteristics which make it unsuitable for other contactor systems (Brierley *et al.*, 1989):

- low mechanical strength
- low density
- small particle size

In stirred tank contactors these characteristics are offset by the suspension of biosorbent in the liquid effluent. Stirred tank contactors take several forms; they may be batch or continuous flow (Fig. 15.1a and b) and may be one-stage or multi-stage. Although multi-stage contactors increase the efficiency of the process there is a concomitant

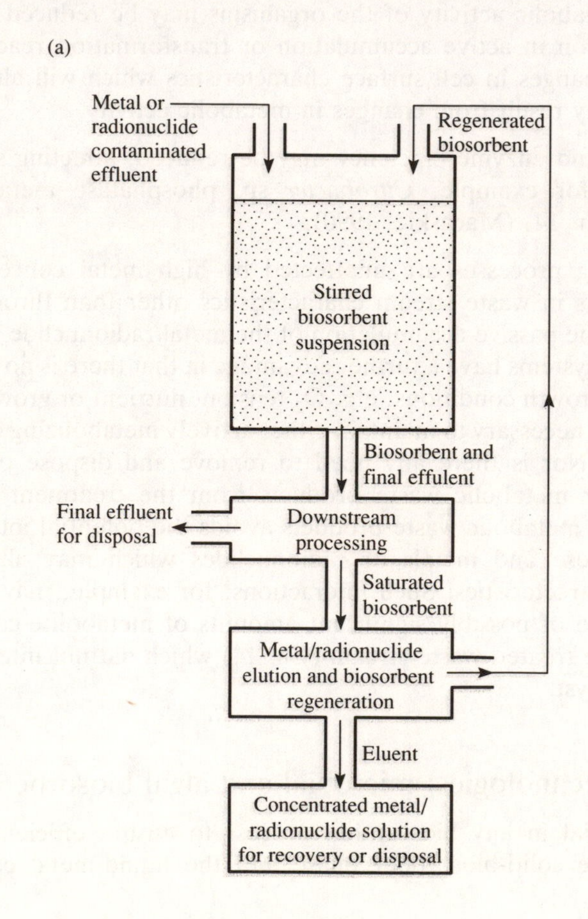

increase in operating cost and capital costs. There is a considerable drawback associated with stirred tank reactors which adds significantly to process costs. Metal-saturated biosorbent must be separated from the liquid effluent before the bound metal can be eluted from the surface and the biosorbent regenerated. Conventional techniques used in the separation include filtration, centrifugation or sedimentation.

Recently, there has been interest in more novel techniques for the removal of metal- or radionuclide-loaded microbial biosorbent from suspensions. These techniques are biomagnetic separation and monoclonal antibody technology (Ashley and Roach, 1990). Biomagnetic separation techniques are applicable to both systems employing active and passive mechanisms of microbial biosorbent accumulation and have been fully described by Ellwood *et al*. (1992). It is possible for microorganisms to acquire significant magnetic moments on accumulation of metals. This may permit their subsequent separation from the liquid phase by high gradient magnetic separation (HGMS). Ashley and Roach (1990) have suggested that it may be appropriate to link biomagnetic separation procedures with monoclonal antibody technology. Monoclonal anti-

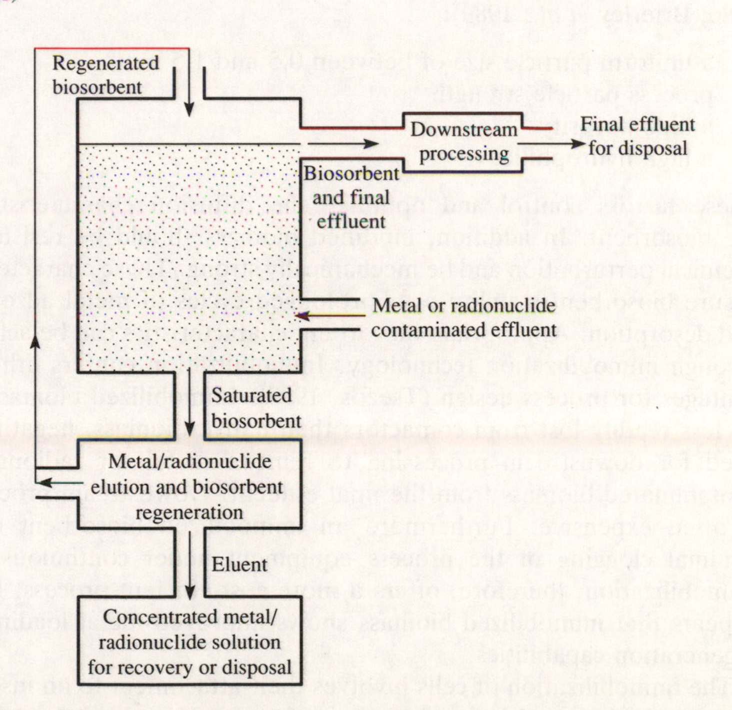

Fig. 15.1. Diagrams to show (a) a stirred tank batch contactor and (b) a continuous countercurrent contactor.

bodies are produced from a single source or clone of cells. They are highly specific antibodies which recognize only one kind of antigen. It is thought that application of separation procedures using monoclonal antibodies together with biomagnetic procedures would permit selective recovery of specific radionuclides or heavy metals from mixed waste streams. This is particularly desirable if the metals are either economically or strategically valuable. It is envisaged that more than one biosorbent would be involved in the sorption process, biosorbent A and biosorbent B. These two biosorbents should be specific for different metals or radionuclides, i.e. 1 and 2, giving metal-loaded biomass A1 and B2. If the treatment mixture is exposed to immobilized monoclonal antibodies for biosorbent A then A1 would be retained in the system. The final effluent would contain only B2 which could be concentrated by HGMS. Subsequently A1 would be eluted and concentrated separately. Thus, metals 1 and 2 would have been effectively separated (Ashley and Roach, 1990). It would of course be possible to develop this type of separation system for more than two biosorbent–metal complexes.

The alternative processes for contacting the biosorbent with the liquid effluent require that the native biomass should be modified to a form reminiscent of activated charcoal or ion exchange resin. That is the modified biomass should have the following characteristics (Tsezos, 1986; Brierley *et al.*, 1989):

- a uniform particle size of between 0.5 and 1.5 mm
- process particle strength
- a high porosity
- a high hydrophilicity

These factors control and optimize the diffusion characteristics of the biosorbent. In addition, modified biomass should be resistant to chemical perturbation and be mechanically strong. These characteristics ensure biosorbent stability over prolonged cycles of metal adsorption and desorption. Appropriate biosorbent characteristics can be achieved through immobilization technology. Immobilization confers other advantages for process design (Tsezos, 1990). Immobilized biomass may be less readily lost from contactors than native biomass, negating the need for downstream processing to remove metal- or radionuclide-contaminated biomass from the final effluent. Downstream processing is often expensive. Furthermore an immobilized biosorbent causes minimal clogging of the process equipment under continuous flow. Immobilization, therefore, offers a more cost-efficient process. It also appears that immobilized biomass shows improved metal loading and regeneration capabilities.

The immobilization of cells involves their attachment to an insoluble matrix which may be the microorganism itself, a biopolymer or a synthetic polymer (Linko and Linko, 1983). One immobilization technique used in biosorbent preparation is the chemical cross-linking of cells. There are several examples of treating natural biomass powder to produce immobilized biosorbents. AMT-BIOCLAIM™ has developed a commercial process for producing granulated biosorbent which is hydrated and porous, with an average diameter of 1 mm from a variety of bacterial species (Brierley *et al.*, 1986). For example, a granulated *Bacillus* product is produced from concentrated biomass which is treated with cross-linking agents such as gluteraldehyde and ground or extruded to give appropriate particle size characteristics (Brierley, 1990). At present granulated *Bacillus* is used in various configurations of column contactors by Advanced Mineral Technologies, Inc., Golden, CO (Brierley *et al.*, 1989). Fungal biomass can also be modified to produce a biosorbent with suitable characteristics for a column contactor by stiffening and cross-linking hyphae before granulation. Stiffening can be achieved by treating the native biomass with high molecular weight polypeptidic compounds, e.g. gelatin. Stiffened fungal biomass can then be cross-linked through the action of chemicals such as aldehydes which are polymerizable. The biosorbent is, then, mechanically granulated (Nemec *et al.*, 1977).

An alternative to this form of immobilization and biosorbent granulation is the immobilization of native biomass by interaction with a matrix, either through entrapment, adsorption, encapsulation or covalent bonding. This form of immobilization is not only appropriate to dead biomass, but also for living or resting cells (see below). There are certain process limitations, however, for this type of immobilization particularly with regard to diffusion characteristics. The size of the immobilization particle determines diffusion limits. Ideally the matrix should be dominated by the biosorbent, i.e. there should be a minimal amount of matrix material (Brierley, 1990a). The advantages of low matrix content were clearly indicated by Tzesos and Deutschmann (1990). These workers found that uranium biosorption efficiency increased for immobilized *Rhizopus arrhizus* as matrix content and particle size decreased. It is possible that diffusion limitations enforced by immobilization within a carrier matrix may restrict the use of these biosorbent systems to small-scale operations (Macaskie, 1990). Even so this process may still be economically viable for low-volume waste and the extraction of valuable metal.

Biosorbents have been immobilized in a variety of supports including alginate, polyacrylamide and collagen (Brierley *et al.*, 1989; Macaskie and Dean, 1989; Brierley, 1990a). Immobilization matrices vary in their process characteristics, e.g. resistance to hydrostatic pressure, and will require careful selection depending on the contactor system (Nakajima *et al.*, 1982; Bedell and Darnall, 1990). For example, *Chlorella* biosorbents were immobilized in both polyacrylamide and silica gel (Darnall *et al.*, 1986). Of the two immobilization methods, silica gel appeared to give the better characteristics, since it was porous, physically strong and relatively economical to produce.

Immobilized biosorbent can be used in a variety of column contactor configurations such as fixed packed bed, pulsating beds and fluidized beds. In fixed packed bed reactors liquid effluent is percolated through a bed of immobilized biosorbent (Fig. 15.2). This process arrangement results in the upper layers of the biosorbent coming into contact with the effluent first, with lower layers exposed to the metal/radionuclide as the effluent percolates down the column. The top layers of biosorbent become saturated first and the bottom layers last. At the 'breakthrough point' all the biosorbent is saturated and there is a rapid rise in the metal content of the treated effluent. The effluent is then switched to a fresh active column while the saturated column is regenerated by eluting (Ch. 14) the metal/radionuclides from the biosorbent in preparation for subsequent adsorption/desorption cycles. This is the basic process design for column contactors, however, there are a variety of modifications to this system. For example, in pulsating beds the metal-loaded effluent flows up through the column. In this column design, saturated biosorbent at the bottom of the column can be replaced by fresh biosorbent to allow continuous operation of the reactor. In fluidized beds effluent is pumped upwards through the column at relatively high flow velocities so that the biosorbent granules are in

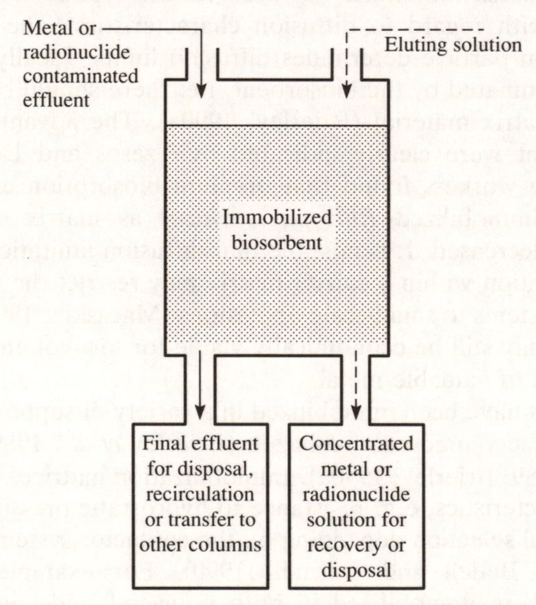

Fig. 15.2. Diagram to show a fixed packed bed contactor with alternate accumulation and elution cycles.

continuous circulation and mixing, i.e. fluidized. All these types of column contactors have been employed for the treatment of metal- and radionuclide-containing liquid effluent (Brierley *et al.*, 1989).

The efficiency of the biosorption process is not only determined by the nature of the biosorbent (Ch. 14), but also by the contactor design. White and Gadd (1990) investigated the efficiency of four kinds of contactors (packed bed with downwards flow, packed bed with upwards flow, stirred bed, and airlift) for the treatment of a thorium simulated waste stream using several different fungal biosorbents. Not only did treatment efficiency vary with biosorbent, but it was apparent that there were major differences in efficiency with reactor design. Poor mixing resulted in low removal of thorium in static bed and stirred bed contactors, while the airlift contactor removed 90–95% of the thorium in the effluent over long treatment periods with a well defined breakthrough point. The advantage of this contactor design was primarily through ensuring the greatest contact between the biosorbents and thorium by efficient circulation within the column.

Efficiency of non-living microbial biosorbents

Immobilized bacterial, fungal and algal biosorbents have been used in several different process designs, some showing a considerable efficiency. The granulated *Bacillus* developed by AMT-BIOCLAIM™ (see above) has an impressive range of characteristics which make it particularly useful in effluent treatment for environmental protection and the minimization of pollutant discharge. This product will simultaneously and non-selectively remove several different heavy metals and radionuclides, e.g. U, Pb, Cd, Ni, Hg, Cr, Zn and Cu, from an effluent. This occurs efficiently regardless of the initial metal concentration, i.e. between <10 ppm and 100s ppm, in the effluent giving biomass loadings >10% of the weight of the *Bacillus* granules. Metal removal from the effluent is more than 99% efficient for dilute waste streams (10–100 mg metal l^{-1}) giving final effluent metal concentrations as low as 10–50 ppb (Hutchins *et al.*, 1986; Brierley *et al.*, 1989; Brierley, 1990a). Typical values for metal accumulation per gramme of granulated *Bacillus* are 1.9 mM Cd, 2.4 mM Cu and 2.9 mM Zn (Brierley, 1990a). It also appears that this granulated biosorbent does not sorb the divalent cations Ca^{2+} and Mg^{2+} and as a result will be unaffected by water hardness. The granulated biosorbent has been used in packed fixed bed (small volume treatment, <15 l min^{-1}) and expanded fluid bed and dispersed bed contactors (large volumes >35l min^{-1}) (Brierley *et al.*, 1989; Brierley, 1990a, b).

Granulated products formed from stiffening and cross-linking of fungal hyphae appear efficient with a high biosorbent loading capacity.

Two patented immobilized fungal biosorbents have uranium sorption capacities of 102.5 mg g^{-1} and 90.5 mg g^{-1} (Brierley, 1990b). Fungal biosorbents immobilized by attachment and entrapment in reticulated foam have also proved efficient at metal effluent treatment. For example, *Aspergillus oryzae* biosorbent removed 90% of the cadmium present in an artificial waste stream. The residence time of the effluent in the column contactor was 5 minutes (Kiff and Little, 1986). Matrix immobilized *R.arrhizus* had a loading capacity of 50 mg U mg^{-1} and removed 100% of the uranium present in an effluent containing <300 mg l^{-1} (Tsezos, 1986).

Non-viable algal biosorbents, e.g. *Chlorella vulgaris*, have also been prepared for use in effluent treatment through immobilization (Nemec *et al.*, 1977; Bedell and Darnall, 1990). Algal biomass immobilized in a silica matrix (AlgaSORB™ produced by Bio-Recovery Systems, Inc.) is available as a commercial product. This biosorbent has been found to efficiently remove a considerable range of metal cations of different valencies and solution chemistries, e.g. Ag^{+}, Co^{2+}, Cr^{3+}, Cr^{6+}, from effluent. The biosorbent is reusable after appropriate regeneration treatment for a great many adsorption/desorbtion cycles (Bedell and Darnall, 1990).

Process technology: biopolymers and exoproduct sorption

The use of biopolymers and microbial exoproducts in metal effluent treatment occurs in both living systems and non-living systems. One of the commonest, metal sorption by bacterial exopolysaccharides, is often associated with a living process, in particular within activated sludge (see below). There is, however, potential to use extracellular complexing agents or products produced from dead plant, animal and microbial biomass in non-living systems, e.g chitin-based compounds and bacterial siderophores (Ch. 14). Some of these are low molecular weight compounds, e.g. siderophores, and require immobilization in order to achieve suitable operating characteristics, e.g. strength, etc., for efficient use in column contactors.

Immobilization of low molecular weight exoproducts is most appropriately performed by attaching the metal-complexing compound to a carrier material. Covalent attachment may be direct to the carrier or through an intermediate linkage and usually requires activation of the carrier. Activation involves the insertion of functional groups on the carrier suitable to enter into covalent bonding. It is achieved by chemical treatment, e.g. organosilane reagents can introduce functional groups on controlled pore glasses. The selection of the type of carrier for metal-binding compounds is based on similar criteria to that for preparation of whole cells and is fully described in Holbein (1990). A range of carriers can be used including natural organic polymers, e.g. cellulose and alginate, and synthetic organic polymers, e.g. nylon and polyacrylamide. Inert inorganic carriers such as controlled pore glass have also been found suitable for immobilization of metal-complexing compounds (Holbein, 1990).

Recent work on complexing agents of biological origin has underlined their promise for application in metal effluent treatment (Ch. 14). Unfortunately little progress has been made in developing processes at industrial level, with the work largely restricted to laboratory-scale studies. There are some commercially available metal-binding agents, however. Siderophores have been prepared as an immobilized product (Compositions™) with high affinity for specific metals. Mercury is specifically removed from waste by this product which has the ability to reduce the mercury loading in effluent from 8 ppm to 1 ppb (Holbein, 1990).

Living biological processes for metal effluent treatment

Heavy metal and radionuclide effluent treatment processes that use living organisms can be divided into two general categories. The first are engineered systems with broadly similar process designs to those used for non-living sorbents, i.e. column contactors and immobilized cells. These may be based on individual populations of microorganisms. The commonest processes using viable microorganisms for metal removal from liquid waste are the biological treatment systems of sewage treatment plants. Trickling filters and activated sludge both possess complex microbial communities efficient at removal of metal from liquid. The second category encompasses the use of living complex communities, consisting of microbial, algal and plant biocatalysts together with associated animal communities in aquatic and wetland habitats. These can be either naturally occurring or man-made artificial enclosures or impoundments. Even if specific varieties of plants and other organisms are initially cultivated in these artificially produced habitats, the subsequent invasion of naturally occurring organisms will be unpredictable and the finally established community will depend on a range of variables largely beyond control.

Engineered microbial living systems

Treatment systems which utilize viable microorganisms as the biocatalyst can be classified into two types:

- immobilized living cells
- biological sewage treatment

Immobilized growing or resting cells Viable bacteria and fungi have frequently been immobilized for application in metal and radionuclide effluent treatment. The techniques for immobilization of active micro-organisms involve entrapment, encapsulation, covalent bonding and adsorption. The adsorption of living cells to inert supports and the development of a biofilm is a commonly used technique. The process of bacterial adsorption to the inert supports, kinetics of biofilm formation and the subsequent steady state processes within biofilms is, however, complex (Characklis and Marshall, 1990). There are certain characteristics of biofilms that require some consideration in process design. Biofilms are dynamic with the processes of attachment, growth (if supplied with nutrients, etc.) and detachment occurring simultaneously and to different extents depending on conditions. The detachment of metal- or radionuclide-loaded cells from a biofilm raises the potential loss of a valuable resource or release of pollutant. Downstream processing is likely to be required, adding substantial costs to a treatment plant.

In addition, thick biofilms can be diffusion limited (Characklis and Marshall, 1990), reducing the efficiency of contact between the biosorbent and the heavy metals or radionuclides in liquid effluent, thereby lowering removal capacity. The suggested advantage of growing biofilms is in the extension of the sorptive life of the biosorbent through their continual replenishment by growth. Individual cells may become saturated but the biofilm as a whole will continually produce new biosorbent material (Macaskie and Dean, 1989; Macaskie, 1990). This advantage may be largely offset by the requirement for downstream processing and diffusion limitation in thick biofilms. Not all growing cells will be on the surface of the biofilm; a number will be within its matrix.

Imaginative equipment design may turn the disadvantage of cell desorption to an advantage and can overcome the problem of diffusion limitation. For example, Shumate and Stranberg (1985) used an immobilized mixed bacterial culture for nitrate removal, i.e. denitrification, and uranium removal from effluent. The bacteria were immobilized on anthracite particles contained within a fluidized bed contactor. Anthracite particles with excess cell mass accumulated in the upper portion of the fluidized bed and were continuously removed and passed through a vibrating screen before returning to the fluidized bed. This meant that excess biofilm was removed, maintaining optimum biofilm thickness within the contactor. The excess cells were also utilized in the process. They were transferred to a stirred tank contactor and used in uranium biosorption.

The maintenance of conditions within the contactor appropriate for bacterial growth, i.e. nutrients, disposal of metabolic products, temperature control, etc., have considerable engineering and economic implications. It may also be the case that growth of the biofilm and activity of the bacteria may be affected by the constituents of the waste stream. The heavy metals or radionuclides themselves are often toxic, as may be other waste stream constituents. Effluent pH may also inhibit microbial growth. It is these features that may ultimately limit the use of biofilms in metal effluent treatment.

Even if a biofilm can continually retain high sorptive capacities by growth, the system will eventually require regeneration. This may be difficult in the case of living cells. Not all sorption may be associated with the cell surface, but internal uptake may occur. The elution (Ch. 14) of surface-bound metal may also be complicated by diffusion limitations into the biofilm from the bulk liquid phase and within thick biofilms. This may make regeneration impossible and require the disposal of degradable biomass contaminated with heavy metals or radionuclides. Controlled disposal will therefore be necessary (Hutchins *et al.*, 1986). Certainly, the recovery of valuable or strategically important metals/radionuclides from biofilms will be difficult. Even if metal desorption can be successfully achieved the biomass may not be reusable. Undoubtedly, however, the use of microbial biofilms will permit the exploitation of metal or radionuclide removal processes which are associated with metabolically active cells including precipitation reactions and transformation reactions (Ch. 14).

A range of materials have been used as inert supports for biofilms. These include solid smooth substrata, e.g. glass, metal and plastic surfaces, solid irregular substrata such as coke and sand, and porous solids such as foams and porous glass (Macaskie and Dean, 1989). Ideally these supports should be of appropriate strength, have a large surface area and permit high effluent flow rates through a contactor. Their arrangement within a contactor and their porosity should be such as to avoid clogging through microbial growth. These criteria are not easily met and a range of solid substrata for biofilm development and application to metal effluent treatment have been shown to have limitations (Table 15.1). Several different contactor designs can be used for biofilm-based systems including fixed bed, fluidized bed and airlift configurations.

A variety of metals have been treated with biofilm-based systems and it is not uncommon for other pollutants to be treated at the same time. This dual treatment potential has obvious advantages in the design of pollution control processes. It may be limited, however, by deleterious effects of the pollutants on the processes. For example, polyvinyl chloride granule immobilized *Pseudomonas fluorescens* (cell loading 0.1–0.4 gm dry wt gm^{-1} plastic) was used for simultaneous nitrate removal and metal removal. The efficiency of metal removal was good at a flow rate of 1500 ml hr^{-1} with 1.0 mg l^{-1} $Pb(NO_3)_2$ reduced to 0.05–0.1 mg l^{-1} and $ZnSO_4$ from 10 to 5 mg l^{-1}. However,

Table 15.1 Limitations of selected biofilm substrata for metal effluent treatment

Substratum	*Limitations*	*Flow rate giving half maximal activity (column vol hr^{-1})*
Smooth solid substrata		
Stainless steel (0.4 × 0.04 mm wide)	Expensive Too heavy for large bioreactor	2.2
Glass helices ('Fenske')	Fragile	Not tested
Porous substrata		
Wood shavings (2 mm wide)	Antimicrobial compounds present requiring removal by methanol wash	1.4
Pumice sponge	Sensitive to shear force	1.3
Porous glass ('Raschig rings')	Pores blocked by biofilm growth 'Active' biosorption sites restricted to annuli	2.6
Coke	Poor biofilm development	2.1
Reticulate foam (20 pores in.$^{-1}$)	Sparse growth in interior of foam bed	7.0

Based on results for *Citrobacter* phosphatase system summarized in Macaskie (1990).

copper was highly toxic to the *Ps. fluorescens* and denitrification was affected (Tengerdy, 1981). Even so, there are commercial processes which simultaneously treat several different pollutants. A rotating disc biofilm-contacting system capable of treating >5 × 10^6 gallons of waste liquid daily has been developed for treatment of gold mining and milling waste. The microbial biofilm degrades cyanide, thiocyanate and ammonia, and heavy metals are removed through biosorption (Hutchins *et al.*, 1986). A *Citrobacter* sp. has been immobilized in a variety of ways including biofilm development on foams, stainless steel wire, solid glass and wood shavings (Macaskie, 1990) to treat heavy metal and radionuclide effluents. The process relies on the action of a cell surface associated phosphatase liberating HPO_4^{2-} which precipitates heavy metals extracellularly (Ch. 14). The process has been found to be highly efficient for removal of cadmium, lead and strontium from effluent, i.e. removal from flow of between 85% and 95% of the metals independent of metal concentration.

Living microorganisms can also be immobilized through entrapment and encapsulation within a support such as polyacrylamide gel. Such immobilization techniques have been used less often for viable microorganisms than dead microorganisms. Immobilization in this manner does permit the regeneration of the living biosorbent for subsequent sorption/desorption cycles. It can produce biosorbent highly efficient in metal removal, even given the associated limitations of immobilization, i.e. diffusion limitation, found also for dead biosorbents. For example, *Citrobacter* sp. immobilized in polyacrylamide gel showed high efficiency of removal of cadmium, copper, lead and uranium. The biosorbent could be regenerated and reused for long periods (Macaskie, 1990). Similarly, a polyacrylamide immobilized *Streptomyces* sp., selectively (in the order $UO_2^{2+} \gg\gg Cu^{2+} > Co^{2+}$) and efficiently removed metal from effluent, could be regenerated without affecting particle stability for five sorption/desorption cycles (Nakajima *et al.*, 1982; Nakajima and Sakaguchi, 1986).

Biological sewage treatment A commonly used living microbial system in the treatment of metal-containing waste liquid is found in the trickling filters and activated sludge systems of sewage treatment plants (described in detail in Horan (1990)). These engineered systems are normally preceded by primary sedimentation which removes particulates, including insoluble metals, together with a large amount of soluble metal. Up to 60% of the total effluent metal load can be removed by sedimentation.

Sedimentation is followed by the biologically-based processes of trickling filters or activated sludge (Ch. 7). A trickling filter consists of a reactor filled with permeable media over which develops a microbial biofilm. Efficiency of the systems depends on an even distribution of effluent over the surface of the filter and on adequate aeration within the filter. There is a mixed community of microorganisms present within the biofilms of trickling filters. The composition of the community includes heterotrophic and autotrophic bacteria, algae, fungi and protozoa, and varies with depth within the filter, primarily because of differences in sewage composition as it passes down the filter (Horan, 1990). Activated sludge reactors can be of several configurations including batch reactors and plug flow reactors (Horan, 1990). The microorganisms are present as a mixed community in suspended growth as flocs. The microbial community consists primarily of heterotrophic and autotrophic bacteria together with ciliate, amaeboid and flagellate protozoa (Horan, 1990).

The removal of metals in trickling filters and activated sludge appears to be broadly comparable and can be rapid and efficient. However, efficiency varies with the metal type, e.g. Cu, Pb, Cr and Zn 50% removal $\gg\gg\gg$ Ni, Mn and Co (Sterritt and Lester, 1986). The key components in binding the heavy metals in these processes are the extracellular polymers produced by the bacteria, particularly by *Zoogloea ramigera* (Ch. 14). Several factors affect the efficiency of metal removal by activated sludge. These include pH, the presence of other contaminating cations, e.g. Al prevented Cu removal (Norberg and Persson, 1984), and the age of the sludge. Optimum removal is obtained if the sludge is under nine days old. Saturation of *Z.ramigera* polymer binding sites occurs at 10 mg metal l^{-1} with little removal of metal below concentrations of 1 mg l^{-1}. In general, the more soluble the metals, the lower the efficiency of metal removal (Brierley *et al.*, 1989). It is possible to desorb metal from *Z.ramigera* biosorbent through acid treatment without deleterious effects on the metal binding capacity of the biomass (Norberg and Persson, 1984). Ultimately the viability of the organisms within activated sludge can be affected by metal toxicity, which eventually influences the capacity of activated sludge to remove metals.

Engineered and natural ecosystems

The use of complete ecosystems for the treatment of liquid effluent containing heavy metals or radionuclides has been found to be efficient and reliable. There are, however, considerable difficulties associated with their use (Brierley *et al.*, 1989). Many of these obstacles are similar to those found for living microbial systems, although some are unique to the artificial and natural ecosystem processes. Those restrictions common to both microbial systems and ecosystems include the following.

1 The effect of waste stream toxicity, induced either by the metals present or by other waste stream components. This may make the maintenance of living ecosystems a demanding problem and may make the continued growth and hence replenishment of the biocatalyst plants, algae and microorganisms difficult to achieve. In practise most of the existing systems are applied to effluent with metal concentrations below toxic levels.

2 A basic requirement in the design of contactors for metal effluent treatment is the efficient mixing of the waste with the biocatalyst biomass. This may be difficult to guarantee in ecosystem treatment processes, particularly natural systems. It may be possible, however, in artificially designed ecosystems to ensure appropriate flow and mixing of the waste through the insertion of baffles, etc.

There are a number of problems unique to the use of whole ecosystems.

1 There will inevitably be a natural loss through death of algal and plant biomass. Even perennial plants will naturally shed material. This presents two problems. First, the necessity to remove and safely dispose of a large amount of potentially metal- or radionuclide- contaminated and degradable biological material which consists of a large amount of water (Brierley *et al.*, 1989). Although a similar problem exists with microbial living systems if metal desorption is impossible, in an ecosystem-based system, collection of the biomass is more problematic. Secondly, there is potential for silting of channels caused by dead and decaying debris inhibiting efficient contact between the waste liquid and biocatalysts. In those systems based on meanders and impoundments this may demand some dredging of the channel beds. If this is the case then sedimentary material contaminated with metals will have to be disposed of safely. The sediment appears to be a key site of metal removal from the liquid phase (Table 15.2). The metal is, however, often present as precipitates and may be largely unavailable for biological uptake.

2 The seasonal nature of plant, algal and microbial growth and activity in complete ecosystems poses another problem. If the mechanism of metal removal from the waste is an active process, e.g. microbial transformations (Ch. 14), then the rate and efficiency of the process may vary with season. In addition, if the phytoplankta are key in the removal process, their input will vary considerably between seasons, i.e. spring/summer algal blooms. The extent of plant growth is also seasonally variable and a system may sustain large losses of annual plants that will not be replenished until the following growing season.

3 These ecosystems, whether artificially produced or naturally occurring, may be difficult to contain within delineated areas. In natural conditions ecosystems are often in a state of flux with spatial boundaries between habitats changing. This type of variation may result in the treatment ecosystem expanding or even contracting.

4 Both artificially designed and natural ecosystems for metal waste treatment have a major disadvantage in that there will undoubtedly be associated invertebrate and vertebrate animal populations. This raises the very real problem of animal exposure to potentially unacceptable levels of metal contamination (see Ch. 13).

The natural ecosystems (or natural-setting systems) such as lakes and wetlands, and the artificially built meanders and impoundments, however, have been successfully employed for the treatment of metal waste (Table 15.2). They have commonly been applied to effluents arising from mining and milling operations. The mechanisms (Ch. 14) and organisms involved in the process are mixed (Table 15.2) (Brierley, *et al.*, 1989).

The whole ecosystem approach to metal effluent treatment requires considerable and prudent management, for example, harvesting and removal of dead biomass at selected times of the year, etc. Appropriate

Table 15.2 The mechanisms and efficiency of natural and artificial ecosystems in metal effluent treatment

Ecosystem	*Source of waste*	*Organisms*	*Mechanisms*	*Efficiency*	*Reference*
Lake (natural)	Mine and smelter wastes	Algae Phytoplankta Bacteria Sulphate reducing bacteria (SRB)	Algal biosorption followed by death and sedimentation. Precipitation of metal sulphides (Zn, Cd, Cu, Fe) by SRBs. Microbial conversion of mercury to dimethyl mercury	Good	Jackson (1978)
Wetlands/ bogs (natural and constructed)	Acid mine drainage	Algae Cyanobacteria (unidentified) Mosses *Sphagnum* *Polytrichum* Higher plants *Typha* (cattails) *Scirpus* (bulrushes) *Carex* (sedges) Bacteria Numerous, unidentified especially SRBs.	Biosorption metal sulphide precipitation through action of SRBs (anaerobic conditions) Fe + Mn precipitation by bacterial oxidation (aerobic conditions	Good e.g. Mn 69–90% in 4 months Fe 60% removal in 4 months (efficiency declines during winter months)	Erickson *et al.* (1987)

Table 15.2 Continued.

Ecosystem	*Source of waste*	*Organisms*	*Mechanisms*	*Efficiency*	*Reference*
Meanders (man-made switchable channels)	Lead mine and mill	Algae *Chlorella, Oscillatoria* *Cladophora, Spirogyra* *Rhizodonium, Hydrodictyon* Higher plants *Potomogeton* (pond weed) *Typha* (cattail)	Biosorption to algae and plants (no consideration of microbial role)	99% efficiency removal Fe,Pb, Cu, Ni and Cd	Gale (1986)
Algae-ponds (man-made)	Uranium mine and mill	Algae *Spirogyra, Chara* *Oscillatoria* Bacteria SRBs including *Desulfovibrio* and *Desulfotomaculum*	(After primary and secondary conventional treatment, physical entrapment and biosorption by algae. No indication of SRB role though likely	U 86% Se 96% Mo 65% efficiency of removal	Ashley and Roach (1990)

management (Ch. 10) will maintain the ecosystem in an active state, capable of sufficient levels of regeneration for metal treatment.

References

ASHLEY, N.V. and ROACH, D.J.W. (1990) Review of biotechnology applications to nuclear waste treatment. *J. Chem. Technol. Biotechnol.*, **49**, 381–394.

BEDELL, G.W. and DARNALL, D.W. (1990) Immobilization of non-viable, biosorbent algal biomass for recovery of metal ions. In Volesky, B. (ed.) *Biosorption of Heavy Metals*, pp. 313–326. CRC Press, Boca Raton.

BRIERLEY, C.L. (1990a) Metal immobilization using bacteria. In Ehrlich, H.L., Brierley, C.L. (eds) *Microbial Mineral Recovery*, pp. 303–323. McGraw-Hill, New York.

BRIERLEY, C.L., BRIERLEY, J.A. and DAVIDSON, M.S. (1989) Applied microbial processes for metals recovery and removal from wastewater. In Beveridge, T.J., Doyle, R.J. (eds) *Metal Ions and Bacteria*, pp. 359–383. John Wiley, USA.

BRIERLEY, J.A. (1990b) Production and application of a *Bacillus*-based product for use in metals biosorption. In Volesky, B. (ed.) *Biosorption of Heavy Metals*, pp. 305–311. CRC Press, Boca Raton.

BRIERLEY, J.A., BRIERLEY, C.L. and GOYAK, G.M. (1986) AMT-BIOCLAIM™: a new wastewater treatment and metal recovery technology. In Lawrence, R.W., Branion, M.R., Ebner, H.G. (eds) *Fundamental and Applied Biohydrometallurgy*, pp. 291–304. Elsevier, Amsterdam.

CHARACKLIS, W.G. and MARSHALL, K.C. (eds) (1990) *Biofilms*. Wiley Interscience.

DARNALL, D.W., GREENE, B., HOSEA, M., MCPHERSON, R.A., HENZL, M. and ALEXANDER, M.D. (1986) Recovery of metals by immobilized algae. In Thompson, R. (ed.) *Trace Metal Removal from Aqueous Solution*, pp. 1–24. Special Publication No. 61, The Royal Society of Chemistry, London.

ELLWOOD, D.C., HILL, M.J. and WATSON, J.H.P. (1992) Pollution control using microorganisms and magnetic separation. In Fry, J.G., Gadd, G.M., Herbert, R.A., Jones, C.W., Watson-Craik, I.A. (eds) *Microbial Control of Pollution*, pp. 89–112. Society General Microbiology, Cambridge University Press.

ERICKSON, P.M., GIRTS, M.A. and KLEINMANN, R.L.P. (1987) Use of constructed wetlands to treat coal mine drainage. *Proc. 90th Nat. West. Mining Conf.* Colarado Mining Association, Denver.

GALE, N.L. (1986) The role of algae and other microorganisms in metal detoxification and environmental clean-up. In Ehrlich, H.L.

Holmes, D.S. (eds) *Workshop on Biotechnology for the Mining, Metal-Refining and Fossil Fuel Processing Industries. Biotechnol. Bioeng, Symp., No. 16*, p. 171. Wiley, New York.

HOLBEIN, B.E. (1990) Immobilization of metal-binding compounds. In Volesky, B. (ed.) *Biosorption of Heavy Metals*, pp. 327–338. CRC Press, Boca Raton.

HORAN, N.J. (1990) *Biological Wastewater Treatment Systems. Theory and Operation*. John Wiley and Sons, Chichester.

HUTCHINS, S.R., DAVIDSON, M.S., BRIERLEY, J.A. and BRIERLEY, C.L. (1986) Microorganisms in reclamation of metals. *Ann. Rev. Microbiol.*, **40**, 311–336.

JACKSON, T.A. (1978) The biogeochemistry of heavy metals in polluted lakes and streams at Flin Flon, Canada, and a proposed method for limiting heavy metal pollution of natural waters. *Environ. Geol.*, **2**, 173.

KIFF, J.R. and LITTLE, D.R. (1986) Biosorption of heavy metals by filamentous fungi. In Eccles, H., Hunt, S. (eds) *Immobilisation of Ions by Biosorption*, pp. 71–80. Ellis Harwood, Chichester.

LINKO, P. and LINKO, Y.-Y. (1983) Applications of immobilized microbial cells. *Appl. Biochem. Bioeng.*, **4**, 53–151.

MACASKIE, L.E. (1990) An immobilized cell bioprocess for the removal of heavy metals from aqueous flows. *J. Chem. Technol. Biotechnol.*, **49**, 357–379.

MACASKIE, L.E. and DEAN, A.C.R. (1989) Microbial metabolism, desolubilization and deposition of heavy metals: metal uptake by immobilized cells and application to the detoxification of liquid wastes. In Mizrahi, A. (ed.) *Biological Waste Treatment*, pp. 159–201. Alan R. Liss Inc., New York.

NAKAJIMA, A. and SAKAGUCHI, T. (1986) Selective accumulation of metals by microorganisms. *Appl. Microbiol. Biotechnol.*, **24**, 59–64.

NAKAJIMA, A., HORIKOSHI, T. and SAKAGUCHI, T. (1982) Recovery of uranium by immobilized microorganisms. *Eur. J. Appl. Microbiol. Biotechnol.*, **16**, 88–91.

NEMEC, P., PROCHAZAKA, H., STAMBERG, K., KATZER, J., STAMBERG, J., JILEK, R. and HULAK, P. (1977) Process of treating mycelia of fungi for retention of metals. US Patent 4,021,368.

NORBERG, A.B. and PERSSON, H. (1984) Accumulation of heavy metals by *Zoogloea ramigera*. *Biotechnol. Bioeng.*, **26**, 239–246.

SHUMATE, S.E. and STRANBERG, G.W. (1985) Accumulation of metals by microbial cells. In Moo-Young, M., Robinson, C.N., Howell, J.A. (eds) *Comprehensive Biotechnology*, pp. 235–247. Pergamon Press, New York.

STERRITT, R.M. and LESTER, J.N. (1986) Heavy metal immobilization by bacterial extracellular polymers. In Eccles, H., Hunt, S. (eds)

Immobilisation of Ions by Biosorption, pp. 201–218. Ellis Horwood, Chichester.

TENGERDY, R.P., JOHNSON, J.E., HOLLO, J. and TOTH, J. (1981) Denitrification and removal of heavy metals from waste water by immobilized microorganisms. *Appl. Biochem. Biotechnol.*, **6**, 3–7.

TSEZOS, M. (1986) Adsorption by microbial biomass as a process for the removal of ions from process or waste solutions. In Eccles, H., Hunt, S. (eds) *Immobilisation of Ions by Biosorption*, pp. 201–219. Ellis Horwood, Chichester.

TSEZOS, M. (1990) Engineering aspects of metal binding by biomass. In Ehrlich, H.L., Brierley, C.L. (eds), *Microbial Mineral Recovery*, pp. 323–339. McGraw-Hill, New York.

TSEZOS, M. and DEUTSCHMANN, A.A. (1990) An investigation of engineering parameters for the use of immobilized biomass particles in biosorption. *J. Chem. Technol. Biotechnol.*, **48**, 29–39.

WHITE, C. and GADD, G.M. (1990) Biosorption of radionuclides by fungal biomass. *J. Chem. Technol. Biotechnol.*, **49**, 331–343.

Further reading

BRIERLEY, C.L., BRIERLEY, J.A. and DAVIDSON, M.S. (1989) Applied microbial processes for metals recovery and removal from wastewater. In Beveridge, T.J., Doyle, R.J. (eds) *Metal Ions and Bacteria*, pp. 359–382. John Wiley and Sons, New York.

EHRLICH, H.L. and BRIERLEY, C.L. (eds) (1990) *Microbial Mineral Recovery*. McGraw-Hill, New York.

GADD, G.M. (1992) Microbial control of heavy metal pollution. In Gadd, G.M., Herbert, R.A., Jones, C.W., Watson-Craik, I.A. (eds), *Microbial Control of Pollution*, pp. 59–89. 48th Symposium, Society for General Microbiology, Cambridge University Press, Cambridge.

VOLESKY, B. (ed.) (1990) *Biosorption of Heavy Metals*. CRC Press, Boca Raton.

Chapter 16

FUTURE PROSPECTS

Industrialization and the associated development of urban life has resulted in the introduction of a wide range of chemicals into the environment which now threaten the balance of the planet's biosphere. Many of the pollution problems evident today are derived from new chemical substances and the products derived from them. A recent European Economic Community inventory listed more than 100 000 different substances, 30 000 of which were considered to be a threat to the environment because they were bioaccumulative, recalcitrant and/or toxic. This has led governments and their agencies to prepare proscribed chemicals listings which form the basis for control of environmental pollution.

Recently the increasing public pressure, government legislation and international agreements have stimulated the introduction of effective waste treatment policies by industry. Concerns over the effects of our previous waste management strategies based on dilution, burial or incineration has provided the motivation for the pollution control industries to look at alternative technologies. With these developments has come a realistic consideration of the biotechnologies for pollution control. Whilst these technologies are still in their infancy, with only a limited number of successful full-scale applications, where they have been applied, they are proving to be effective alternate or adjunct technologies, providing economically acceptable pollution control strategies.

Only a decade ago 'bioremediation' was a new, often misunderstood and overstated technology. Today there are over 200 companies in the United States which offer bioremediation services and there is evidence to suggest that the biotechnologies are becoming the favoured approach to site remediation. It is hoped that these considerations are based on sound scientific principles and a realization of the limitations of the technology. If this is the case then the market predictions of \$200 million to \$1 billion for bioremediation technologies should be realized by the beginning of the next century.

The control of industrial wastes at source, prevention rather than cure, has gained momentum as industrial concerns have started to regard end-of-pipe discharges as products of the manufacturing processes, rather than an effluent. This has in part been brought about by economic pressure for efficient manufacturing practices designed to reduce production costs by recycling catalysts and synthons (chemical precursors) and also by the public desire to see not just products, but also the manufacturing processes that make the products being more environmentally friendly.

Many existing extraction and manufacturing processes are intrinsically polluting. To a large extent the technologies involved in remediation and the control of pollution have been developed to address this problem. The success of the last decade would lead one to predict that with further development and a rational approach to biocatalyst development and bioreactor design, the application of pollution control biotechnologies will continue to expand. However, perhaps the more exciting possibilities for biotechnological development are in their application to 'clean technologies'.

Industrial processes based on modern biotechnology do not rely on the use of hazardous conditions or toxic chemicals, the wastes generated are usually produced at lower volumes than in traditional process plants and are biodegradable. Biotechnological processing can also be used to modify process streams during manufacture either to change the nature of synthons as they enter the process or to remove unwanted chemical contaminants in product streams.

An example of synthon biotransformation is the use of the chiral-selective nature of the activity of some enzymes. Such systems can be used to resolve racemic solutions of synthons before they are added to the chemical reaction vessel. Resolution into a single isomeric form of the synthon enables subsequent production of chiral, rather than racemic products. This is of particular interest when the product is a pharmaceutical or an agrochemical, where only one of the two chiral forms is biologically active. Products based on only the active form can be employed at half the concentration of the racemic product and still achieve the same level of efficacy. This then reduces the side-effects, allergic responses or environmental pollution. The use of biocatalysts in chiral resolution is relatively inexpensive and has made the production of chiral products economically feasible.

As market forces have made environmental acceptability of the products a positive influence on manufacturing processes through sales figures, the environment-friendly 'green' label has become a significant selling point. Hence, there is also the potential for the introduction of biotechnologies as in-process unit operations for environmental biotransformations leading to enhancement of existing products of the chemical industry. For instance, many chemical syntheses utilize halogenated compounds as solvents or synthons. By their very nature these chemicals lead to environmental problems not only in waste treatment but also because they contaminate the end products of the process.

For product enhancement, the decision on the type of reactor will be largely dependent on the nature of the product and the offending contaminants and also on the degree of acceptability of the presence of the biocatalytic biomass in the product stream. The reactors may be simple stirred tank systems operated in a batch or continuous mode dependent on the nature of the rest of the process. This could then be associated with a downstream operation to remove the biomass from the product stream if necessary, or at least to inactivate the biocatalyst. It may involve the application of biofilm reactors, especially if the biotreatment can be disassociated from the growth phase of the biofilm. In which case the amount of biomass contaminating the product would be significantly reduced. If the treatment required only involved simple or few biocatalytic steps the biocatalyst could be presented in the form of an enzyme, in a free, or more likely, an immobilized state.

Techniques such as these will obviate the need for a cure of pollution problems caused by industrial effluents and reduce pollution caused by the use of products contaminated with environmental chemicals.

The biotechnological concepts and processes described in this book appear to offer a realistic opportunity to remediate and control a significant number of pollutants. However, a plea must go out to all protagonists of their own biotechnological answers to specific pollution problems—be realistic, do not make exaggerated claims for the efficiency or efficacy of the technology, and base claims on research and development at the laboratory *and* process or field scale. In addition, environmental biotechnologists, in common with all protagonists of other biotechnologies, should make a major effort to educate governments and the public in a totally open manner, so as to avoid the fate of other technologies which have foundered for want of public acceptance.

INDEX